FRICTION STIR WELDING AND PROCESSING V

FRICTION STIR WELDING AND PROCESSING V

Related titles include:

- *Friction Stir Welding and Processing IV*
- *Friction Stir Welding and Processing III*
- *Materials Science & Technology 2008 (MS&T'08)*

HOW TO ORDER PUBLICATIONS

For a complete listing of TMS publications, contact TMS at (724) 776-9000 or (800) 759-4TMS or visit the TMS Knowledge Resource Center at http://knowledge.tms.org:

- Purchase publications conveniently online and download electronic publications instantly.
- View complete descriptions and tables of contents.
- Find award-winning landmark papers and webcasts.

MEMBER DISCOUNTS

TMS members receive a 30 percent discount on TMS publications. In addition, members receive a free subscription to the monthly technical journal *JOM* (both in print and online), free downloads from the Materials Technology@TMS digital resource center (www. materialstechnology.org), discounts on meeting registrations, and additional online resources to name a few of the benefits. To begin saving immediately on TMS publications, complete a membership application when placing your order or contact TMS:

Telephone: (724) 776-9000 / (800) 759-4TMS

E-mail: membership@tms.org or publications@tms.org

Web: www.tms.org

FRICTION STIR WELDING AND PROCESSING V

Proceedings of a symposia sponsored by
the Shaping and Forming Committee of
the Materials Processing & Manufacturing Division of
TMS (The Minerals, Metals & Materials Society)

Held during TMS 2009 Annual Meeting & Exhibition
San Francisco, California, USA
February 15-19, 2009

Edited by

Rajiv S. Mishra
Murray W. Mahoney
Thomas J. Lienert

A Publication of

A Publication of **The Minerals, Metals & Materials Society (TMS)**
184 Thorn Hill Road
Warrendale, Pennsylvania 15086-7528
(724) 776-9000

Visit the TMS Web site at
http://www.tms.org

ISBN Number 978-0-87339-737-7

If you are interested in purchasing a copy of this book, or if you would like to receive the latest TMS Knowledge Resource Center catalog, please telephone (724) 776-9000, ext. 270, or (800) 759-4TMS.

TABLE OF CONTENTS
Friction Stir Welding V

Friction Stir Welding V

Session I

Session II

Session III

Session IV

Session V

Session VI

About the Editors

Rajiv Mishra is currently a Curators' Professor of Metallurgical Engineering in the Department of Materials Science and Engineering at the University of Missouri-Rolla. He is also the UMR Site Director of the NSF I/UCRC for Friction Stir Processing. His highest degree is Ph.D. in Metallurgy from the University of Sheffield, UK (1988). He has received a number of awards which include: the Firth Pre-doctoral Fellowship from the University of Sheffield, the Brunton Medal for the best Ph.D. dissertation in the School of Materials from the University of Sheffield in 1988, the Young Metallurgist Award from the Indian Institute of Metals in 1993, Associate of the Indian Academy of Sciences in 1993, the Faculty Excellence Awards from the University of Missouri-Rolla in 2001, 2002, 2003, 2004, 2005, 2006 & 2007. He has authored or co-authored 170 papers in peer-reviewed journals and proceedings and is principal inventor of three U.S. patents.

Murray W. Mahoney has a B.S. (UC Berkeley) and M.S. (UCLA) both in physical metallurgy. Mr. Mahoney, consultant, has more than 40 years experience in physical metallurgy and related disciplines. Most recently, his work has centered on developing friction stir welding and friction stir processing. This work has led to the introduction of friction stir welding to join metals considered unweldable and advance FSW to higher temperature alloys. In addition, developments in friction stir processing have advanced superplasticity to very thick section structural Al alloys, enhanced room temperature formability, and improved mechanical and corrosion properties in cast Al and Cu alloys. These studies have resulted in a more complete metallurgical understanding of joining fundamentals, microstructural evolution, formability, and corrosion sensitivity associated with friction stir welding and processing. Mr. Mahoney has authored or co-authored over 90 publications and been awarded over 20 U.S. patents.

 **Dr. Thomas J. Lienert** is currently a Technical Staff Member at Los Alamos National Laboratory. He has held previous positions at the University of South Carolina, Edison Welding Institute and Sandia National Laboratories. Dr. Lienert holds Ph.D. and M.S. Degrees in Materials Science and Engineering and a B.S. Degree in Welding Engineering all from The Ohio State University. His current research interests include laser materials processing and friction welding, including friction stir welding. Dr. Lienert has served as Chairperson for the Joining Critical Technology Sector for ASM International and is a member of the Technical Programming Board for ASM International. Dr. Lienert is also Chairman of the AWS C6 Committee. He is a Principal Reviewer for the Welding Journal and is an active reviewer for Metallurgical and Materials Transactions. Dr. Lienert has also been an organizer for several symposia for ASM International, TMS and AWS. He is a member of Tau Beta Pi Engineering Honors Society and Alpha Sigma Mu Metallurgy/Materials Science Honors Society. Dr. Lienert has published over 25 papers in refereed journals and conference proceedings, and was the author of the chapter on Selection and Weldability of Aluminum Metal-Matrix Composites in the ASM Metals Handbook on Welding, Brazing and Soldering. He has also made over 60 presentations at technical conferences. Dr. Lienert has been awarded both the Charles H. Jennings Award and the McKay-Helm Award from AWS for papers published in the Welding Journal.

FRICTION STIR WELDING AND PROCESSING V

Session I

Session Chair

Rajiv S. Mishra

Friction Stir Welding and Processing V
Edited by: Rajiv S. Mishra, Murray W. Mahoney, and Thomas J. Lienert
TMS (The Minerals, Metals & Materials Society), 2009

A MICROSTRUCTURE – PROCESSING RELATIONSHIPS IN FRICTION STIR PROCESSING (FSP) OF NiAl BRONZE

Terry R McNelley[1], Srinvasan Swaminathan[2], Jianqing Su[1] and Sarath Menon[1]

[1]Naval Postgraduate School
Department of Mechanical and Astronautical Engineering
Monterey, CA 93943-5146
[2]GE Global Research
Bangalore, Karnataka 560066 India

Keywords: Friction stir processing; Multi-pass processing; Microstructure; Strain distribution; NiAl bronze

Abstract

The evolution of SZ and thermomechanically affected zone (TMAZ) microstructures during single-pass and multi-pass FSP by rectangular and spiral raster processes will be summarized. Microstructures produced by thermomechanical simulations will be compared to those produced during FSP. The distortion of microstructure constituents in thermomechanical simulations may be applied to estimation of SZ and TMAZ strain distributions in the initial FSP pass. Recrystallization in the primary α constituent is initiated at κ_{iv} (Fe$_3$Al) particles prior to their dissolution during straining.

Introduction

An allied process of friction stir welding[1] (FSW), friction stir processing (FSP) is emerging as a novel metal working technology that can provide localized modification and control of microstructures in near-surface layers of processed components[2-4]. In FSP, a cylindrical, wear-resistant tool consisting of a smaller diameter pin with a concentric, larger diameter shoulder is rotated and forced into a surface of the work piece. As the tool pin penetrates, a combination of frictional and adiabatic heating softens the material so that tool rotation induces a stirring action and flow of material about the pin. Severe but localized plastic deformation results in formation of a stir zone (SZ) while adjacent regions experience only moderate straining and comprise the thermomechanically affected zone (TMAZ). The tool shoulder prevents upward flow of material and serves to forge the SZ as it comes in contact with the work piece surface; large areas may then be processed by traversing the tool in a pattern on the work piece surface.

Traversing patterns in FSP have included repeated linear traverses that are offset from one another by a characteristic step-over distance. Raster and rectangular or circular spiral patterns have also been employed wherein successive traverses are, again, offset by a characteristic step over distance. In FSP, the step-over distance is an important process parameter in addition to the tool rpm and traversing rate. When the step-over distance is smaller than the minimum diameter of the tool pin successive passes will overlap one another and FSP inherently becomes a mult-stage process.

FSP has been employed to homogenize and refine microstructures in both cast and wrought metals including alloys of Al[5-10] and Mg[11,12] and higher melting alloys of Cu[13], Fe[14] and Ti[15]. Benefits of FSP in cast metals include the elimination of porosity and local conversion of the as-cast microstructure to a wrought condition in the absence of macroscopic shape change. Cast NiAl bronze alloys are used for components in a wide range of marine applications due to good combinations of corrosion resistance strength, toughness, friction coefficients and non-sparking behavior. Many cast components are large and resulting slow cooling rates contribute to coarse microstructures and reduced physical and mechanical properties. In such circumstances NiAl bronze materials may not be readily heat treatable and, so, FSP represents an alternative means of selectively strengthening the surfaces of such components. However, the full benefit of FSP will only be realized when tool design and process parameters are selected to achieve uniform deformation and homogeneous microstructures throughout the SZ. The present work will illustrate a comparison of SZ microstructures after an initial FSP pass and following multi-pass FSP. The SZ microstructure is inhomogeneous after the initial pass, with some locations exhibiting features suggestive of only modest deformation (von Mises strains ≤3.0). Thermomechanical simulations have been conducted and analyzed to assess the distribution of local strains in the vicinity of the tool during the initial pass.

Experimental

Two plates, each approximately 300mm × 152mm × 42mm in thickness, were sectioned from a large NiAl bronze casting. Composition data for the alloys of these plates are included in Table I. Details of the alloy constitution of as-cast material as well as the effect of FSP on microstructure have been given previously. The FSP was conducted using a tool fabricated in Densimet® and

Table I. Compositions (in wt. pct.) of the NiAl bronze alloy in this investigation

Element	Cu	Al	Ni	Fe	Mn	Si	Pb
Min-max	(min)79.0	8.5-9.5	4.0-5.0	3.5-4.5	0.8-1.5	0.10(max)	0.03(max)
Nominal	81	9	5	4	1	-	-
Alloy	81.2	9.39	4.29	3.67	1.20	0.05	<0.005

having a pin in the shape of a truncated cone 12.7mm in length, a base diameter of 15mm, and tip diameter of 6.3mm. The pin also had a step-spiral thread feature. All processing was conducted at 1200rpm and a traversing rate of 51mm min⁻¹.In order to examine the microstructure corresponding to an initial pass, a single linear FSP traverse was made. The traverse was approximately 250mm in length and centered along the long dimension of one of the plates. Multi-pass FSP was conducted on the other plate using a rectangular spiral pattern. Processing was initiated with a 200mm traverse centered along the long dimension of the plate. A spiral pattern was followed with the advancing side of the tool to the outside of the pattern and with a step-over distance of 4.5mm. Separately, cylindrical samples 11.1mm in diameter and 25.4mm in length were prepared and subjected to hot compression in a Gleeble® thermomechanical simulator. For the current investigation, a sample was heated to 900°C, held for 2mins, then compressed to a true strain of 0.58 at a displacement rate of 200mm s⁻¹ and, finally, cooled to ambient at ~7°Cs⁻¹.

Samples for optical microscopy were prepared either by sectioning on a transverse plane or on the plane of the plan view after FSP of the plates. A sample from the hot compression test was sectioned to reveal the plane containing the compression axis. Transverse views of the SZ in both of the FSP plates were sectioned from locations near the middle of the processed region and correspond to a steady state. The plan view sample was sectioned from the pin tool extraction site for the single traverse at the mid-depth of the stir zone. Optical microscopy sample preparation procedures have been given previously. Etched samples from the hot compression material were also examined using secondary imaging in a Topcon S510 scanning electron microscope operating at 15kV.

Results

The Initial FSP Pass

A montage of images illustrating the microstructure of a transverse section through the SZ after a single FSP pass is shown in Fig. 1. The tool pin profile is superimposed on the image in Fig. 1; the advancing side (tool rotation and travel directions are the same) is to the right, and retreating side (indicated by Rt in the image; tool rotation and travel directions are opposite) is to the left. The SZ microstructure is internally inhomogeneous and the surrounding microstructures reflect

Fig. 1. A montage of images showing the microstructure of a transverse section through the SZ after a single FSP pass; the tool pin profile is superimposed on the image.

asymmetric deformation from the advancing to the retreating side of the tool. The advancing side interface with the base metal is distinct while the interface is more diffuse under the tool and on the retreating side. The heating portion of the FSP thermomechanical cycle results in the eutectoid reversion reaction $\alpha + \kappa_{iii} \rightarrow \beta$ in locations where the temperature, T, is $\geq 800°C$, the eutectoid temperature in the quasi binary formed by the α and κ_{iii} phases in the Cu-Al-Ni ternary[16-20]. The dark-etching constituent in the TMAZ is the cooling transformation product of this β. This transformation product embrittles the heat affected zone of fusion welds as well as the TMAZ following FSP of as-cast NiAl bronze. The light-etching constituent is the undissolved primary α. The α and β have deformed in a compatible manner in many locations around the SZ periphery and TMAZ; the distortion of these constituents results in a layer appearance of the structure and suggests that they are pulled upward along the Rt side from under and then across the upper portion of the SZ as they experience moderate deformation. In contrast the central portion of the SZ appears to be much more highly refined and details of the structure are not discernable at this magnification. The layered arrangement of the primary α and β transformation product is apparent at higher magnifications in the micrographs of Fig. 2(a) and (b), which are from the location denoted Rt in Fig. 1. Recrystallization is apparent within the primary α in the microsgraph of Fig. 2(b), although this constituent remains separate and distinct in this location. In the central regions of the SZ, shown in Fig. 2(c), the layered structure is no longer apparent, suggesting that the $\alpha + \beta$ mixture became fragmented, likely due to much larger local strains during FSP.

Multi-Pass FSP

A montage of images from the SZ in the plate subjected to multi-pass FSP is shown in Fig. 3(a) wherein pin tool profiles are superimposed for two successive passes. These profiles show that all SZ locations experience deformation from at least two passes, and more than two passes in

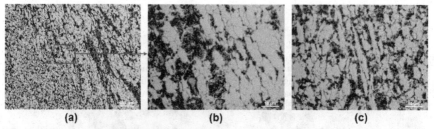

(a)   (b)   (c)

Fig. 2. Higher magnification micrographs of the region denoted Rt in Fig. 1 are shown in (a) and (b), illustrating the layered nature of the microstructure. A more homogeneous microstructure is apparent in the center of the SZ shown in (c).

the upper regions of the SZ. In general, the microstructure is more uniform throughout the SZ volume although some locations in the lower SZ exhibit flow lines indicative of moderately deformed material from beneath the SZ that is being pulled upward by the pin tool rotation. The micrograph in Fig. 3(b) is from the location indicated in Fig. 3(a) and shows a representative example of this SZ at a higher magnification. At this location the microstructure comprises 5 - 8μm primary α grains and discrete, irregularly shaped regions of β transformation products. Thus the additional deformation induced by overlapping passes increases strains and homogenizes and refines the structure throughout the SZ.

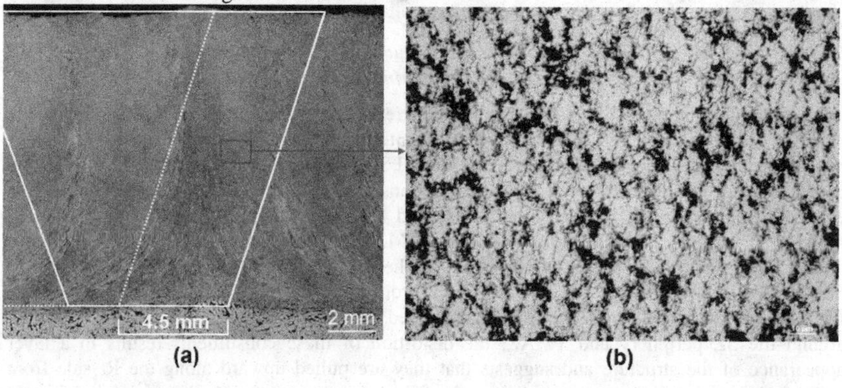

(a)   (b)

Fig. 3. A montage of images of the stir zone for multi-pass FSP is shown in (a) and overlap the the pin tool profiles is illustrated by the solid and dotted lines. A microstructure that is representative of the SZ is shown in (b) and consists of refined grains of α and dispersed β transformation products.

Estimation of Strains during FSP

Quantitative analysis of single-pass microstructures has been employed to estimate peak SZ temperatures and their distribution[16-20]. Values obtained from these estimates are 900 – 1000°C and direct thermocouple measures have subsequently confirmed the validity of these estimates. While numerous micrographs showing TMAZ and SZ grain distortion have appeared in the literature there have been no reported attempts to quantify the corresponding strains. An example illustrating strains and strain gradients in the vicinity of the tool extraction site for material subjected only to the initial FSP pass in this investigation is shown in Fig. 4. This montage of images was obtained from a plan view of a location at the mid-depth of the SZ. Ahead of the tool (the upper micrograph) the onset of straining is reflected in the shearing distortion of the primary and eutectoid constituents in a direction tangential to the surface of the tool and in the direction of tool rotation. In this location the distortion extends outward a distance of ~400μm, and a distinct gradient in strain is also evident over this distance. The grain distortion on either side of the tool extends outward much farther (~1mm) than it does ahead of the tool. The reflects the backward tilt of the tool by ~3° such that the influence of the tool shoulder is greater to the sides of the tool than it is ahead of the tool. At this depth in the SZ, the microstructure behind the tool corresponds to the very fine grain structure evident at the mid-depth of the SZ in Fig. 1.

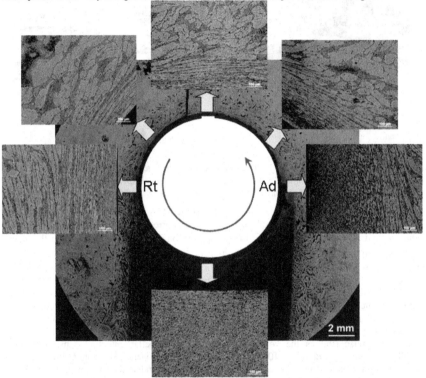

Fig. 4. A montage of images from the tool extraction site for a single FSP pass; these images were acquired for the plan view and from the mid-depth of the stir zone, and illustrate the varying gradient in strain around the tool pin.

The two constituent (primary α and β) apparently deform compatibly during FSP and retain their identity to moderate strains and during subsequent cooling. Fig. 5 shows two secondary images from scanning electron microscopy examination of a NiAl bronze sample heated to 900°C and then compressed to a nominal strain of 0.54 in a Gleeble thermomechanical simulator. At a higher magnification in Fig. 5(a), recrystallization accompanies deformation. This is apparent in the presence of annealing twins formed after the recrystallzation event as well as in the decoration of grain boundaries by precipitation of fine particles, which likely are the κ_{iv} (Fe_3Al)

phase, upon cooling after deformation. At a lower magnification, the flattening of the originally ellipsoidal primary α grains is evident. These grains are embedded in transformation products of the β, which also deformed during the compression at 900°C. As indicated in the schematic

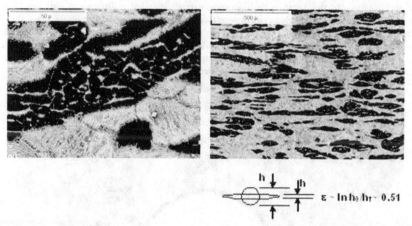

$$\varepsilon \sim \ln h_0/h_f \sim 0.51$$

Fig. 5. Secondary SEM images from a sample deformed to a strain of 0.54 by Gleeble thermomechanical simulation and showing recrystallization in the primary α in (a) and that the distortion of the constituent may employed to estimate strains in the TMAZ.

below the microsgraph, a value of strain may be estimated by analysis of the distortion of the primary α; the value obtained at this location ($\varepsilon \cong 0.51$) corresponds closely to the compressive strain applied to the sample ($\varepsilon_{app} \cong 0.54$). An application of such measurements to the strain field ahead of the pin tool is illustrated in Fig. 6. Direct temperature measurements from a thermocouples embedded in a location incorporated into the SZ of an NiAl bronze plate subjected to FSP are shown at the left. The portion of the overall plot of temperature as a function of time that corresponds to the extent of region ahead of the tool examined metallographically is shown in this plot. Quantitative microscopy measurements of the volume fraction of β transformation products as well as strain estimates from the distortion of the primary α are summarized in the plot to the right in Fig. 6. The volume fraction of β is ~0.2 in as cast NiAl bronze (this is the volume fraction of the lamellar eutectoid constituent after equilibrium cooling at rates of ~10^{-3} °Cs^{-1}). As the tool approaches, the volume fraction of β transformation products increases as β forms upon heating and is then retained at the higher cooling rates characteristic of FSP (10 – 100°Cs^{-1}). Simultaneously, the estimated strain values increases, reflecting the distortion of the primary α constituent. Clearly, the onset of straining has the effect of greatly accelerating the forward transformation α + κ_{iii} → β; this likely reflects the generation of excess vacancies during the severe deformation of the material[21-23]. Local strain estimates by this method suggests that von Mises equivalent strains up to about 3.0 are encountered at a distance of ~100μm ahead of the tool pin. In future work this approach will be taken to estimation of local strains at several locations around the tool in a attempt to characterize the entire deformation field during FSP.

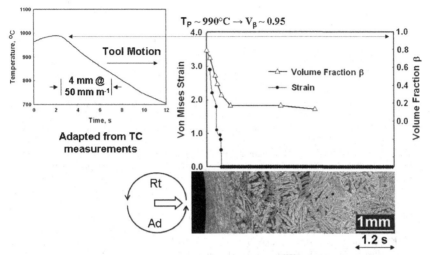

Fig. 6. Direct thermocouple measurements of SZ temperature[17] are plotted at the upper left; corresponding evaluations of volume fraction of β are plotted at the right[16,18,19] and compared to measurements of the strain from distortion of the primary α. Concurrent straining accelerates the reversion of the eutectoid so that the volume fraction of β approaches the equilibrium value (0.95) for the measured local peak temperature(990°C).

Conclusions

The following conclusions may be drawn from this work:

- Straining is non-uniform in the initial pass of FSP and SZ microstructures are inhomogeneous. Some locations may reflect local von Mises strains that are ≤1.0.
- With a small step-over distance such that the pin tool profile overlaps on successive pass a uniform and homogeneous SZ microstructure can be achieved.
- The strain distribution about the tool pin is complex and reflects effects of both the tool pin and tool shoulder.
- The distortion of the primary α may be employed as a means to estimate local strains in locations surrounding the tool.
- Estimates of the strain from shape change of microstructure constituents in FSP of NiAl bronze suggest that the local von Mises strain values are ≤3.0 at many SZ locations.
- Control of step-over distance in multi-pass FSP is required to impart uniform and homogeneous strains and obtain optimum SZ microstructures and mechanical properties.

References

1. W.M. Thomas, E.D. Nicholas, J.C. Needham, M.G. Murch, P. Templesmith and C.J. Daws: G.B. Patent Application No. 9125978.8, December, 1991; U.S. Patent No. 5460317, October, 1995
2. R.S. Mishra: *Advanced Materials and Processes*, 2003, vol. 161(10), pp. 43-46
3. R.S. Mishra, Z.Y. Ma and I. Charit: *Mater. Sci. Engng. A*, 2003, vol. A341, pp. 307-10

4. Z.Y. Ma, R.S. Mishra and M. W. Mahoney: in *Friction Stir Welding and Processing II*, K.V. Jata, M.W. Mahoney, R.S. Mishra, S.L. Semiatin and T. Lienert, eds., TMS, Warrendale, PA, 2003, pp. 221-30
5. R.S. Mishra, M.W. Mahoney, S.X. McFadden, N.A. Mara and A.K. Mukherjee: *Scripta Mater.*, 2000, vol. 42, pp. 163-68
6. R.S. Mishra and M.W. Mahoney: in *Superplasticity in Advanced Materials – Proceedings of ICSAM2000*, N. Chandra, ed., Materials Science Forum, Trans Tech Publications, Switzerland, 2001, vol. 357-59, pp. 507-14
7. Z.Y. Ma, R.S. Mishra and M.W. Mahoney: *Acta Mater.*, 2002, vol. 50, pp. 4419-30
8. I. Charit and R.S. Mishra: *Mater. Sci. Eng. A*, 2003, vol. A359, pp. 290-96
9. Z.Y. Ma, R.S. Mishra M.W. Mahoney and R. Grimes: *Mater. Sci. Engng. A*, 2003, vol. A351, pp. 148-53
10. Y.S. Sato, M. Urata and H. Kokawa: *Metal. Mater. Trans. A*, 2002, vol. 33A, pp. 625-635
11. S.H.C. Park, Y.S. Sato and H. Kokawa: *Scripta Mater.*, 2003, vol. 49, pp. 161-6
12. D. Zhang, M. Suzuki and K. Maruyama: *Scripta Mater.*, 2005, vol. 52, pp. 899-903
13. H.S. Park, T. Kimura, T. Murakami, Y. Nagano, K. Nakata and M. Ushio: *Mater. Sci. Eng. A*, 2004, vol. A371, pp. 160-169
14. Y.S. Sato, T.W. Nelson and C.J. Sterling: *Acta Mater.*, 2005, vol. 53, pp. 637-45
15. A.P. Reynolds, E. Hood and W. Tang: *Scripta Mater.*, 2005, vol. 52, pp. 491-4
16. K. Oh-ishi, A. P. Zhilyaev, R. Williams and T. R. McNelley: in *Friction Stir Welding and Processing III*, K.V. Jata, M.W. Mahoney, R.S. Mishra and T.J. Lienert, eds., TMS, Warrendale, PA, 2005, pp. 107-14
17. T.R. McNelley, K. Oh-ishi and A. P. Zhilyaev: in *Processing and Manufacturing of Advanced Materials – Proceedings of THERMEC'2006*, T. Chandra, K. Tsuzki, M. Militzer and C. Ravichandran, eds., Materials Science Forum, Trans Tech Publications, Switzerland, 2007, vols. 539 – 43, pp. 3745-50
18. K. Oh-ishi and T.R. McNelley: *Metall. Mater. Trans. A*, 2004, vol. 35A, pp. 2951-61
19. K. Oh-ishi and T.R. McNelley: *Metall. Mater. Trans. A*, 2005, vol. 36A, pp. 1575-85
20. K. Oh-ishi, A.P. Zhilyaev and T.R. McNelley: *Metall. Mater. Trans. A*, 2006, vol. 37A, pp. 2239-51
21. J.L. Robbins, O.C. Shepard and O.D. Sherby: *J. Iron Steel Inst.*, 1964, vol. 202, pp. 804-7
22. O.D. Sherby, B. Walser, C.M. Young and E.M. Cady: *Scripta Metall.*, 1975, vol. 9, pp. 569-74
23. B. Walser and O.D. Sherby: *Metall. Trans. A*, 1979, vol. 10A, pp. 1461-71

Friction Stir Welding and Processing V
Edited by: Rajiv S. Mishra, Murray W. Mahoney, and Thomas J. Lienert
TMS (The Minerals, Metals & Materials Society), 2009

MICROSTRUCTURAL EVOLUTION DURING FRICTION STIR WELDING OF NEAR-ALPHA TITANIUM

R. W. Fonda and K. E. Knipling

Naval Research Laboratory
Code 6356, 4555 Overlook Ave., SW, Washington, DC 20375, USA

Keywords: titanium, grain structure, texture

Abstract

The microstructural evolution occurring around the tool was investigated in friction stir welds of the near-α alloy, Ti-5111. Specifically, the purpose of this investigation was to determine how the material ahead of the tool was influenced by the rotating tool to produce the refined grain structure observed adjacent to the tool and in the tool wake. This involved characterizing the base plate microstructure to show the original β grain structure and its decomposition to form specific combinations of α lath orientations. The microstructure and texture of the final refined grain structure near the tool and in the deposited weld is also discussed.

Introduction

Friction stir welding (FSW) is a solid-state joining process that was developed in the early 1990s by TWI [1]. This technique, which uses a rotating, non-consumable tool to weld the workpieces together by "stirring" together the surrounding material, was initially developed for use on aluminum alloys because of the lower temperatures and stresses required to weld those alloys and the ready availability of tool materials to perform the welding. Most of the subsequent development and commercial applications of FSW have similarly been focused on aluminum alloys. However, there is also substantial interest in developing FSW for higher strength alloys such as titanium and steels. Such applications require tools that can retain their strength at much higher temperatures and may also require welding machines that can withstand the higher loads needed for some of these alloys. Although FSW of these high strength alloys has seen much less development than FSW of aluminum alloys, FSW of many high strength alloys in titanium and steel has been demonstrated and research on the welding of these alloys is increasing [e.g., 2–14].

There has been little analysis of the microstructures and textures that are produced during FSW of titanium and steel alloys, and even less on the evolution of those microstructures and textures. This is partly due to the predominant focus of this technique on aluminum alloys, but is also due in part to the complexity that is introduced into these analyses by the allotropic phase transformations that often occurs during welding. Most titanium and steel alloys have a different crystal structure at high temperature from what is present at room temperature, and friction stir welding often induces this phase transition to occur. This can lead to complex microstructural and crystallographic changes as the weld is cooled down from the welding temperature that can be difficult to deconvolve from the evolution that occurs during the welding process.

In this study, we intend to extend some of the more recent research on the evolution and development of grain structure and crystallographic texture in aluminum alloys [15–21] to the understanding of the comparable processes within the near-α titanium alloy, Ti 5111. This alloy was developed to exhibit a high toughness, good weldability, and good stress-corrosion cracking resistance, and is primarily considered for marine applications that require a superior toughness and corrosion resistance [22].

Experimental

The weld examined in this study is a bead-on-plate (no seam) weld in ½" (12.7 mm) thick 5111 titanium. The Ti plate was prepared at Timet, provided by Concurrent Technologies Corporation and then welded at the Edison Welding institute with a tungsten-based alloy tool at a welding speed of 2 inches per minute (0.85 mm/s) and 140 rpm, corresponding to a 360 µm tool advance per revolution. The tool used in this welding had a truncated conical geometry with a narrow shoulder and contained no threads, flats, or other features. The small size of the shoulder was selected to facilitate an even distribution of heating during the welding process because of the poor thermal conductivity of titanium.

The weld prepared using a stop-action technique wherein the tool was extracted from the plate immediately upon completion of the weld and the weld end was quenched with cold water. This process was intended to preserve the microstructure surrounding the welding tool as a representation of the actual microstructure present around the tool during the welding process. A plan-view cross section through the plate mid-thickness at the weld end, see Figure 1, was prepared by standard metallographic techniques, with final polishing accomplished using a solution of 20% hydrogen peroxide (30%) and 80% colloidal silica solution. The resultant surface was

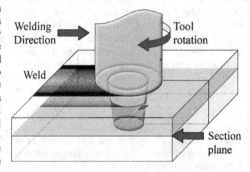

Figure 1. Schematic of the friction stir welding process showing the plan view sectioning plane used in this study and the location of Figure 2 on that sectioning plane.

analyzed in an FEI Nova 200 NanoLab dual column field-emission gun scanning electron microscope and focused ion beam (FEGSEM/FIB) operating at 18 kV and equipped with the HKL Channel 5 system.

Results and Discussion

Base Plate Microstructure

In order to study the microstructural evolution that occurs during friction stir welding, it is essential to understand the initial microstructure of the base plate. All microstructural evolution derives from this initial microstructure and needs to be interpreted relative to that starting point.

The microstructure of the Ti-5111 alloy consists of very large prior-β grains that can reach 8 mm in diameter (see Figure 2). These prior-β grains are subdivided into regions that contain different α lath orientations. Some of these regions are indicated by white rectangles in Figure 2. Careful examination of these regions reveals that they each contain three specific orientations of α laths that are characterized by a 60° separation between basal {0001}$_α$ planes and a coincidence of the close-packed ⟨1 1 -2 0⟩$_α$ directions. Furthermore, a particular α lath orientation is only observed in one of these regions in each prior-β grain.

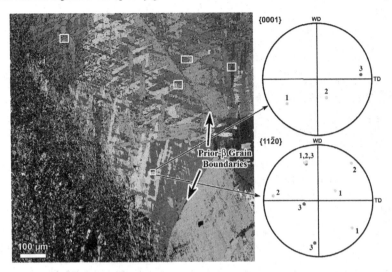

Figure 2. Plan view EBSD map of the initial grain structure ahead of the FSW tool and equal angle pole figures from the indicated region. Tool exit hole is towards the lower left, and tool rotation is clockwise.

The α laths within each prior-β grain are each related to the original β grain orientation by the Burgers orientation relationship (OR) [23], where close-packed planes {1 1 0}$_β$ ‖ (0 0 0 1)$_α$ and close-packed directions ⟨1 1 1⟩$_β$ ‖ ⟨1 1 -2 0⟩$_α$ are parallel. This is the OR most commonly observed between the α and β phases in titanium [24]. From a geometric standpoint, the Burgers OR is theoretically able to generate 12 different α variants from one parent β orientation. The three α variants observed in each region of Figure 2 are the ones that share a common ⟨1 1 -2 0⟩$_α$ direction, indicating that these are strain accommodation variants that relieve the transformation strain that each variant generates perpendicular to their common ⟨1 1 -2 0⟩$_α$ direction.

Initial Effects of Friction Stir Welding on the Microstructure

The microstructure evolves as it approaches the rotating tool and is exposed to the increasing thermal and deformation gradients generated by the tool. The microstructural evolution occurs across a narrow (~400 μm) transition region between the unaffected base plate and the fine, equiaxed α grains near the tool, as shown in Figure 3. The initial stages of this evolution appear as a gradual refinement of the α lath structure. During this lath refinement, new crystallographic variants of α are also introduced with different lath orientations than the original variants.

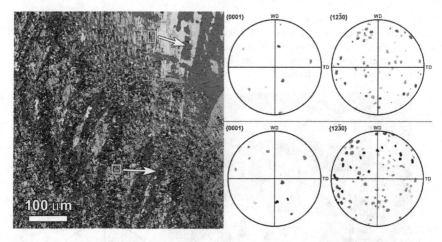

Figure 3. Plan view EBSD map of the microstructural evolution occurring through the transition region between the base plate (upper right) and the refined grain structure near the tool (lower left) and pole figures in equal angle projection showing the production of additional α variants through this transition region.

The pole figures in Figure 3 illustrate the crystallographic evolution that occurs along with this grain refinement. At the outer edge of the deformation, three additional primary orientations appear in the {0001}$_\alpha$ pole figure, and some minor orientations also begin to appear (see Figure 3, top pole figures). A gradual counter-clockwise rotation is apparent along with the α lath refinement as this material becomes influenced by the deformation field induced by the rotating tool. There is a discontinuity in the rotation at the outer edge of the banded region where the laths abruptly rotate approximately 10° and then retain the same orientation across the banded region (Figure 3, bottom pole figures). The same six crystallographic variants remain dominant throughout these grain rotation and refinement processes, with some additional variants appearing in minor amounts.

Other regions of the weld exhibit a more complete development of α variants that better illustrate the genesis of the observed α lath orientations. Pole figures from a region of highly refined α laths further towards the retreating side of the weld exit hole are shown in Figure 4. The symmetry of these pole figures illustrates a symmetry between the different α lath variants that demonstrates that these α lath orientations are all derived from the original β grain orientation. The six primary {0001} orientations are distributed with the angular separations of {110} poles in a bcc structure, indicating an adherence to the Burgers OR. A similar distribution of orientations can be discerned in Figure 3, although it is complicated by a greater orientation spread and the appearance of some additional variants. The Burgers OR also maintains a parallelism between the ⟨1 1 -2 0⟩$_\alpha$ direction and the ⟨1 1 1⟩$_\beta$ direction, which is reflected in the three-fold symmetry evident in the {1 2 -3 0}$_\alpha$ pole figure of Figure 4, but also present in Figure 3. The four-fold symmetry of the corresponding ⟨1 0 0⟩$_\beta$ directions is evident in the {1 0 -1 2}$_\alpha$ pole figure. Analysis of these pole figures reveals the presence of 12 predominant α variants. There are 6 orientation variants corresponding to an alignment of a (0 0 0 1)$_\alpha$ plane with one of the 6 bcc {1 1 0}$_\beta$ planes. Each of those plane-matching variants contains two rotational variants separated by about 10.5° to align a ⟨1 1 -2 0⟩$_\alpha$ close-packed direction with one of the two ⟨1 1 1⟩$_\beta$ orientations in that {1 1 0}$_\beta$ plane. Thus most of the new α lath variant generation, and likely

14

most of the intervariant transformation, appears to arise from a local transformation to the parent β grain structure that is enabled by the high temperatures, possibly aided by the strain, of the FSW process. The subsequent decomposition of that bcc parent structure produces α laths according to the Burgers OR. Since these transformations occur through a single β grain orientation that is the same as the orientation of the original β parent grain, it is likely that small remnants of retained β were preserved between the α laths that can serve as nuclei for the allotropic α to β transformation during welding. Otherwise, multiple β grains, each of which satisfies the Burgers OR with the α laths, would be likely to form.

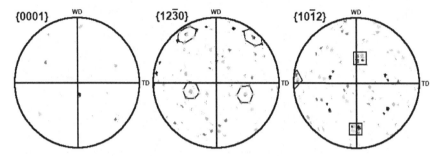

Figure 4. Pole figures in equal angle projection illustrating the cubic symmetry exhibited by the hcp α crystal variants after refinement of the lath structure and production of new α lath variants.

<u>Texture of refined α laths in the stir zone</u>

The predominant deformation during FSW, particularly in regions close to the tool, is expected to be simple shear, as confirmed in previous FSW studies of aluminum alloys [16, 18, 20, 21, 25, 26]. Analysis of the deformation texture that occurs around the tool during friction stir welding of this near-α titanium alloy, however, is complicated by the allotropic phase transformation. The temperatures achieved during friction stir welding of titanium alloys often exceed the transus temperature, transforming the material around the tool to the high-temperature β phase. Subsequent cooling after passage of the welding tool transforms virtually all of this material back to α, preventing a direct observation of the microstructures and textures that existed around the tool during welding. Instead, it is necessary to use the characteristics of the resultant α phase to infer the microstructure and texture that existed around the tool during welding indirectly.

The texture of this Ti-5111 friction stir weld was determined at a number of locations around the tool. Two specific locations — on the retreating side of the weld and in the deposited weld nugget — are shown in Figure 5 and typify the results obtained from all the regions examined. The left-hand pole figures, Figure 5(a), have the same relative orientation with respect to the specimen (i.e., plan view with the welding direction upward). There is a strong texture in both sets of pole figures. The basal $\{0\ 0\ 0\ 1\}_\alpha$ planes are tilted ~35° from ND in the direction away from the welding tool and the $\langle 1\ 1\ -2\ 0 \rangle_\alpha$ directions are aligned tangent to the tool, which is parallel to the presumed direction of maximum shear.

The pole figures can be rotated into the shear deformation frame of reference, with a horizontal shear direction (SD) and a vertical shear plane normal (SPN), as shown in Figure 5(b). The geometry of the tool can be used to determine the orientation of the SD and SPN adjacent to the tool at those locations. The SD can be assumed from the tool tangent. Aligning the pole figures with this presumed shear direction aligns the predominant $\langle 1\ 1\ -2\ 0 \rangle_\alpha$ direction with the horizontal axis of the pole figure. The shear plane is parallel to the surface of the truncated conical tool, which has a taper angle of ~27° (see tool schematic in Figure 5). Rotating the pole figure to account for this taper aligns a $\{0\ 0\ 0\ 1\}_\alpha$ orientation near the ND. Thus, the close-packed $\{0\ 0\ 0\ 1\}_\alpha$ basal plane appears to be perpendicular to the shear plane and the close-packed $<1\ 1\ -2\ 0>_\alpha$ directions are parallel to the SD. The pole figures shown in Figure 5(b) reflect this local orientation of the texture.

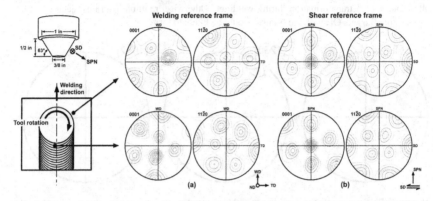

Figure 5. Schematics of the FSW tool geometry and the end of the weld, indicating the locations adjacent to the exit hole from which the $\{0\ 0\ 0\ 1\}_\alpha$ and $\{1\ 1\ -2\ 0\}_\alpha$ pole figures shown in parts (a) and (b) (in stereographic projection) were obtained. In (a), the pole figures have the same orientation as in the weld schematic. In (b), the pole figures have been rotated to align the presumed shear plane normal (SPN) with the vertical direction and the shear direction (SD) with the horizontal direction. The contours shown in the pole figures are multiples of 0.5 times random density.

In bcc metals, simple shear deformation produces partial fibers belonging to $\{hkl\}\langle111\rangle$ and $\{110\}\langle uvw\rangle$, as established by modeling and experimental studies of deformation textures in torsion tests [e.g., 27–28]. The texture components of these ideal simple shear orientations are shown in the $\{110\}_{bcc}$ pole figure, Figure 6(a). Figure 6(b) displays these $\{110\}_{bcc}$ ideal shear orientations from Figure 6(a) superimposed on the $\{0001\}_\alpha$ pole figure from the deposited weld (bottom of Figure 5(b)). This Figure reveals an extremely good agreement between the experimental $\{0001\}_\alpha$ texture and the ideal $\{110\}_{bcc}$ shear texture, which is expected since these close-packed planes are parallel according the Burgers OR. Figure 6 demonstrates that the observed $\{0001\}_\alpha$ texture is inherited directly from the shear texture of the high-temperature β phase, and closely matches the $D\{112\}\langle111\rangle$ texture component.

There have been two previous FSW studies on the texture produced in titanium alloys that should be compared to these results. Reynolds et al. [8] studied friction stir welds of a β titanium alloy and observed excellent agreement between the observed shear texture in the weld nugget (after a rotation of the pole figures) and the shear textures of bcc tantalum reported by Rollet and Wright [29]. Mironov et al. [30] measured the stir zone texture developed in friction stir welds of an α-β titanium alloy, Ti-6Al-4V, and found that the retained β phase exhibited a $J\{\bar{1}\ \bar{1}\ 0\}\langle\bar{1}\ 1\ \bar{2}\rangle$ simple shear texture that presumably developed from $\{110\}\langle111\rangle$ slip. The authors, however, admitted that the small amount of retained β in that sample limited the statistics supporting this result.

Mironov et al. [31] also measured the texture developed from friction stir processing of pure iron and consistently observed, from several locations within the stir zone, a $D_2(11\bar{2})[111]$ simple shear texture. Moreover, this $D_2(11\bar{2})[111]$ texture is the one most commonly observed during equal channel angular extrusion of bcc iron [32,33]. The similar texture produced during the current examination of friction stir welding of Ti-5111, friction stir processing of pure iron, and the equal channel angular extrusion of bcc iron further supports the conclusion that the texture observed near the FSW tool and in the deposited weld is directly derived from a simple shear texture of the high-temperature β phase.

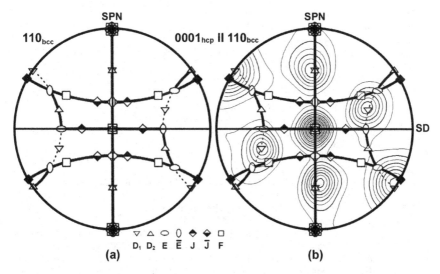

Figure 6. (a) Schematic {110} pole figure in stereographic projection showing the main simple texture component orientations and fibers associated with simple shear deformation of bcc metals (after Li et al. [31]). (b) Superposition of the simple shear texture components on the {0001} pole figure of the deposited weld (rotated into the shear reference frame) from Figure 5(b).

Conclusion

The microstructural evolution occurring during friction stir welding of a near-α titanium alloy, Ti-5111, was discussed. The microstructure transitions over ~400 μm from the unaffected base plate, which consists of course α laths within prior-β grains, to a fine equiaxed α microstructure near the tool. In the unaffected baseplate, the α laths are related to the original β grain orientation by the Burgers OR, which aligns close-packed planes $\{1\,1\,0\}_\beta \parallel (0\,0\,0\,1)_\alpha$ and close-packed directions $\langle 1\,1\,1 \rangle_\beta \parallel \langle 1\,1\,-2\,0 \rangle_\alpha$. Prior-β grains are typically subdivided into regions that each contain three α variants that share a common $\langle 1\,1\,-2\,0 \rangle_\alpha$ direction, indicating that these are strain accommodation α variants. Closer to the tool, there is as a gradual refinement of the α laths, which also rotate slightly in response to the shear deformation generated by the tool. Additional α variants are also generated through this region that are derived from the original β grain orientation. Adjacent to the FSW tool, the high temperatures and extensive shear deformation results in refined grains of α exhibiting a $\{0001\}_\alpha$ basal texture closely matching the

17

$D_2(11\bar{2})[111]$ simple shear texture observed in high-temperature torsion and equal channel angular extrusion (ECAE) of bcc iron. This indicates that the observed α texture in the stir zone is inherited directly from a simple shear texture of the high-temperature β phase.

References

1. W.M. Thomas, E.D. Nicholas, J.C. Needham, M.G. Murch, P. Templesmith, and C.J. Dawes, "Friction Stir Butt Welding", Int. Patent App. PCT/GB92/02203 and GB Patent App. 9125978.8, Dec. 1991. U.S. Patent No. 5,460,317, Oct. 1995.
2. W.M. Thomas, P.L. Threadgill, E.D. Nicholas, "Feasibility of Friction Stir Welding Steel", *Sci Tech Weld Join*, **4**, 365-372 (1999).
3. A.P. Reynolds, W. Tang, M. Posada, and J. DeLoach, "Friction Stir Welding of DH36 Steel", *Sci Tech Weld Join*, **8**, 455-460 (2003).
4. P.J. Konkol, J.A. Mathers, R. Johnson, and J.R. Pickens, "Friction Stir Welding of HSLA-65 Steel for Shipbuilding", *J Ship Production*, **19**, 159-164 (2003).
5. M. Posada, J. DeLoach, A. P. Reynolds, R. W. Fonda, and J. P. Halpin, "Evaluation of Friction Stir Welded HSLA-65", *Proc 4th Int Symp on FSW*, TWI Ltd., S10A-P3 (2003).
6. A.J. Ramirez and M.C. Juhas, "Microstructural Evolution in Ti-6Al-4V Friction Stir Welds", *Mat Sci Forum*, **426-4**, 2999-3004 (2003).
7. T.W. Nelson, J.-Q. Su, and R.J. Steel, "Friction Stir Welding of Ferritic Steels", *Proc 14th Int Offshore and Polar Eng Conf*, 50-54 (2004).
8. A.P. Reynolds, E. Hood, and W. Tang, "Texture in Friction Stir Welds of Timetal 21S", *Scripta Mater*, **52**, 491-494 (2005).
9. Y.S. Sato, T.W. Nelson, and C.J. Sterling, "Recrystallization in Type 304L Stainless Steel During Friction Stirring", *Acta Mater*, **53**, 637-645 (2005).
10. W.B. Lee, C.Y. Lee, W.S. Chang, Y.M. Yeon, S.B. Jung, "Microstructural Investigation of Friction Stir Welded Pure Titanium", *Mat Sci Lett*, **59**, 3315-3318 (2005).
11. S.H.C. Park, Y.S. Sato, H. Kokawa, K. Okamoto, S. Hirano, M. Inagaki, "Microstructural Characterisation of Stir Zone Containing Residual Ferrite in Friction Stir Welded 304 Austenitic Stainless Steel ", *Sci. Technol. Weld. Joining*, **10**, 550–56 (2005)
12. A.L. Pilchak, M.C. Juhas, and J.C. Williams, "Microstructural Changes due to Friction Stir Processing of Investment-cast Ti-6Al-4V", *Met Mat Trans A*, **38A**, 401-408 (2007).
13. S. Mironov, Y. Zhang, Y.S. Sato, and H. Kokawa, "Development of Grain Structure in β-Phase Field During Friction Stir Welding of Ti-6Al-4V Alloy", *Scripta Mater*, **59**, 27-30 (2008).
14. S. Mironov, Y. Zhang, Y.S. Sato, and H. Kokawa, "Crystallography of Transformed β Microstructure in Friction Stir Welded Ti-6Al-4V Alloy", *Scripta Mater*, **59**, 511-514 (2008).
15. J.-Q. Su, T.W. Nelson, R. Mishra, and M. Mahoney, "Microstructural Investigation of Friction Stir Welded 7050-T651 Aluminium", *Acta Mater*, **51**, 713-729 (2003).
16. R.W. Fonda, J.F. Bingert, and K.J. Colligan, "Development of Grain Structure During Friction Stir Welding", *Scripta Mat*, **51**, 243-248 (2004).
17. J.A. Schneider and A.C. Nunes, Jr, "Characterization of Plastic Flow and Resulting Microtextures in a Friction Stir Weld", *Met Mat Trans B*, **35B**, 777-783 (2004).
18. P.B. Prangnell and C.P. Heason, "Grain Structure Formation During Friction Stir Welding Observed by the 'Stop Action Technique'", *Acta Mater*, **53**, 3179-3192 (2005).
19. R.W. Fonda and J.F. Bingert, "Precipitation and Grain Refinement in a 2195 Al Friction Stir Weld", *Met Mat Trans A*, **37A**, 3593-3604 (2006).

20. R.W. Fonda, J.A. Wert, A.P. Reynolds, and W.Tang, "Friction Stir Welding of Single Crystal Aluminum", *Sci Tech Weld Join*, **12**, 304-310 (2007).
21. R. W. Fonda, K. E. Knipling, and J. F. Bingert, "Microstructural Evolution Ahead of the Tool in Aluminum Friction Stir Welds", *Scripta Mater*, **58**, 343-348 (2007).
22. Titanium Metal Corporation: TIMETAL 5111, Report # TMC-0170, Titanium Metal Corporation, Denver, Co, (2000).
23. W. G. Burgers, "On the Process of Transition of the cubic-body-centered Modification into the hexagonal-close-packed Modification of Zirconium," *Physica*, **1**, 561–586 (1934).
24. G. Lutjering and J. C. Williams, Titanium, Springer-Verlag, pp. 27–31 (2003).
25. Y.S. Sato, H. Kokawa, K Ikeda, M. Enomoto, S. Jogan, and T. Hashimoto, "Microtexture in the Friction-Stir Weld of an Aluminum Alloy", *Met Mat Trans A*, **32A**, 941-948 (2001).
26. D.P. Field, T.W. Nelson, Y. Hovanski, and K.V. Jata, "Heterogeneity of Crystallographic Texture in Friction Stir Welds of Aluminum", *Met Mat Trans A*, **32A**, 2869-2877 (2001).
27. F. Montheillet, M. Cohen, and J. J. Jonas, "Axial Stresses and Texture Development During the Torsion Testing of Al, Cu and α-Fe," *Acta Metall*, **32**, 2077–2089 (1984).
28. J. Baczynski and J. Jonas, "Texture development during the torsion testing of alpha-iron and two IF steels," *Acta Mater*, 44, 4273–4288 (1996).
29. A. D. Rollett and S. I. Wright, in Texture and Anisotropy, edited by U. F. Kocks, S. N. Tome, and H. R. Wenk (Cambridge University Press, 1998), p. 202.
30. S. Mironov, et al., "Development of Grain Structure in β-Phase Field during Friction Stir Welding of Ti-6Al-4V Alloy," *Scripta Mater*, **59**, 27–30 (2008).
31. S. Mironov, Y. S. Sato, and H. Kokawa, "Microstructural Evolution during Friction Stir-Processing of Pure Iron," *Acta Mater*, **56**, 2602–2614 (2008).
32. S. Li, I. J. Beyerlein, and M. A. M. Bourke, "Texture Formation during Equal Channel Angular Extrusion of fcc and bcc Materials: Comparison with Simple Shear," *Mater Sci Eng A*, **394**, 66–77 (2005).
33. A. A. Gazder, et al., "Microstructure and Texture Evolution of bcc and fcc Metals Subjected to Equal Channel Angular Extrusion," *Mater Sci Eng A*, **415**, 126–139 (2006).

19

Friction Stir Welding and Processing V
Edited by: Rajiv S. Mishra, Murray W. Mahoney, and Thomas J. Lienert
TMS (The Minerals, Metals & Materials Society), 2009

PHYSICAL SIMULATION OF FRICTION STIR PROCESSED TI-5111

M. Rubal[1], J. Lippold[1], M. Juhas[2]

[1]Welding Engineering Program, The Ohio State University;
1248 Arthur E. Adams Dr.; Columbus, OH 43221, USA
[2]Dept. of Material Science and Engineering, The Ohio State University;
2041 College Rd.; Columbus, OH 43210, USA

Keywords: Friction Stir Processing, Titanium, Hot Torsion

Abstract

Friction stir processing (FSP) of Ti-5111 was performed above and below the beta transus temperature allowing for investigation of the microstructural evolution in both conditions. Each processed panel was instrumented with thermocouples to record the thermal histories in the stir zone and adjacent heat-affected zone. Single sensor differential thermal analysis (SS-DTA) was used to determine the beta transus temperature during processing. The FSP microstructures were characterized using optical and scanning electron microscopy, while the microtextures of the FSP regions were compared using electron backscatter diffraction (EBSD). FSP produced extreme grain refinement in both processing conditions – reducing the 200-500 µm base material grains to 1-20 µm. The microstructures observed in the FSP panels were simulated using a Gleeble® 3800. The strain and strain rate data may be used to verify FSP modeling programs of titanium to reduce the parameter selection phases of future friction stir projects.

Introduction

Ti-5111 (5Al-1Sn-1Zr-1V-.8Mo) is a near-α titanium alloy with an equilibrium beta transus of approximately 980 °C [1]. The US Navy developed this alloy as a lower cost alternative for Ti-6Al-4V for structural applications in ships and submarines [2]. The desirable properties of Ti-5111 include intermediate strength in combination with excellent toughness, corrosion and room temperature creep resistance [2,3].

Friction stir processing (FSP), a modification of friction stir welding (FSW), can be used to reduce the grain size of titanium alloys. This refined structure improves the mechanical properties, including increasing the yield stress and resistance to fatigue crack initiation [4]. A previous FSW study was conducted on Ti-5111 by Fonda et al. at the Naval Research Laboratory. A microstructural analysis revealed the welds contained only three zones: the stir zone (SZ), base material (BM) and a narrow region of deformation between the two [5]. This deformed region, which experiences only partial recrystallization, is referred to as the transition zone (TZ) [6]. The stir zone microstructure is significantly affected by the phase field in which processing occurs, specifically if deformation is introduced above or below the beta transus temperature. This was explored in Ti-6Al-4V FSP by Pilchak et al. [6]. A goal of this study is to determine the effect of processing above and below the beta transus temperature on the microstructures observed in Ti-5111 FSP panels.

Single sensor differential thermal analysis (SS-DTA), developed at the Ohio State University, is one of several techniques capable of evaluating phase transformations during welding and processing. SS-DTA is unique from classical differential thermal analysis (DTA) in that it does not require a reference sample; instead, a software-calculated reference curve is fit to the *in situ* collected thermal data. Deviations from the reference curve indicate the start and finish temperatures of a phase transformation [7,8]. Both classical DTA techniques and dilatometry have been used to confirm the accuracy of SS-DTA [9].

Electron backscatter diffraction (EBSD) is an important tool in determining the degree of microtexture present in friction stir welds. Several texture studies have been conducted on titanium FSW and FSP panels, including the near-α Ti-5111, β alloy TIMET 21S and the α-β alloy Ti-6Al-4V. Fonda et al. observed shear deformation and deformation twinning of the α phase in Ti-5111 FSW near the edges of the weld, while a strong texture associated with the Burgers orientation relationship occurred near the tool [10]. A shear texture was detected in TIMET 21S FSW by Reynolds et al. which could be rotated to correspond to a previously published BCC torsion texture. [11]. Finally, a study on Ti-6Al-4V FSP by Pilchak et al. revealed a moderate intensity (~5x) transformation texture in the α phase governed by the Burgers relationship when processed above the beta transus and a shear texture when processed sub-transus [12].

The Gleeble® is a thermo-mechanical simulator, first developed by Nippes et al. and currently produced by DSI, Inc. [13]. The 3800 model has an optional torsion mobile conversion unit (MCU) capable of imposing high strains and strain rates on a sample while maintaining a steady temperature gradient. Previous studies on steels have proved that the microstructures observed within the FSW regions can be simulated using Gleeble® torsion tests [14,15]. Determining the strain, strain rate and temperature combinations capable of producing a particular microstructure could be used to validate titanium FSW and FSP models.

Experimental Procedures

Beta-annealed plates of Ti-5111 were friction stir processed under two conditions: above and below the beta transus temperature. The processing was performed at the Edison Welding Institute (EWI) using an Army-supplied, position-controlled GTC AccuStir machine. Variable penetration tools (VPT) of a refractory-based material with different dimensions were used to achieve the necessary heat inputs. Table I contains the processing parameters for the above and below transus conditions. Argon shielding was used during processing to avoid oxidation of the panels.

Table I. FSP Parameters for Above and Below Transus Conditions

FSP Condition	Tool Rotation	Travel Speed	
	rev/min	in/min	mm/sec
Above Transus	450	3	1.3
Below Transus	300	7	3.0

Each panel contained eight thermocouples imbedded at various depths along the centerline. These thermocouples recorded the thermal histories experienced in the SZ and heat affected zone (HAZ) at a sampling rate of 1000 sec^{-1}. SS-DTA analysis was performed on the *in situ* collected data to determine the on-heating and on-cooling beta transus temperatures.

22

The processed zones were cut into transverse sections, mounted in conductive Bakelite and ground through 800 grit silicon carbide paper. A colloidal silica vibratory polisher was used to achieve the final .02 μm finish. Kroll's reagent (2 ml HF – 10 ml HNO_3 – 88 ml H_2O) was selected to etch the optical and scanning electron microscopy samples. The SZ and TZ for both conditions were analyzed using an FEI Quanta scanning electron microscope (SEM). Electron backscatter diffraction (EBSD) was performed on unetched samples to compare the microtextures.

Hot torsion tests were performed on the Ti-5111 base material using a Gleeble® 3800 thermo-mechanical simulator equipped with a torsion system mobile conversion unit (MCU). The temperature, strain and strain rate were varied to simulate the microstructures observed in the SZ and TZ of the processed panels.

Results and Discussion

Single Sensor Differential Thermal Analysis

The *in situ* collected thermal histories were analyzed using SS-DTA. The software fit a reference curve to the data using a 3-term 2nd order polynomial on-heating and a 5-term Rosenthal equation on-cooling. A deviation between the reference curve and thermal history is caused by the heat of reaction from the α-β phase transformation. The maximum and minimum deviation (δT) indicates the start and finish temperatures of the transformation.

The sub-transus processed panels did not exhibit any thermal effects associated with the α-β phase transformation. The data collected during above-transus processing, however, contained a positive thermal effect on-heating, which could correspond to the beta transus or the superposition of the transus and dynamic recrystallization. This effect occurred while the tool was directly over the thermocouple, though, and the signal noise prevented accurate analysis. Two cooling curves collected during above-transus processing produced smooth thermal histories that exceeded the beta transus. Figure 1 shows the analyzed cooling curve for a thermocouple located within the stir zone.

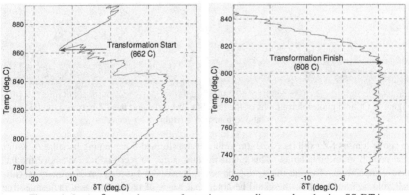

Figure 1. *In situ* β-to-α phase transformation on-cooling, analyzed using SS-DTA

The transformation start temperature in Figure 1 corresponds to the non-equilibrium beta transus. The on-cooling transformation start temperature during processing was determined to be 862 °C at one thermocouple location within the SZ and 845 °C for the other analyzed thermocouple. These are depressed by more than 100 °C compared to the published equilibrium transus. Thus, the effect of non-equilibrium heating and extreme deformation experienced during FSP can significantly lower the $\beta \rightarrow \alpha$ transformation. The difference in the two analyzed transformation temperatures can be attributed to the varying amount of strain experienced by each thermocouple and the effect of the plasticized material sticking or slipping past the tool.

SS-DTA analysis of a furnace test on Ti-5111 base material heated at 5 °C/min and slow cooled produced a clear indication of the on-heating transformation at 1006 °C and the on-cooling $\beta \rightarrow \alpha$ transformation at 960 °C. Both are similar to the published beta transus of approximately 980 °C. The super-heating and under-cooling associated with the non-equilibrium α-β transformation was also observed by Elmer et al. during welding of Ti-6Al-4V [16].

Optical and Scanning Electron Microscopy

Three regions were observed within the processed panels: base material (BM), transition zone (TZ) and stir zone (SZ) which confirms the findings of the FSW study performed by Fonda et al. on Ti-5111 [5].

The β-annealed BM exhibits a fully lamellar microstructure with a prior-β grain size of 200 to 500 μm. The small volume fraction of β is present as ribs between the α laths. The TZ is characterized by distortion of the α lamellae and the beginning of recrystallization near the SZ. Figure 2 contains optical photomicrographs of the BM and TZ in the panel processed above the beta transus.

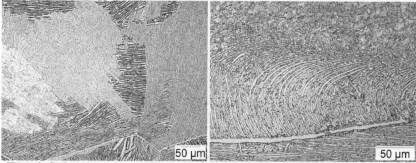

Figure 2. Optical photomicrographs of the lamellar BM (left) and the TZ between the BM and above-transus SZ (right)

The above-transus SZ exhibits 10-20 μm grains of transformed β containing lamellar α, with α at the prior-β grain boundaries. This microstructure is due to recrystallization above the beta transus temperature. The sub-transus SZ, however, consists of 1 μm equiaxed primary α with a small volume fraction of β. It should be noted that bands of refractory-based inclusions from tool wear were observed within the sub-transus SZ. Backscattered electron SEM photomicrographs of the SZ for both processing conditions are in Figure 3.

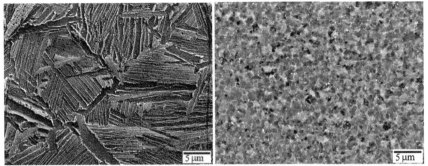

Figure 3. Backscattered electron SEM photomicrographs of the lamellar α grains in the above-transus SZ (left) and equiaxed α grains in the sub-transus SZ (right)

Electron Backscatter Diffraction

The microtextures associated with the SZ for both processing conditions were determined using EBSD. All scans were performed on transverse sections of the FSP panels on the retreating side of the SZ. The TD direction in the pole figure corresponds to the advancing side and RD is the tool travel direction. Figure 4 contains the α and β pole figures for the above and below-transus SZ, displayed as equal area projections.

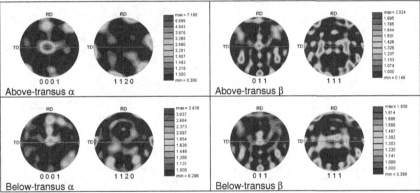

Figure 4. Equal area projection pole figures describing the retreating-side SZ microtexture for the above- and below-transus FSP panels

The α pole figure for the above-transus SZ shows a transformation texture with one main intensity on the basal plane. The maximum intensity associated with this pole figure is more than double the others. This is due to the limited orientations of α transforming from β, as dictated by the Burgers orientation relationship. The β pole figure for the above-transus processing condition reflects a deformation texture. Both pole figures for the below-transus SZ exhibit deformation textures as no transformation from β-to-α occurred. The similarity between the β pole figures of both conditions should be noted.

Gleeble® Torsion Simulation

A Gleeble® 3800 equipped with a torsion MCU was used to simulate the microstructure of the above-transus SZ. Thermal histories obtained during processing were input into the test program and the strain and strain rate were varied. The torsion samples were quenched with argon both on the sample surface and through the hollow interior to achieve the cooling rates experienced during processing.

The below-transus SZ cannot be simulated using Gleeble® torsion tests due to the high flow stress of the material at the processing temperature. The simulation samples sheared at the center prior to recrystallization.

In the samples used to simulate the above-transus SZ, strain localization was observed at testing temperatures below 1000 °C, which prevented the determination of a strain gradient as in the studies performed by Norton and Sinfield [14,15]. However, this localization allowed for full recrystallization of the sample microstructure. Figure 5 shows backscattered electron SEM photomicrographs of a torsion sample and the SZ of the above-transus FSP panel. The torsion sample was tested at a peak temperature of 940 °C and experienced one full revolution at a speed of 200 rev/min.

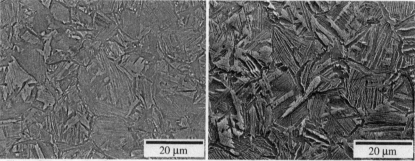

Figure 5. Backscattered electron SEM photomicrographs of a torsion simulation sample (left) and the SZ of the above-transus FSP panel (right)

Both the torsion sample and the above-transus SZ of the FSP panel contain lamellar α grains with α at the prior β grain boundaries. However, the α lamellae and grain boundary α are coarser in the SZ microstructure. This is due to detachment of the control thermocouple during testing and the torsion sample cooling faster than the FSP panel. The grain size for both is approximately 10-20 μm.

Once a method is developed for determining the strain experienced in the sample, the data collected from the simulations can be used to verify a model and reduce the parameter selection phase for titanium FSW and FSP.

Conclusions

After analysis of the Ti-5111 FSP panels and the initial Gleeble® 3800 simulations, the following conclusions were made.

1. The on-cooling β→α transformation start temperature during processing was determined using SS-DTA. The non-equilibrium and highly-strained processing conditions depressed the transformation by over 100 °C, compared to the equilibrium transus.

2. The FSP panels contained only three microstructural regions: the base material, transition zone and stir zone. The above-transus stir zone was characterized by 10-20 μm lamellar α grains and the below-transus stir zone exhibited 1 μm equiaxed α grains.

3. Processing above the beta transus caused a transformed-α and deformed-β microtexture. The intensity of the α microtexture was higher due to the Burgers orientation relationship. The sub-transus stir zone exhibited deformation textures for both the α and β phases.

4. Initial Gleeble® 3800 torsion simulations were able to match the microstructure of the above-transus stir zone. However, a method for determining the strain experienced in the sample must be established.

Acknowledgements

The authors would like to thank the Office of Naval Research, Cognizant Program Officer Johnnie Deloach, for support of this project. Our gratitude is also expressed to Brian Thompson and Seth Shira of the Edison Welding Institute for performing the friction stir processing. Special appreciation goes to Prof. Jim. Williams and Adam Pilchak of the Department of Materials Science and Engineering at the Ohio State University for helpful discussions on microtexture and titanium metallurgy.

References

1. E. J. Czyryca et al., "Titanium Alloy Ti-5111 for Naval Applications" (Paper presented at the RINA International Conference - Advanced Marine Materials: Technology and Applications, London, 9-10 October 2003), 41-49.

2. D. Baxter et al., "Joining of Titanium Alloy 5111" (Paper presented at the 10th World Conference on Titanium, Hamburg, 13-18 July 2003), 651-658.

3. J. Been et al., "Using Ti-5111 for Marine Fastener Applications," *JOM*, 51 (6) (1999), 21-24.

4. Gerd Lütjering and James C. Williams, *Titanium* (Berlin: Springer, 2003), 191.

5. R. W. Fonda et al. "Microstructural Evolution in Ti 5-1-1-1 Friction Stir Welds," *Friction Stir Welding and Processing IV*, ed. R. S. Mishra et al. (Warrendale, PA: TMS, 2007), 295-302.

6. A. L. Pilchak et al., "Microstructural Changes Due to Friction Stir Processing of Investment-Cast Ti-6Al-4V," *Met. and Mat. Trans. A*, 38 (2) (2007), 401- 408.

7. B. T. Alexandrov et al., "Methodology for In-situ Investigation of Phase Transformations in Welded Joints" (IIW Doc. IX-2114-04, International Institute of Welding, 2004).

8. B. T. Alexandrov et al., "A New Methodology for Studying Phase Transformations in High Strength Steel Weld Metal" (Paper presented at the 7th International Trends in Welding Research Conference, Pine Mountain, Georgia, 16-20 May 2005).

9. B. T. Alexandrov et al., "Single Sensor Differential Thermal Analysis of Phase Transformations and Structural Changes during Welding and Postweld Heat Treatment," *Welding In the World*, 51 (11/12) (2007), 48-59.

10. R. W. Fonda et al., "Microstructure of Ti 5111 Friction Stir Welds" (Presentation at the 136th TMS Annual Meeting, Orlando, Florida, 25 February – 1 March 2007).

11. A. P. Reynolds et al., "Texture in Friction Stir Welds of Timetal 21S," *Scripta Materialia*, 52 (6) (2005), 491-494.

12. A. L. Pilchak et al., "A Comparison of Friction Stir Processing of Investment Cast and Mill-annealed Ti-6Al-4V," *Welding in the World*, 52 (9/10) (2008), Research Supplement.

13. E. F. Nippes et al., "An Investigation of the Hot Ductility of High Temperature Alloys," *Welding Journal*, 4 (1955), 183s-196s.

14. S. J. Norton, "Ferrous Friction Stir Weld Physical Simulation" (Ph.D. Thesis, The Ohio State University, 2006).

15. M. F. Sinfield et al., "Physical Simulation of Friction Stir Weld Microstructure of a High-Strength, Low Alloy Steel (HSLA-65)" (Paper presented at the 7th International Symposium on Friction Stir Welding, Awaji Island, Japan, 20-22 May 2008).

16. J. W. Elmer et al., "Phase Transformation Dynamics During Welding of Ti-6Al-4V," *Journal of Applied Physics*, 95 (12) (2004), 8327-8339.

Friction Stir Welding and Processing V
Edited by: Rajiv S. Mishra, Murray W. Mahoney, and Thomas J. Lienert
TMS (The Minerals, Metals & Materials Society), 2009

FATIGUE CRACK GROWTH IN FRICTION STIR WELDED Ti-5111

P.S. Pao, R.W. Fonda, H.N. Jones, C.R. Feng, and D.W. Moon

Naval Research Laboratory, Washington, DC 20375, USA

Keywords: Friction stir welding, Fatigue crack growth, Microstructure, Titanium, Residual stresses.

Abstract

The effects of weld microstructure and weld speed on the fatigue crack growth kinetics of friction stir welded Ti-5111 were investigated. The FSW welds in Ti-5111 consist of very fine recrystallized grains, in contrast to coarse basketweave grains in the base plate. Fatigue crack growth rates are significantly lower and fatigue crack growth thresholds are significantly higher through the weld than those in the base plate. As the weld speed increases, the fatigue crack growth rates are progressively higher and fatigue crack growth thresholds lower through the weld. However, after stress-relief annealing, such differences in fatigue crack growth kinetics among different weld speeds no longer exist. Fatigue crack growth rates through post stress-relieved welds are higher than those in the base metal. The observed fatigue crack growth responses are discussed in terms of differences in crack tip microstructure, compressive residual stresses, crack closure, and crack deflection.

Introduction

Ti-5Al-1Sn-1Zr-1V-0.8Mo-0.1Si (Ti-5111) is a near-alpha titanium alloy in the 700 MPa yield strength class. Ti-5111 was developed as a lower cost substitute for Ti-6Al-2Nb-1Ta-0.8Mo (Ti-6211) with better producibility. Ti-5111 offers high strength, excellent corrosion resistance, good fracture toughness and stress-corrosion cracking resistance and is a good candidate for many structural applications in marine environments. Ti-5111 structures are typically joined by fusion welding techniques such as electron-beam, laser-beam, and gas-metal arc welding. Because of the high reactivity of molten titanium, oxygen can readily enter the weld. Interstitial oxygen, in relatively low concentration, has been known to significantly lower the stress-corrosion cracking resistance of titanium alloys [1]. Thus, conventional titanium fusion welding is typically performed in vacuum or with an inert gas shielding.

Friction stir welding (FSW), which was introduced by TWI in 1991, has emerged as a promising alternative to conventional fusion welding [2]. FSW is a solid-state welding process in which joining is achieved by the combined action of plastic deformation and frictional heat. Since the material to be joined stays below its melting point during FSW, the extent of oxidation and oxygen pick up is significantly reduced. Besides low oxygen pickup, FSW also offers many advantages over conventional fusion welding such as low weld distortion, easy automation, low defect population, environmental friendliness, lower cost, and the ability to join materials that are deemed unweldable by conventional fusion techniques. FSW has been gaining acceptance in the aerospace and transportation industries for joining aluminum alloys for application such as the Delta rocket and ship structures, primarily because the tools are readily available for these soft and low melting point alloys. Because titanium alloys have higher strength and toughness, the tooling required for FSW is much more demanding and challenging.

Friction stir welding has been shown to produce significantly different weld region microstructures when compared to the base plate [3-9]. The FSW weld nugget region in aluminum alloys, steels, and titanium alloys typically consists of very fine equiaxed grains. Previous investigations have indicated that fatigue crack growth is affected by grain size, with fatigue crack growth rates significantly higher through fine grain materials [10,11]. Residual stresses developed during FSW have been shown to be substantial and can improve fatigue crack growth resistance [3,7,8]. However, the effect of FSW weld microstructure and residual stresses on fatigue crack growth kinetics of Ti-5111 have not been systematically investigated.

In this paper, the microstructures, microhardness, and fatigue crack growth kinetics through FSW Ti-5111 weld region produced under various welding speeds were studied and were compared to those of base metal.

Experimental Procedures

Both 12.7 mm and 6.3 mm thick Ti-5111 plates were used in this study. Single-pass FSW butt joints were prepared at the Edison Welding Institute using a tungsten based tool. The rotational speed was 225 rpm. The weld speed was 51 mm/min for the 12.7 mm thick plate (Weld 1). For the 6.3 mm thick plates, two weld speeds, 25.4 mm/min (Weld 2) and 102 mm/min (Weld 3), were used. Transverse cross section microhardness maps were obtained for these welds. Optical microscopy was used to examine microstructures in the weld nugget and base metal.

While the two 6.3 mm thick Ti-5111 welds, Weld 2 and Weld 3, were defect-free, radiographic examinations indicated the presence of wormholes in the 12.7 mm thick Ti-5111 Weld 1. Most of these wormholes were located near the bottom of the plate. As such, 3.2 mm of material was machined off from both the top and the bottom of Weld 1 to remove these defects. Thus, only the middle 6.3 mm-thick portion of Weld 1 plate was used for fatigue crack growth studies.

5.8 mm thick, 50.8 mm-wide compact tension (CT) fracture mechanics type specimens were used in this study. The notch direction and the crack growth direction were parallel to the welding direction. For added constraint, 5% side grooves were machined on the specimen surfaces. The notch of the CT specimen was positioned at the center of weld. For comparison, CT specimens with the notch located in the base plate region far from the weld were machined from 6.3 mm thick Ti-5111 plate (W3 plate). To determine the effect of weld residual stresses on fatigue crack growth, half of the as-welded and base plate CT specimens were stress-relief annealed in vacuum at 1100 °C for 2 hours, followed by a rapid cooling in flowing argon to ambient temperature. All fatigue crack growth tests were carried out in ambient air (20°C and 42% relative humidity) at a cyclic frequency of 10 Hz and a stress ratio, R, of 0.1. A compliance technique was used to continuously monitor the fatigue crack growth. Post-fatigue fracture surfaces were examined by scanning electron microscopy (SEM) and the average fracture roughness was determined by an optical profilometer.

Results and Discussion

Microstructural Analysis and Microhardness Map

The microstructural characteristics and microhardness maps for Ti-5111 FSW W1, W2, and W3 are similar and only W1 is discussed in detail here. The transverse cross section macrograph and the microhardness map across the mid-thickness of the weld nugget shown in Figs. 1 and 2 are taken from the bead-on plate weld on the same 12.7 mm thick Ti-5111 plate using identical welding parameters and tools as those of Weld 1. The low magnification transverse cross section of 12.7 mm thick Ti-5111 FSW weld is shown in Fig. 1. Fig.1 reveals only two distinct regions, a fine-grained weld nugget region surrounded by a coarse grained base plate region. The base plate consists of very large grains, particularly in the center of the plate where grains as large as 8 mm were present. The large grains in the base plate clearly reveal the boundary between the base plate and the weld nugget. Unlike FSW aluminum alloys and FSW steels where the transition from base plate microstructure to weld nugget microstructure is gradual and is comprised of several zones such as the thermomechanical-affected zone (TMAZ) and heat-affected zone (HAZ), the transition in FSW titanium is abrupt as shown in Fig.1 and no apparent transitional zone is observed. A previous study on the microstructural evolution in Ti-5111 confirmed that the transition from the base plate to weld nugget was abrupt with a transition zone width on the order of only several hundreds micrometers [5]. Because of this unique "two-zone" microstructure, fatigue crack growth studies in FSW Ti-5111 are concentrated on comparing fatigue crack growth kinetics through the fine-grained FSW nugget region to that through the base plate.

Figure 3a shows a typical optical micrograph taken from the base plate. It consists of coarse Widmanstatten structure with many colonies in which α plates align in the same orientation. This base plate microstructure suggests that the plate has been hot rolled above the β transus followed by an α/β anneal. Most of the prior β grains are several millimeters in size with some exceeding 8 mm. Figure 3b shows the fine equiaxed microstructure in the FSW nugget region. In contrast to the coarse grains found in base plate, the average grain size in the FSW nugget region is only about 30 μm. Fine α plates, often in the same orientation, precipitate in the prior beta grains and the alpha phase decorating the grain boundaries. This microstructure clearly indicates that the temperature excursion during friction stir welding has exceeded the beta transus temperature. Furthermore, the very fine equiaxed grain structure observed in the nugget region suggests that the time spent above the beta transus is short, as beta grains would otherwise coarsen rapidly.

The microhardness map and microhardness scan across the mid-thickness of the transverse section of the 12.7 mm thick Ti-5111 FSW weld is shown in Fig. 2. As shown in Fig. 2, the microhardness across the coarse grain base plate region varies between 260 and 360 HVN, while the fine equiaxed grain nugget region exhibits more uniform microhardness in the range of 300 to 320 HVN. Thus, unlike FSW HSLA-65 steel which shows a significant overmatch in the weld nugget, and FSW Al 7050 which shows a considerable undermatch, the difference in microhardness between the base plate and weld nugget of Ti-5111 is small [3,8].

Thus, the major microstructural characteristics in FSW Ti-5111 is the conversion of coarse Widmanstatten structure of the base plate to fine equiaxed grains in the nugget.

31

Fig. 1 Transverse cross section of the Ti-5111 FSW weld.

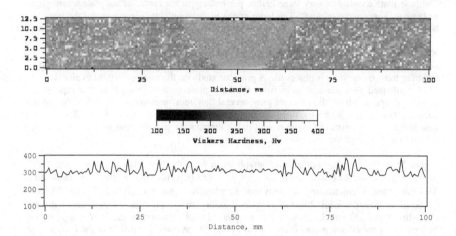

Fig. 2 Microhardness map and microhardness scan across the mid-thickness of the Ti-5111 weld.

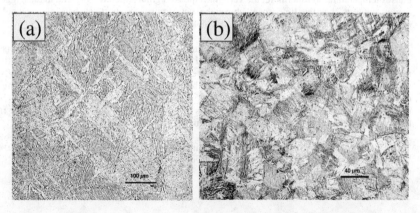

Fig.3 Optical micrographs from (a) the base plate and (b) weld nugget of FSW W1 plate.

Fatigue Crack Growth

The fatigue crack growth kinetics through the Ti-5111 FSW weld nugget regions and through the base plate are shown in Fig. 4. For comparison, Fig. 4 also includes previous fatigue crack growth rate data from 25 mm thick Ti-5111 plate that was not subjected to welding conditions. As shown in Fig. 4, irrespective of the plate thickness, the fatigue crack growth rate through the weld nugget is highest in the fastest weld speed plate W3, followed by middle weld speed plate W2, and is lowest in slowest weld speed plate W1. The fatigue crack growth threshold stress intensity, $\bullet K_{th}$ below which the crack will not propagate, is lowest in fastest weld speed plate W3 and is highest in slowest weld speed plate W1. The difference in fatigue crack growth rates through the weld nugget regions of W1, W2, and W3 plates progressively diminishes as the stress intensity, $\bullet K$, increases. Above a $\bullet K$ of 20 MPa $\bullet$m, fatigue crack growth rates through these weld nuggets are comparable. The fatigue crack growth rates through the base plate and the weld nugget regions of W3 plate are comparable and are substantially higher than those through the weld nugget regions of W1 and W2 plates.

The fatigue crack growth rates obtained from the 25.4 mm thick Ti-5111 plate, at similar $\bullet K$ levels, are higher than those from the base plate region of W3 plate used in this study. The 25.4 mm thick Ti-5111 plate used in a previous fatigue study did not experience welding conditions and, thus, is free from welding-induced residual stresses [12]. It is believed that, during FSW of the 6.3 mm thick Ti-5111 plate, the high clamping loads and the thermal excursion may be enough to introduce significant, though small, residual stresses in the base plate region. The presence of these small residual stresses in the base plate region of 6.3 mm thick W3 plate can effectively reduce the crack tip driving force and result in slightly lower fatigue crack growth rates and higher fatigue crack growth thresholds, when compared to those from the residual stress-free 25.4 mm thick Ti-5111 plate used in previous study.

The better fatigue crack growth performance through the weld nugget regions, as manifested by the lower fatigue crack growth rates and higher fatigue crack growth thresholds, is believed to be caused by the presence of compressive residual stresses in the weld plates. The presence of compressive residual stresses and thus the better fatigue crack growth resistance through the FSW weld nugget regions also have been reported for FSW HSLA-65 steel and FSW aluminum alloys [3,7,8]. To verify the presence of residual stresses in FSW Ti-5111 plates and to determine the extent of their influence on fatigue crack growth, CT fatigue specimens made from the FSW nugget regions as well as from the base plate region were stress-relief annealed. The fatigue crack growth kinetics obtained from stress-relief annealed CT specimens is compared in Fig. 5.

After stress-relief annealing, the fatigue crack growth kinetics obtained from the base plate region of W3 plate, as shown in Fig. 5, are identical to that obtained from the residual stress-free 25.4 mm thick Ti-5111 plate used in the previous study. This suggests that residual stresses were indeed present in the base plate region of plate W3. Relieving these residual stresses causes the fatigue crack growth response of the base plate region of Ti-5111 FSW weldments to approach that of the residual stress-free Ti-5111.

Figure 6 directly compares the fatigue crack growth kinetics, before and after residual stress-relief annealing, through the weld nugget of slowest weld speed W2 plate. The residual stresses have considerable beneficial effect, as the fatigue crack growth rates of as-welded material are significantly lower than those from the stress-relieved material. Furthermore, the nugget region fatigue crack growth threshold decreases from about 12 to 3 MPa$\bullet$m following the stress-relief annealing. Of course, after the stress-relief annealing, this beneficial effect is gone.

33

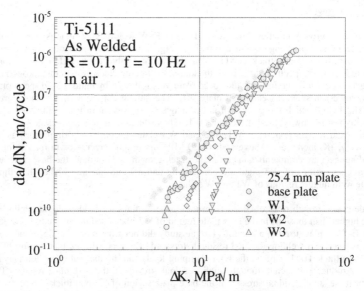

Fig. 4 Fatigue crack growth kinetics obtained from as-welded Ti-5111.

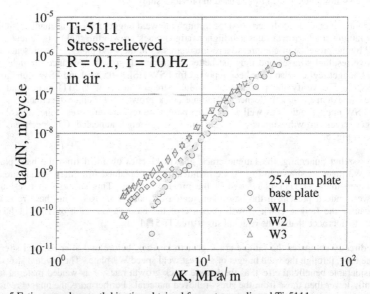

Fig. 5 Fatigue crack growth kinetics obtained from stress-relieved Ti-5111.

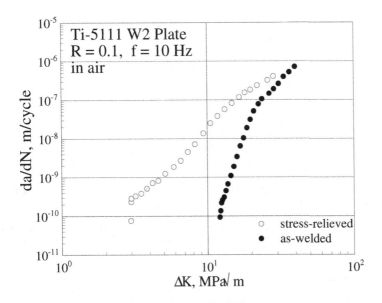

Fig. 6 Comparison of through weld nugget fatigue crack growth kinetics in as-welded and residual stress-relieved Ti-5111 W2 plate.

The fatigue crack growth rate curves for the stress relieved weld nuggets all shift to the left of that obtained from the residual stress-free base plate, as shown in Fig. 5. In stress-relieved weld nugget regions, the fatigue crack growth rates are higher and the fatigue crack growth thresholds are significantly lower when compared to those from the annealed base plate. Since both weld nugget and base plate CT specimens are annealed and contain no or little residual stresses, the difference in fatigue crack growth response, as shown in Fig. 5, can be attributed only to the difference in crack tip microstructure. As shown in Fig. 1, the nugget region contains very fine grains that are about two orders-of-magnitude smaller than the coarse grains found in the base plate. This difference in crack tip grain size results in a significant difference in fracture surface morphology and fracture surface roughness, particularly in the near-threshold region.

Figure 7 shows near-threshold fracture surface morphologies taken from weld nugget and base plate. As shown in Fig. 7a, the fracture surface obtained from the coarse-grained base plate is very rough and the fracture path is highly tortuous as compared to the weld nugget's much smoother fracture surface appearance and straighter fracture path shown in Fig. 7b. Average fracture roughness, as measured by optical profilometer, for base plate and weld nugget regions are 15 μm and 2 μm, respectively. Rougher fracture surfaces can cause roughness-induced crack closure and prematurely close the crack during the unloading cycle. This would result in the reduction of crack tip driving force, •K, and lower fatigue crack growth rates, as is the case for the coarse-grained base plate shown in Fig. 5. In addition, the base plate's tortuous fracture paths would also cause crack deflection and further reduce the effective crack tip driving force

Fig. 7 Near-threshold fracture surface morphology taken from stress-relieved W3 plate, (a) base plate region and (b) weld nugget region.

and fatigue crack growth rates [13]. Thus, the weld nugget regions' higher fatigue crack growth rates and lower fatigue crack growth thresholds can be attributed directly to their fine grain sizes, which result in diminished crack closure and a low degree of crack deflection.

The fatigue crack growth rate curve taken from the weld nugget before residual stress-relief annealing, as shown in Fig. 4, progressively shifts to the left as weld speed increases. That is, at comparable •K, the fatigue crack growth rate through the weld nugget is the highest in the fastest weld speed W3 plate (102 mm/min) and is the lowest in slowest weld speed W1 plate (25.4 mm/min). The fatigue crack growth threshold goes the other way and is lowest in the fastest weld speed plate and highest in the slowest weld speed plate. Such a fatigue crack growth kinetics dependence on weld speed is believed to be caused by the difference in residual stresses, which is the result of different heat inputs during welding. Slower weld speeds correspond to hotter welding conditions, which could increase the residual stresses. Large residual stresses present in the slow weld speed plate can lower the effective •K and reduce the fatigue crack growth rates. When such residual stresses are removed by stress-relief annealing, then the difference in fatigue crack growth kinetics among plates with various weld speeds would be greatly diminished, as demonstrated in Fig. 5.

It is interesting to note that, prior to residual stress-relief annealing as shown in Fig. 4, the fatigue crack growth kinetics obtained from the nugget region of the fastest weld speed Ti-5111 W3 plate is identical to that from the base plate region of the same plate. The weld nugget and base plate have very dissimilar grain sizes and must also have very different residual stress distributions. However, the fatigue crack growth response in FSW Ti-5111 is primarily dictated by two parameters, crack tip microstructure and residual stress distribution. On the one hand, the weld nugget region's fine grain size would result in smoother fracture surface morphology, low crack closure level, and a low degree of crack deflection, all of which would enhance fatigue crack growth. On the other hand, the high compressive residual stresses present in the FSW weld nugget would reduce the effective •K and lower the fatigue crack growth rates. It is coincidental that the effects of grain size and residual stresses, which have opposite influences on fatigue crack growth in FSW Ti-5111 cancel each other out and result in an apparent identical fatigue crack growth kinetics, as observed for fatigue crack growth through the weld nugget and base plate of W3 plate.

Conclusions

1. The weld nugget consists of a single weld region with a grain size significantly smaller than that of the base plate.
2. The fatigue crack growth kinetics through the weld nugget is strongly affected by the presence of residual stresses and the crack tip microstructures.
3. In the as-welded condition, the fatigue crack growth rates are significantly lower and fatigue crack growth thresholds are significantly higher through the weld than those in the base plate.
4. As the weld speed increases, the fatigue crack growth rates become progressively higher and fatigue crack growth thresholds lower through the weld. However, after stress-relief annealing, such differences in fatigue crack growth kinetics among different weld speeds greatly diminishes.
5. Fatigue crack growth rates through post stress-relieved welds are higher than those in the base metal. This can be attributed to the weld nugget's lower crack closure level and lower degree of crack deflection.

Acknowledgements

The authors would like to acknowledge financial support from the Naval Research Laboratory under the auspices of the Office of Naval Research. We also would like to thank Ernie Czyryca of Naval Surface Warfare Center at Carderock for providing information on stress-relief.

References

1. P.S. Pao, M.A. Imam, H.N. Jones, R.A. Bayles, and C.R. Feng, "Effect of Oxygen on Stress-Corrosion Cracking and Fatigue Crack Growth of Ti-6211," in the Proceedings of the 11th World Conference on Titanium, The Japan Institute of Metals, pp. 279-282 (2007).
2. W.M. Thomas, E.D. Nicholas, J.C. Needham, M.G. Nurch, P. Temple-Smith, and C.J. Dawes, "Friction Stir Butt Welding," International Patent Application No. PCT/GB92/02203, GB Patent Application No. 9125978.8 (1991), and U.S. Patent No. 5,460,317 (1995).
3. P.S. Pao, R.W. Fonda, H.N. Jones, C.R. Feng, and D.W. Moon, "Friction Stir Welding of HSLA-65 Steel," in Friction Stir Welding and Processing IV (Eds. R.S. Mishra, M.W. Mahoney, T.J. Lienert, and K.V. Jata), pp. 243-251, The Minerals Metals, and Materials Society, 2007.
4. M. Posada, J. DeLoach, A.P. Reynolds, R.W. Fonda, and J.P. Halpin, "Evaluation of Friction Stir Welded HSLA-65," in Proceedings of the Fourth International Symposium on Friction Stir Welding, TWI Ltd., S10A-P3 (2003).
5. R.W. Fonda, K.E. Knipling, C.R. Feng, and D.W. Moon, "Microstructural Evolution in Ti-5111 Friction Stir Welds," in Friction Stir Welding and Processing IV (Eds. R.S. Mishra, M.W. Mahoney, T.J. Lienert, and K.V. Jata), pp. 295-301, The Minerals Metals, and Materials Society, 2007.
6. C.G. Rhodes, M.W. Mahoney, W.H. Bingel, R.A. Spurling, and C.C. Bampton, "Effects of Friction Stir Welding on Microstructure of 7075 Aluminum," *Scripta Materialia*, 36 (1997), 69-75.
7. K.V. Jata, K.K. Sankaran, and J.J. Ruschau, "Friction-Stir Welding Effects on Microstructure and Fatigue of Aluminum Alloy 7050-T7451," *Met Mat Trans A*, 31A (2000), 2181-2192.
8. P.S. Pao, S.J. Gill, C.R. Feng, and K.K. Sankaran, "Corrosion-fatigue crack growth in friction stir welded Al 7050," *Scripta Materialia* 45 (2001), 605-612.
9. J. Corral, E.A. Trillo, Y. Li, and L.E. Murr, "Corrosion of Friction-Stir Welded Aluminum Alloys 2024 and 2195," *J Mat Sci Letters*, 19 (2000), 2117-2122.
10. P.S. Pao, H.N. Jones, S.F. Cheng, and C.R. Feng, "Fatigue Crack Propagation in Ultrafine Grained Al-Mg Alloy," International Journal of Fatigue, Vol. 27, 10-12, p. 1164(2006).

11. T. Hanlon, Y.N. Kwon, and S. Suresh, "Grain Size Effects on the Fatigue Response of Nanocrystalline Metals," *Scripta Materialia*, 49 (2003), 675-680.

12. P.S. Pao, S.J. Gill, M.A. Imam, C.R. Feng, and R.A. Bayles, "Stress-Corrosion Cracking and Corrosion Fatigue Resistance in Ti-5111 and VLI Ti-6Al-4V," in Ti-2003 Science and Technology, Wiley-VCH, pp. 2083-2090 (2004).

13. S. Suresh, "Fatigue of Materials," Cambridge University Press (1992).

Friction Stir Welding and Processing V
Edited by: Rajiv S. Mishra, Murray W. Mahoney, and Thomas J. Lienert
TMS (The Minerals, Metals & Materials Society), 2009

FASTER TEMPERATURE RESPONSE AND REPEATABLE POWER INPUT TO AID AUTOMATIC CONTROL OF FRICTION STIR WELDED COPPER CANISTERS

L. Cederqvist[1,2], R. Johansson[3], A. Robertsson[3], and G. Bolmsjö[2]

[1]Swedish Nuclear Fuel and Waste Management Company (SKB)
Box 925, 57229 Oskarshamn, Sweden
[2]Lund University, Division of Design Sciences, Box 118, 22100 Lund, Sweden
[3]Lund University, Department of Automatic Control, Box 118, 22100 Lund, Sweden

Keywords: Friction stir welding, copper, automation, automatic control

Abstract

The thermal boundary conditions change throughout the weld cycle to seal copper canisters resulting in variable power input requirement to keep the welding temperature within the process window. The variable power input requirement together with the lag time of approximately 20 seconds in the tool temperature reading results in problems when trying to adaptively control the tool temperature using software. By adding new thermocouples in the shoulder, the lag time in the temperature responding to power input changes were reduced to approximately 10 seconds. In addition multiple weld cycles show that the power input requirement to maintain the tool temperature in the middle of the process window throughout the 45 minute long weld cycle is repeatable. By using the faster responding thermocouple reading and the known power input requirement, a more accurate and reliable closed-loop control of the tool temperature using software can be developed.

Introduction

The Swedish Nuclear Fuel and Waste Management Company (SKB) will join at least 12,000 lids and bottoms to the extruded 5 cm thick copper tubes containing Sweden's nuclear waste. High quality welds are currently produced thanks to manual changes of the spindle speed by a skilled welding operator to keep the tool temperature within the process window through the non-uniform thermal boundary conditions. However, to not be dependent on a skilled welding operator for 30+ years of production, an automatic welding procedure controlled by software needs to be developed. In addition, correctly programmed software will most probably be more reliable and repeatable than a welding operator.

The reliability of the potential automatic procedure is however limited by the lag time in the tool temperature responding to changes in power input. Currently, the tool temperature (measured with a thermocouple inside the probe) takes approximately 20 seconds to respond to power input changes. Another limitation is the non-uniform thermal boundary conditions throughout the weld cycle, see Figure 1, that results in variable power input requirements [1].

Before software to control the process automatically was to be developed, the welding procedure was optimized with regards to stability and repeatability [2]. A convex scroll shoulder geometry was found to aid the process stability, and the tool geometry now used can be seen in Figure 2.

Figure 1. Sequences in a full-circumferential weld cycle: Figure 2. Tool geometry.
1. acceleration, 2. downward, 3. jointline, 4. overlap, 5. parking.

Experimental Procedure

Due to relatively large lag time for the thermocouple in the probe, efforts were made to reduce the lag time. Two new thermocouples, named shoulder ID and OD (inside and outside diameter), were added at new locations in the shoulder, see Figure 3 [3], which shows shoulder ID, probe and shoulder OD thermocouples on top, middle and bottom, respectively. In addition to the shoulder material (Densimet D176) having much higher thermal conductivity than the probe material (Nimonic 105), 74.0 versus 10.9 W/m·K at 20°C, the location of the new thermocouples are closer the weld metal. In addition to the three thermocouples, an infra-red camera is recording the maximum temperature on the leading side of the shoulder, see Figure 4.

Figure 3. Thermocouple placements.   Figure 4. Infra-red picture of shoulder.

After the input welding parameters were optimized with regards to stability using statistical DOE [2], several full-circumferential welds were made to investigate the repeatability of the welding procedure. Subtle manual changes of the spindle speed were made to keep the probe temperature around the desired value of 850°C, which is in the middle of the process window ranging from 790°C (wormhole defects produced below this temperature) and 910°C (risk of probe fracture above this temperature). The reason for spindle speed being the adjusted parameter is due to previous studies [4] showing it to be the most influential input parameter on the probe temperature, which makes sense since the product of spindle speed and spindle torque is power input.

Results and Discussion

Figure 5 shows how the three thermocouple and the infra-red readings respond to changes in the power input. This is only an example including ¾ of the downward sequence and a few degrees of jointline welding, but it can be seen that the power input changes from negative to positive slope at an x-axis value of 548 seconds. As a result, the infra-red, probe, shoulder ID and shoulder OD readings change to positive slope 0, 21, 10 and 14 seconds later, respectively. The infra-red reading responds immediately, but the reading scatters and also if there is flash produced in front of the shoulder it can severely affect the accuracy of the reading, which makes it unsuitable for use in the automatic software. Of the thermocouples, the shoulder ID reading responds the fastest, and its location is also a plus since it is close to where wormhole defects are produced at low temperatures, which could make the value correlate better with weld quality than the probe temperature value currently used to define the process window.

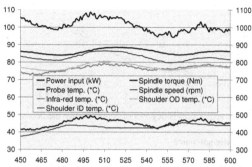

Figure 5. Weld data showing the four temperature readings in the middle, spindle torque on top and power input and spindle speed on the bottom. All values on right y-axis except power input.

Using the most stable input parameters from previous studies [2] several full-circumferential welds were made using subtle manual changes of the spindle speed to keep the probe temperature around 850°C. Figure 6 shows the power input requirement versus the location around the canister (in degrees), and is also showing the different sequences (1 through 5) defined in Figure 1 by vertical lines.

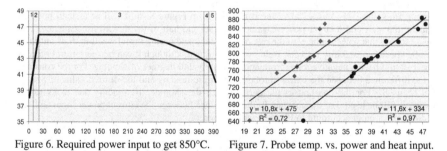

Figure 6. Required power input to get 850°C.   Figure 7. Probe temp. vs. power and heat input.

Figure 7 shows results from the input parameter study [2], where the correlation between the probe temperature at the end of downward sequence and the power and heat input (on the x-axis) was investigated. It can be seen that the heat input (in kJ/mm, power input divided by travel speed) scatters (diamonds) while there is a linear relationship between the probe temperature and the power input (dots). This also shows that the power input requirement is repeatable between weld cycles.

41

In addition to the required power input being repeatable between weld cycles, it appears that the spindle speed changes during the downward sequence have enough repeatability between weld cycles to be able to aid the automatic software. Figure 8 shows how the spindle speed (in rpm) were changed during the 16.6 degrees in the downward sequence to maintain a probe temperature around 850°C.

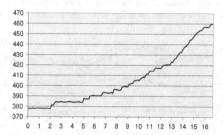

Figure 8. Spindle speed versus degrees of travel during the downward sequence.

First generation automatic software

Using the reduced lag time for the shoulder ID reading, the repeatable power input requirement in Figure 6, and the repeatable spindle speed changes during the downward sequence presented in Figure 8, the development and testing of automatic software can begin. The chosen approach is to use PID-regulator(s) [5] throughout the weld cycle, but also to use feed-forward of spindle speed during the downward sequence. It is possible that feed-forward of spindle speed can be used also during the overlap and park sequences, but it depends on the repeatability (between weld cycles) of the spindle speed changes during these sequences.

Equation 1 shows the planned PID-regulator to be used together with feed-forward of the spindle speed during the weld cycle. The spindle speed is updated every 5 seconds according to the current $\Delta\omega$-value. The $\Delta\omega_{feedforward}$-value is derived from Figure 8, so it will be a small positive value at the start and a larger positive value at the second half of the downward sequence. The $\Delta\omega_{feedforward}$-value will then be zero the rest of the weld cycle during initial welding trials until the spindle speed repeatability between weld cycles is known, although it can be seen from Figure 6 that it should have a negative value from approximately 240 degrees of welding. For the constant k_1 the difference between the current probe temperature reading and the desired value of 850°C will be used. For the constant k_2 the derivative of the shoulder ID reading due to its faster response will be used. For the constant k_3 the difference between the current power input and the desired power input according to Figure 6 will be used.

$$\Delta\omega = \Delta\omega_{feedforward} + k_1 \cdot \left(T_{probe} - 850^\circ C\right) + k_2 \cdot \frac{dT_{shoulderID}}{dt} + k_3 \cdot \left(Power_{current} - Power_{currentlydesired}\right) \quad (1)$$

By determining the time constants (dead and lag time) for the probe temperature and shoulder ID temperature from multiple welds, the k-values for the initial welding trials can be found using a variation on Ziegler-Nichols PID tuning called AMIGO [6].

42

Also using the time constants and data from multiple welds, including data presented in Figures 6 and 7, a dynamic model of the probe temperature as a function of the position around the lid (i.e. thermal boundary condition) and the power input will be developed. The model can then be used in MATLAB Simulink to simulate the regulator (equation 1) on data from various weld cycles and find appropriate spindle speed update rates and k-values throughout the weld cycle. It is most probable that the constants will be varied depending on sequence, for example faster spindle speed update rate during the downward sequence when the process is more transient. It is possible that the most stable and repeatable control is achieved if a cascade loop using only the k_3-part in equation 1 is used at a faster update rate.

Summary

Although the reduced lag time in the shoulder ID reading, the repeatable power input requirement and repeatable spindle speed changes will aid the development of automatic software, a lot of simulations and verifying welding trials are left before a reliable regulator can do a similar or better job that the currently skilled operator. A key to minimizing the number of welding trials (and simulations) will be the repeatability between weld cycles, since high repeatability could make the feed-forward part ($\Delta\omega_{feedforward}$) of the regulator almost unaided control the probe temperature accurately.

References

1. A. Fehrenbacher et al., "Closed-loop control of temperature in friction stir welding," Proceedings from 7th International Friction Stir Welding Symposium, Awaji Island, Japan, May 2008.
2. L. Cederqvist, C.D. Sorensen, A.P. Reynolds and T. Oberg, "Improved stability during friction stir welding of 5 cm thick copper canisters through shoulder geometry and parameter studies," *Science and Technology in Welding and Joining*, submitted Sept. 2008 for review.
3. J.L. Covington, "Experimental and numerical investigation of tool heating during friction stir welding," (Master thesis, Brigham Young University, 2005), 17-18.
4. L. Cederqvist and T. Oberg, "Reliability study of friction stir welded copper canisters containing Sweden's nuclear waste," *Reliability Engineering and System Safety*, 93 (2008), 1491-1499.
5. L. Cederqvist, C.D. Sorensen, and G. Bolmsjo, "Adaptive control of novel welding process to seal canisters containing Sweden's nuclear waste using PID algorithms," Proceedings from 18th International Conference on Flexible Automation and Intelligent Manufacturing, Skovde, Sweden, July 2008.
6. K.J. Astrom and T. Hagglund. Advanced PID control. ISA - The Instrumentation, Systems, and Automation Society, 2006.

43

Friction Stir Welding and Processing V
Edited by: Rajiv S. Mishra, Murray W. Mahoney, and Thomas J. Lienert
TMS (The Minerals, Metals & Materials Society), 2009

THE EFFECTS OF FRICTION STIR PROCESSING ON THE MICROSTRUCTURAL EVOLUTION AND MECHANICAL PROPERTIES OF Ti-6Al-4V ALLOY

Nilesh Kumar, Jeff Rodelas and Rajiv S. Mishra

Center for Friction Stir Processing and Department of Materials Science & Engineering,
Missouri University of Science and Technology, Rolla, MO 65409 USA

Keywords: Ti-6Al-4V, Friction Stir Processing, Mechanical Property, Microhardness,
Microstructure

Abstract

Friction stir processing (FSP) was applied to Ti-6Al-4V alloy to modify the microstructure and improve the mechanical properties. Experiments were carried out at three different tool rotational rates: 1200, 1000, and 800 rpm. Other parameters were kept constant. The material processed at 800 rpm showed a very narrow HAZ. Hardness in the nugget region was higher than the parent material in each case. The sample processed at 1200 rpm showed an improvement of approximately 31% in YS and UTS over as-received material was observed. There was an improvement of 33% and 27% respectively in the samples processed at 1000 rpm and 800 rpm. The greatest improvement in strength after FSP was 1234 MPa, compared to 923 MPa in as-received material. This improvement was observed with no compromise in the ductility of the material (24% elongation for parent and 22-26% elongation for FSPed samples).

Introduction

FSP is an offshoot of friction stir welding (FSW) which was invented in 1991 at The Welding Institute (TWI) by Thomas et al. [1]. FSP was developed by Mishra et al. [2,3] on the principles of FSW and it has proved to be a powerful processing technique for microstructural homogenization and surface modification [4] by the incorporation of second phase particles such as SiC, TiC, or WC. Application of FSP to various alloys has also shown enhanced superplasticity [2] and permitted casting modification [5].

In the past, the research in FSW/FSP has been hindered by extensive tool wear, deformation and tool reactivity during the processing of high softening temperature materials such as steels, Ni-based and Ti-based alloys. Recently, however, new tools have been developed based on refractory metals (W-based alloys [6]), cermets (WC + Co [7]), and PCBN [8] for processing such materials. The development of mechanically and chemically robust tool materials for FSW/P has accelerated the research for high temperature materials.

Ti-6Al-4V is the workhorse material of aerospace industry due to its excellent corrosion resistance, high specific strength and ductility compared to other Ti grades. Only limited studies exist that examine the microstructural modification and mechanical properties of the Ti-6Al-4V alloy during FSW/FSP. Lienert et al. [9, 10] first reported the FSW of Ti-6Al-4V and discussed microstructural evolution based on pseudo-binary section of the Ti-Al-V ternary system at constant Al content. Pilchak et al. [11] have discussed the microstructural changes in investment-cast Ti-6Al-4V alloy as a result of FSP. In addition, Zhang et al. [12] have briefly discussed the effects of various tool rotational rates on the nugget microstructure and discussed mechanical properties. However, there exists only limited knowledge of the microstructural evolution within the nugget for Ti-6Al-4V as a result of varied tool rotational rate.

The present study examines the effects of various tool rotational rates on the nugget profile and determines in each case the microstructural evolution in various parts of the nugget and heat-affected zone (HAZ). Changes in tensile properties and Vickers hardness with varying tool rotational rates were also measured.

Experimental Procedures

A mill-annealed Ti-6Al-4V alloy was used in the present study. A plate of dimensions 0.25"x 2.4"x 3.7" was used. A hybrid bimaterial tool of WC-6wt%Co shoulder-pin and tungsten heavy alloy shank was used to process the material. Tool dimensions are shown schematically in Figure 1.

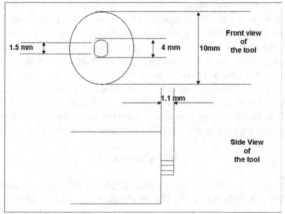

Figure1. A schematic of the FSP tool.

The process parameters used are shown in the Table 1. Experiments were carried out at three different tool rotational rates – 1200, 1000 and 800 rpm. Other parameters such as tool plunge depth, tool tilt, plunge rate, traverse rate, and traverse length were kept constant. At high temperature, Ti and its alloys are prone to oxidation. To circumvent oxidation, FSP experiments were carried out in an argon atmosphere. Transverse cross section of the processed region were polished to 1 µm finish and etched using Kroll's reagent (2% HF + 10% HNO₃). Optical microscopy (OM) and scanning electron microscope[1] (SEM) were used for microstructural characterization.

Mini tensile samples were used for characterization of the mechanical properties of the resulting processed region. The mini-tensile sample was machined from material within the

---

[1] Hitachi S-570 scanning electron microscope.

processed zone. The orientation of the tensile gauge section was transverse to the processing direction. Before testing, tensile samples were polished to 1 µm surface finish. A strain rate of 1 x 10^{-3} s^{-1} was used for all tests. Hardness measurements of the processed region were performed by Vickers microhardness on polished transverse cross-sections. A 0.5 kg load was applied with a dwell time of 10 s for the microhardness measurement.

Table I. Process parameters used during processing of Ti-6Al-4V alloy.

Tool Rotational Rate, rpm	Tool Traverse Speed, ipm	Tool Tilt Angle, °	Plunge Depth, in	Run Length, in
1200	1	3	0.061	1
1000	1	3	0.061	1
800	1	3	0.061	1

Results and Discussions

Figures 2a and 2b show the top surface of the beads made at 1000 and 800 rpm, respectively. Surface appearance of the run made at 1000 rpm is distinctly different from that of the run made at 800 rpm. The colored surface of the bead formed at 1000 rpm indicates that some oxidation occurred during the run despite the protective atmosphere. It can be noted that surface oxidation indicates possibility of O and N pick-up during welding. Addition of these interstitials will impact phase balance and subsequent mechanical properties. The surface of the run made at 800 rpm was free of oxidation and indirectly indicates that the thermal cycle was insufficient to cause noticeable oxidation. This inference was further corroborated with microstructural examination of the processed region. Surface appearance of the 1200 rpm run showed oxidation similar to the 1000 rpm run.

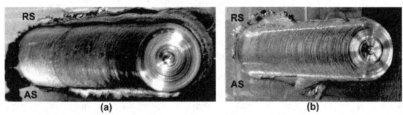

Figure 2. The top surface of the FSP plate. a) Colored surface indicates the oxidation of the material during FSP at 1000 and 1200 rpm run. b) Run made at 800 rpm tool rotational rate showing a clean surface suggestive of minimal oxidation during FSP.

Microstructural Features

Figure 3 shows macro-images of the transverse cross-section of the runs made at 800, 1000 and 1200 rpm. As can be noted from Fig. 3a, there exists a distinct transition region on the advancing side of the run between the stir zone and the parent material. This transition region is unlike the diffuse regions shown in figures 3b and 3c for the runs made at 1000 and 1200 rpm, respectively. As will be discussed later, this is because of a critical difference in the peak temperature for the different rotation rates. The HAZ appears to be absent in run made at 800 rpm.

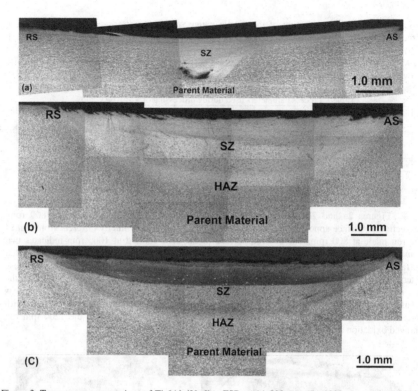

Figure 3. Transverse cross-sections of Ti-6Al-4V alloy FSP at, (a) 800 rpm, (b) 1000 rpm, and (c) 1200 rpm. Run at 800 rpm showing a defective region (black region) on the retreating side of the processed zone.

Figures 4, 5 and 6 show the microstructure of the various regions of the nuggets for the runs made at 800, 1000, and 1200 rpm, respectively. Figure 4a is an image of the parent microstructure taken with a SEM. The parent material has an average grain size of ~20-25 μm. With FSP, grain size was reduced to 0.5-1 μm in the top region of the nugget (figure 4b). The grain size increases from the top to the bottom of the nugget for all conditions. The grain size at the nugget midline is approximately 1-2 μm (figure 4c). Micrographs from the retreating and advancing side of the nugget are shown in the figure 4e and 4f respectively. The equiaxed α morphology and β phase on the grain boundaries suggest the entire nugget region remained under the β transus temperature (950°C) during FSP. Note that the microstructure will be mainly beta phase at 950°C (i.e. β with grain boundary α) and that much of the beta will transform again to alpha on cooling. The volume fraction of the β phase appears to be higher in the nugget region as compared to the parent material, however a quantitative evaluation has not been done. There are two interesting implications of sub-β transus FSP: (a) the microstructure is very fine, and (b) the HAZ is absent.

The material processed at 1000 rpm also shows very fine grain size near the top surface of the stir zone. Figures 5b and 5c show the microstructure in the stir zone. Both the micrographs have similar kind of phases present but they only differ in the scale and amount of the phases present. The presence of grain boundary α and acicular α+β phases suggest that the temperature crossed the β-transus temperature. Figure 5d shows the untransformed α, acicular α+β and coarsened β phases. Figures 5f and 5g show microstructure outside the nugget region on the retreating and advancing sides. The microstructures suggest that temperature in that region was below β-transus temperature. Figure 5e shows the coarsening of β phase in HAZ.

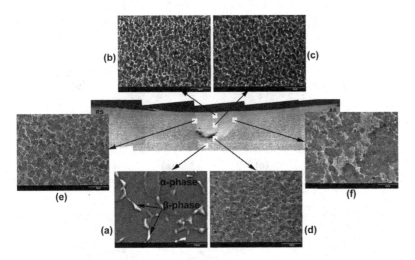

Figure 4. Micrographs (a)-(f) represents various regions of the processed material. (a) Base material, (b)-(f) various regions of stirred zone.

The material processed at 1000 rpm also shows very fine grain size near the top surface of the stir zone. Figures 5b and 5c show the microstructure in the stir zone. Both the AS and RS microstructure exhibit similar phases; however, these phases differ in the size and volume fraction. The presence of grain boundary α and acicular α+β phases suggest the peak FSP temperature in the nugget exceeded the β-transus temperature. Figure 5d shows the untransformed α, acicular α+β, and coarsened β phases at the bottom of the nugget. Figures 5f and 5g show microstructure outside the nugget region on the retreating and advancing sides. The microstructure consisting of equiaxed α with β at the grain boundaries suggests that temperature in that region was below the β-transus temperature. The sub - β transus temperature HAZ (figure 5e) shows a similar microstructure with slightly coarsened β phase.

The microstructure of Ti-6Al-4V processed at 1200 rpm (highest heat input) is shown in Figure 6. Similar to other conditions the grains were smallest at the top of the weld in closest proximity to the tool. Submicron grains were achieved in the top region of the nugget. The figure 6c of the nugget just above the bottom of the nugget shows the continuous α-phase at grain-boundaries and acicular α+β phases. Micrographs 6e and 6f represent the regions outside the nugget from the retreating and advancing sides, respectively.

49

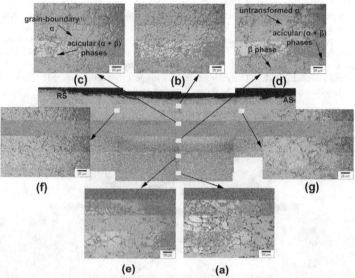

Figure 5. Micrographs (a)-(g) represent various regions of the processed material. a) Base material, (b), (c), (f), and (g) various regions of the stirred zone, (d)-(e) HAZ.

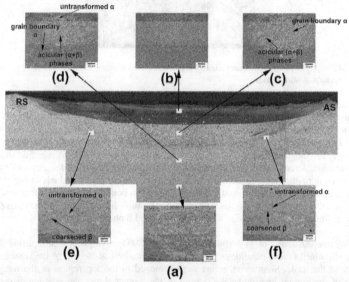

Figure 6. Micrographs (a)-(f) represent various regions of the processed material. (a) Base material, (b)-(c) various regions of the stirred zone, and (d)-(f) HAZ.

Microhardness results of the FSP regions for all the three runs (800, 1000 and 1200 rpm) is shown in figure 7. Hardness profiles for Ti-6Al-4V FSP at 1000 and 1200 rpm are similar. Microhardness of the section for 800 rpm run in the stirred zone is lower than that of other two runs In case of 1000 and 1200 rpm runs, in the stirred zone region, of the hardness is more consistent compared to the parent material away from the weld. This could be attributed to the refined microstructure within the stirred zone. The large scatter in the parent material may be attributed to a volumetric sampling difference of the larger alpha and beta grains in the parent material. Hardness adjacent to the stir region increases in all three cases which is unique to Ti-6Al-4V alloys. In case of Al-alloys, hardness decreases in the HAZ [13]. Increased HAZ hardness has been reported in previous works [12] on the same alloy system by several other authors. Aaronson et al. [14] attributed HAZ hardening to strains caused by the thermal expansion differential between alpha and beta phases. In addition to the development of strain upon cooling, precipitation of α-phase along the prior β-phase grain boundaries may also lead to increased hardness [15].

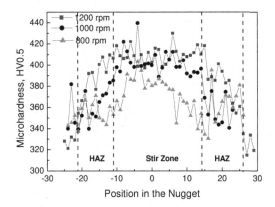

Figure 7. Vickers microhardness across transverse cross-section of the processed material at 800, 1000 and 1200 rpm.

Tensile Properties

Figure 8 shows the engineering stress-strain curve for the base alloy and runs made at 800, 1000 and 1200 rpm. The yield strength (YS), ultimate tensile strength (UTS) and total elongation are also summarized in Table II. Runs made at 1000 rpm showed the highest YS and UTS. FSP Ti-6Al-4V material showed an improvement of 28 to 33% in YS and 25 to 34% in UTS over the parent material. There was minimal change in the ductility of the material after FSP. Material processed at 1000 rpm had the highest strength and highest ductility. From the stress-strain curve it can be inferred that strain hardening exponent is different for base material and FSPed materials. However, changing the tool rotation rate does not have any impact on the strain hardening exponent. It can also be observed that maximum uniform elongation is shown by the sample processed at 1000 rpm. The non-uniform elongation, however, appears to be unchanging for all three conditions. The tensile properties obtained in this work are higher than previously reported values [12,16]. The increased strength properties are likely due to the finer microstructure obtained in this study.

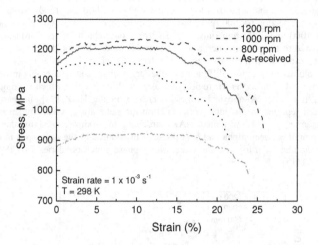

Figure 8. The effects of tool rotational rate on the stress-strain curve of Ti-6Al-4V alloy.

Table II. Tensile properties of Ti-6Al-4V alloy for as-received and friction stir processed conditions.

	YS, MPa	UTS, MPa	% Elongation
Base material	880	923	23.9
800 rpm	1129	1155	21.6
1000 rpm	1168	1234	25.8
1200 rpm	1154	1207	23.2

Conclusions

1. Friction stir processing showed substantial improvement in tensile and hardness of the processed material over base material.
2. Materials processed at 1000 rpm showed the maximum improvement in tensile properties.
3. Temperature crossed the β-transus temperature for runs made at 1000 and 1200 rpm but remained below β-transus temperature for 800 rpm run.

References

1. W.M. Thomas, *Int. Patent 1 PCT/GB92/02203*, 1991.

2. R.S. Mishra et al., "High Strain Superplasticity in a Friction Stir Processed 7075 Al Alloy", *Scr. Mater.*, 42 (2) (1999), 163-168.

3. R.S. Mishra and M.W. Mahoney, "Friction Stir Processing: A new grain refinement technique to achieve high strain rate superplasticity in commercial alloys", *Mater. Sci. Forum*, 357-359 (2001), 507-514.

4. R.S. Mishra, Z.Y. Ma, and I. Charit, "Friction Stir Processing: A Novel Technique for Fabrication of Surface Composite", *Mater. Sci. Eng. A*, 341 (2003), 307-310.

5. W.J. Arbegast and P.J. Hartley, "Method of Using Friction Stir Welding to Repair Weld Defects and to Help Avoid Weld Defects in Intersecting Welds" *U S Patent 6,230,957*, May 15, 2001.

6. T.J. Lienert et al., "Friction Stir Welding Studies on Mild Steel", *Weld J.*, 82 (1) (2003), 1s-9s.

7. T. Yasui et al., "Friction Stir Welding Between Aluminum and Steel with High Welding Speed", *Proceedings of the Fifth International Conference on Friction Stir Welding*, Sept 14-16, 2004 (Metz, France), TWI, paper on CD.

8. C.D. Sorensen, T.W. Nelson, and S.M. Packer, "Tool Material Testing for FSW of High-Temperature Alloys", *Third International Symposium on Friction Stir Welding*, 2001, (Kobe, Japan), paper on CD.

9. T. J. Lienert et al, "Friction Stir Welding of ti-6Al-4V Alloys", *Proceedings from joining of Advanced and Speciality Materials*, Oct. 9-11, 2000, (St. Louis, USA), ASM International, 2000.

10. T. J. Lienert, "Microstructure and Mechanical Properties of Friction Stir Welded Titanium Alloys", *Friction Stir Welding and Processing*, ed. Rajiv S. Mishra and Murray W. Mahoney, (Materials Park, OH: ASM International, 2007), 123-154.

11. A.L. Pilchak, M.C. Juhas, and J.C. Williams, "Microstructural Changes Due to Friction Stir Processing of Investment-Cast Ti-6Al-4V",*Met. Trans. A*, 38A (2007), 401-408.

12. Y. Zhang, "Microstructural Characteristics and Mechanical Properties of Ti-6Al-4V Friction Stir Welds", *Mater. Sci. Eng. A*, 485 (2008), 448-455.

13. P. Prevey, and M. Mahoney, "Improved Fatigue Performance of Friction Stir Welds with Low Plasticity Burnishing: Residual Stress Design and Fatigue Performance Assessment", THERMEC 2003, International Conference on Processing and Manufacturing of Advanced Materials (Leganes, Madrid, Spain), *Mater. Sci. Forum*, 426-432 (2003), 2933-2940.

14. J.W. Elmer et al., "In-situ Observations of lattice expansion and transformation of α and β phases in Ti-6Al-4V", *Mater. Sci. Eng. A*, 391 (2005), 104-113.

15. A.J. Ramirez and M.C. Juhas, "Microstructural Evolution in Ti-6Al-4V Friction Stir Welds", *Materials Science Forum*, 426-432 (2003), 2999-3004.

16. D.G. Sanders et al., "Characterization of Superplastically Formed Friction Stir Weld in Ti-6Al-4V: Preliminary Results", *J Mater. Eng Perform.*, 17 (2) (2008), 187-192.

Friction Stir Welding and Processing V
Edited by: Rajiv S. Mishra, Murray W. Mahoney, and Thomas J. Lienert
TMS (The Minerals, Metals & Materials Society), 2009

TEXTURE AND MICROSTRUCTURAL EVOLUTION DURING LINEAR FRICTION WELDING OF Ti-6Al-4V

E. Dalgard[1], M. Jahazi[2] and J.J. Jonas[1]

[1]McGill University, 3610 University St, Montreal, QC, H3A 2B2, Canada
[2]National Research Council of Canada, IAR-AMTC, 5145 Decelles Ave, Montreal, QC, H4T 1W5, Canada

Keywords: Linear Friction Welding, Ti-6Al-4V, Microtexture

Abstract

The linear friction welding (LFW) behaviour of Ti-6Al-4V was investigated using various processing conditions of frequency (30-50 Hz) and pressure (50-90 MPa). LFW samples were examined using electron backscatter diffraction (EBSD) to relate the texture to the strain and flow stress. Characterization of the welds included analysis of the microstructure of the weld and of the thermomechanically affected zones (TMAZ) in relation to the parent material. The relationship of the transformed structure to the prior grains was examined using EBSD and scanning electron microscopy (SEM).

Introduction

Linear friction welding (LFW) is a solid-state joining process. It involves oscillating one part in a linear fashion against another, which is stationary. No axial symmetry is required and parts can have quite complex geometries involving curves. An LFW process was first patented in 1969 [1] and in the early 1980's The Welding Institute (TWI) demonstrated a working LFW process for metals.

Most LFW development so far has been driven by the aerospace industry's desire to fabricate integrally bladed disks, which offer lower weight and improved performance over existing slotted blade/disk assemblies. LFW can also be used to join dissimilar alloys, which allows component parts to be optimized separately; for example, high cycle fatigue resistant, high temperature alloys can be employed for the blade and lower cycle fatigue resistant materials for the disk [2]. Although manufacturers are investing large amounts of money in researching LFW, very few publications concerning this research are available in the open literature [3].

Titanium alloys are highly prone to texture heterogeneity, resulting in the anisotropy of mechanical properties. The recrystallization dynamics of an LFW joint may be affected by the material's original texture, leading to microstructural changes at the joint. The formability and fatigue life, both important in aerospace applications, are strongly affected by anisotropy. Given the interest in using LFW for both the manufacture and repair of aerospace components, an investigation of the texture behaviour of these alloys is also relevant.

Titanium and its alloys have a strong affinity for the atmospheric gases oxygen, nitrogen and hydrogen and thus care must be taken when welding so that the molten metal does not come into contact with them. LFW, as a solid-state process, eliminates the necessity for a protective environment when welding, since the material does not reach fusion temperatures. Due to the way that heat is generated directly at the interface in friction welding, a high heat density comparable to that developed in laser or e-beam welding can be achieved. This, with the low thermal conductivity of Ti and its alloys, creates a very small heat affected zone (HAZ)[2]. Friction welding processes also create a narrow and extensively deformed TMAZ adjacent to the weld line. As a result, dramatic microstructural and textural variations can be expected in this region [3].

The four stages of LFW
The LFW process can be divided into four distinct phases that occur during welding [4]. The first phase, known as the initial or contact phase, begins with contact between the two pieces in order to initiate the wear of surface asperities. Heat is generated from the solid friction between the two bodies. During this phase, the true contact area between the two bodies increases as roughness is worn away [4]. The second phase, known as the transition or conditioning phase, begins when the large wear particles that were created during the first phase begin to be expelled from the interface. Frictional heat creates a soft plasticized region that is no longer able to support the applied axial load and which begins to deform permanently.

When moving into the third phase, known as the equilibrium or burn-off phase, the flash begins to form [4]. The axial pressure is increased and oscillation continues as in the prior phases. Frictional heat is conducted away from the interface and the plastic zone develops further, extruding material away from the interface to the flash. The last phase is known as the deceleration or forge phase, where the materials are brought to rest after the desired shortening has been attained. Once the materials have been brought to rest and aligned, the axial pressure is increased and the weld is consolidated [4].

Parameters
The four most important process parameters for LFW are as follows:

The **amplitude** of motion of the moving piece with respect to the stationary piece

The **frequency** of oscillation

The **force** applied to the two pieces

The upset, or axial **shortening** distance (time can be specified rather than distance).

Specific Power Input
Vairis proposed an equation for a specific power input parameter, used to characterize the LFW process [5], which combines the important parameters. The expression is based on combining the effective velocity, represented by amplitude times frequency, with a factor added to account for the variable velocity of the sinusoidally oscillating piece, and the pressure applied in the axial direction. The product is a value proportional to the frictionally induced power.

An amendment to the original expression is proposed in the current work, as it appears that the cross-sectional area was employed twice: once in the pressure term (force divided by *area*) and then as the cross-sectional area itself. One of the two is sufficient.

The original formula proposed by Vairis was:

$$w \ (\mathrm{kW/mm^2}) \quad = \quad \frac{a*f*P}{2\pi*A} \tag{1}$$

where a = amplitude
    f = frequency
    P = pressure
and A = cross-sectional area

In the current work:

$$w \ (\mathrm{kW/mm^2}) \quad = \quad \frac{a*f*F}{2\pi*A} \tag{2}$$

where a, f, A are as above, and F = axial force applied.

A minimum value of the specific power input seems to be necessary in order for a successful weld to be produced. This critical value has not as yet been modeled and must be determined empirically for each material (or set of materials) to be welded.

In terms of the effect of the welding parameters on the stress experienced by the material, in previous work it was noted that the maximum residual stress (calculated from the measured residual strain) decreased with an increase in pressure. With an increase in amplitude, the maximum residual stress remained roughly unchanged. No change in residual stress was observed with a change in frequency, power, or time [6].

Previous researchers [3, 7, 8] have noted that the hot working of Ti-6Al-4V at temperatures above the beta-transus temperature, combined with fast cooling, will result in a fine martensitic and/or Widmanstatten microstructure. When processed in the β phase field, the general microstructure of the Ti–6Al–4V alloy consists of only martensite. In the work of Lutjering [9], it was found that the deformation temperature determines the texture type.

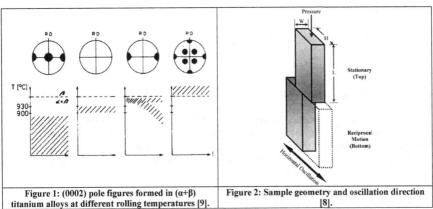

Figure 1: (0002) pole figures formed in (α+β) titanium alloys at different rolling temperatures [9].	Figure 2: Sample geometry and oscillation direction [8].

Deformation imposed in the high alpha fraction temperature range (low temperature) will result in a texture based on the alpha phase, thus a basal/transverse texture as shown in Figure 1. At higher deformation temperatures, where a high fraction of beta is present, the texture that develops is based on the beta phase and the transformation texture upon cooling depends on variant selection (only one of the variants of the Burgers relationship, (110)‖(0002), is selected). This results in a transverse (T) type of transformation texture.

Experimental Procedure

Ti-6Al-4V, an α + β alloy, is the most commonly used alloy in the aerospace industry, and was the primary focus of this study. The main piece of equipment used was the MTS LFW Process Development System (PDS). Table I shows the experimental plan used in this study. These values were based on optima reported in the literature [8]. Samples for welding were machined to a configuration of 13 mm in width, 35 mm in length, and 26 mm in height (Figure 2).

Samples designated "welded parallel to RD" were cut so that the oscillation of the moving piece was parallel to the rolling direction; it was perpendicular on those designated "welded perpendicular to RD". Metallographic and EBSD samples were sectioned from the centres of the weld areas, transverse to the oscillation direction. Conventional polishing procedures were used for optical microscopy and for preliminary preparation of the EBSD samples. Final EBSD polishing was carried out in a vibratory polisher for 12 hours. Etching was done using Kroll's reagent. Back-scatter imaging and EBSD mapping were performed at 30 kV on a Hitachi S-3000N VP-SEM equipped with an Oxford (HKL) EBSD acquisition system.

Table I: Samples prepared

Sample ID	Orientation relative to RD	Amplitude (mm)	Frequency (Hz)	Pressure (MPa)	Upset Dist. (mm)	Specific Power Input (kW/mm^2)	Welding time (s)
#1	Parallel	2	50	50	2	796	1.14
#2	Perpendicular	2	50	50	2	796	1.37
#3	Parallel	2	30	90	2	860	1.49
#4	Perpendicular	2	30	90	2	860	1.71

Results and Discussion

The appearance of the welded samples is presented in Figure 3. Welding perpendicular or parallel to the rolling direction did not seem to influence the appearance of the welded sample, nor did the magnitude of the axial pressure. The as-received material was analyzed using optical microscopy and EBSD to provide a comparison point for the welded specimens. Figure 4 illustrates the microstructure of the as-received material, with bands of different microstructures. The darker bands consist of equiaxed α grains and a transformed β matrix that contains coarse, acicular α. The grains averaged 10 μm in diameter. The lighter bands consist of 5-10 μm α grains surrounded by grain boundary β.

Microstructures
The microstructure of a sample welded with the oscillation direction *parallel* to the rolling direction is illustrated in Figure 5. At high optical magnification (see (a)), the grains are not easily seen along the weld centre using optical microscopy, although faint indications of a martensitic structure can be seen. The small dark areas may represent etch pitting (also visible in the other micrographs). This highly deformed region does not etch as readily as the undeformed

grains; furthermore, the literature [3,8] suggests that these grains are very fine, contributing to the low etching response. In Figure 5(b), an SEM backscattered electron micrograph of the weldline, is presented, in which the martensitic structure can be seen more clearly; here the needles are approximately 5 to 7 μm in length and under 500 nm in width. It is difficult to identify grain boundaries in such a structure. In micrograph (c), representing the structure about 0.375 mm from the weldline, it is apparent that the grains are elongated along the direction of material flow. Finally, in Figure 5(d) 0.725 mm from the weld line,, the undeformed bimodal alpha/transformed beta microstructure can be seen clearly.

| Figure 3: LFW Ti-6Al4V joint showing flash extending towards back and front. | Figure 4: Ti-6Al-4V as received, Etched with Kroll's reagent. |

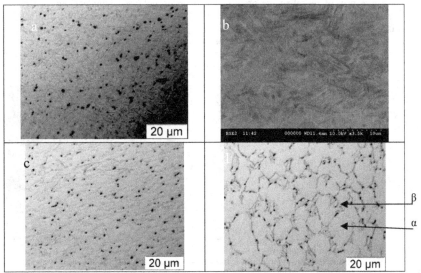

Figure 5: Ti-6Al-4V welded parallel to the rolling direction at 50 MPa axial pressure, 50 Hz frequency; etched with Kroll's reagent. (a) weld centre at high magnification (b) BSE image of weld centre at 3.5 times the magnification in (a) (c) 0.375 mm above weld centre (d) 0.725 mm above weld centre.

Optical micrographs of the sample welded with the oscillation direction *perpendicular* to the rolling direction are presented in Figure 6. Here the same welding parameters were employed as in the previous example. The weld region appears bright due to its resistance to etching, which is attributed to this region being heavily deformed and very fine grained. This phenomenon occurred in the parallel sample as well. An approximate measurement of the weld (unetched) region indicated that it was 0.75 mm across. The deformed region (TMAZ) from the weld line up into the parent metal is illustrated in Figure 6(b). In Figure 6(c), similar to Figure 5(a), faint grain lines in the weld centre indicate the presence of a martensitic structure. The microstructures (not presented) for this sample at 0.325 mm and 0.725 mm from the weld, in turn, resemble those of the parallel-welded sample presented in Figure 5(c) and (d).

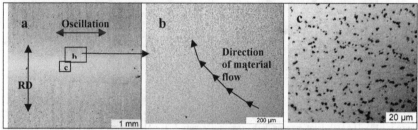

Figure 6: Ti-6Al-4V welded perpendicular to the rolling direction at 50 MPa, 50 Hz; etched with Kroll's reagent. (a) Weld line (low mag) (b) Weld line (c) Weld line (high mag).

An alternative way of revealing the grain structure, using polarized light without an etch, is displayed in Figure 7T. Here, the fine-grained weld region is approximately 0.5 mm across, which is approximately 50% smaller than the weld region in sample #1, welded at lower pressure and a higher frequency. The transition zone between the fine recrystallized grains in the weld zone and the larger adjacent grains is illustrated in Figure 7(b). The grain size in the weld appears to average 1-3 μm and 5-10 μm in the immediately adjacent area.

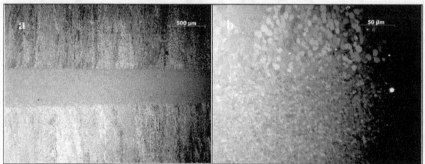

Figure 7: Ti-6Al-4V welded parallel to the rolling direction at 90 MPa axial pressure and 30 Hz frequency; polarized light, unetched. (a) Banded structure next to the weld line and fine grains in the weld.(low mag) (b) Border between recrystallized weld line grains and large grains adjacent to the weld.

Microhardness measurements

It can be seen from Figure 8 and Figure 9 that the weld region is somewhat harder than the surrounding TMAZ. The TMAZ hardness is comparable to that in the as-received material. The gray shaded bar represents the standard deviation of the hardness measured in the as-received material. Comparing the samples welded at low and high pressure, the higher pressure can be seen to produce a narrower weld zone (Figure 9) with a sharp, well-defined drop at +/- 1 cm, in agreement with the microstructural evidence, while the lower pressure sample (Figure 8) has a lower peak hardness and a much less abrupt drop.

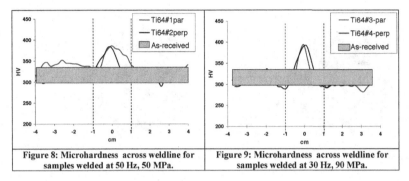

Figure 8: Microhardness across weldline for samples welded at 50 Hz, 50 MPa.	Figure 9: Microhardness across weldline for samples welded at 30 Hz, 90 MPa.

Microtexture measurements

Orientation data were obtained from the as-received and welded samples. The material required an extensive and sensitive polishing procedure, as electropolishing is not recommended in order to preserve any beta that may be found in the structure. In fact, very few beta grains were detected and indexed in any case. The pole figures below refer solely to the orientations of the alpha (HCP) grains.

The as-received texture, Figure 10, has a strong basal component in the direction normal to the rolling direction, as predicted in the literature. Metals or alloys with c/a ratios less than the ideal 1.633, such as titanium, tend to form rolling textures with the basal poles tilted plus or minus 20 to 40 degrees from the normal toward the transverse direction and [10-10] poles aligned with the rolling direction [10].

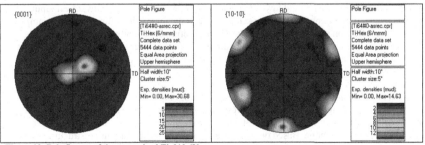

Figure 10: Pole figures of the as-received Ti-6Al-4V.

From Figure 11, it can be seen that the texture of the as-received material has been completely changed by LFW-ing parallel to the rolling direction. Here the data are taken from the thermo-mechanically affected zone near the weld centre. Note that the basal pole has been rotated by approximately 90° about the transverse direction, as have the prismatic poles. Similar remarks apply to Figure 12, which represents the texture of material LFW'd *perpendicular* to the rolling direction. In this case, the basal and prismatic poles have again been rotated by 90° about the transverse direction; the prismatic poles have also been shifted by 90° about the rolling direction with respect to those displayed in Figure 11b, as expected.

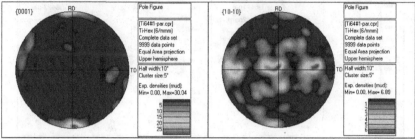

Figure 11: Pole figures of Ti-6Al-4V welded parallel to RD, near weld centre.

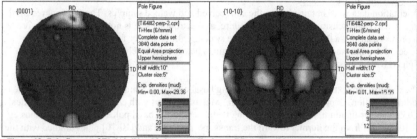

Figure 12: Pole figures of Ti-6Al-4V welded perpendicular to RD near weld centre.

Welding at higher pressure (and lower frequency), with an overall higher power/heat input, results in a similar microtexture for the sample welded parallel to the rolling direction, sample #3, but the reflection peaks appear to be sharper, indicating a more pronounced texture, see Figure 13. This agrees with the findings of Lütjering [9], who observed that, while the texture type depends on the deformation temperature, the texture intensity depends on the degree of deformation.

The final set of pole figures presented, in Figure 14, does not appear to relate to the previous textures observed. Since the first pole figures were acquired over a larger area than this final set, it is possible that only this set represents the real weld centre and not the surrounding thermo-mechanically affected zone. Based on the appearance of the weld centre microstructure, this region is highly martensitic and may have a less pronounced deformation texture than the areas immediately adjacent to the weld.

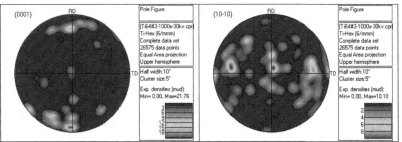

Figure 13: Pole figures of Ti-6Al-4V#3 welded parallel to RD, near weld centre.

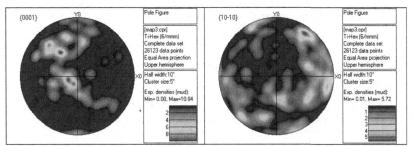

Figure 14: pole figures of Ti-6Al-4V#4 welded perpendicular to RD at weld centre.

Summary and Conclusions

The as-received material contained a banded structure of macrozones approximately 50 to 150 μm in width. One band consisted of equiaxed alpha with transformed beta grains containing lamellae of alpha and beta. The other contained alpha grains with grain boundary beta.

A martensitic structure was present in the centre of the welded specimens, as revealed with Kroll's etch under the optical microscope. This was confirmed with the aid of backscattered electron images from the SEM. The needles were approximately 5 to 7 μm in length and under 0.5 μm in width.

Deformed grains are visible at some distance from the weld centre line, indicating that the total width of the recrystallized, fine-grained weld zone is 0.75 mm for the samples welded at 50 MPa, 50 Hz and 0.5 mm for those welded at 90 MPa, 30 Hz. Equiaxed grains are visible at greater distances from the weld centre line, indicating a total width of the TMAZ of 1.5 mm for the samples welded at 50 MPa, 50 Hz. Samples welded at higher pressure and lower frequency, but overall higher specific power input, had narrower weld regions.

The hardness values and microstructures of the present samples are in agreement with previous findings for material welded under similar conditions. Welding parallel versus perpendicular to the rolling direction does not significantly influence either the hardness or the weld microstructure.

The pole figures show that the as-received material has a typical rolled titanium alloy texture, with strong basal texture components in the directions normal and transverse to the rolling direction. The textures obtained from the welded material oscillated *parallel* to the rolling direction reveal that close to the weld line and in the TMAZ, this texture (described in terms of the basal pole and prismatic poles) is rotated 90°. In the samples oscillated *perpendicular* to the rolling direction, there is an additional shift of 30° towards the transverse direction.

The pole figures generated from orientation data for the sample welded at higher specific power input, perpendicular to the rolling direction, do not match those obtained at lower specific power input. This may be due to a smaller area, closer to the weld centre, being analyzed for this data set, as opposed to the larger region including the TMAZ for the previous sets.

Acknowledgments

The authors are grateful to Dr. S. L. Semiatin of Air Force Research Laboratory, Dayton, Ohio, who generously provided the Ti-6Al-4V used in this study.

References

1 P. Threadgill, Knowledge Summary, web published by TWI, http://www.twi.co.uk/j32k/index_our.xtp.

2 H. Wilhelm, R. Furlan, and K.C. Moloney, "Linear friction bonding of titanium alloys for aero-engine applications", Titanium '95: Science and Technology, Institute of Materials, London, (1996) pp. 620-27.

3 M. Karadge, M. Preuss, C. Lovell, P.J. Withers and S. Bray, "Texture development in Ti-6Al-4V linear friction welds", *Materials Science and Engineering* A 459 (2007) pp. 182-191.

4 A. Vairis and M. Frost, "Modelling the linear friction welding of titanium blocks", *Materials Science and Engineering* A292 (2000) pp. 8–17.

5 A. Vairis, "High frequency linear friction welding" (PhD Thesis, University of Bristol, 1997).

6 M.M. Attallah, "Influence of linear friction welding parameters on the residual stress development in Ti-6246", Trends in Welding Research 8, Georgia (2008).

7 R. Ding, Z.X. Guo, and A. Wilson, "Microstructural evolution of a Ti-6Al-4V alloy during thermomechanical processing", Materials Science and Engineering A327 (2002).

8 P. Wanjara and M. Jahazi, "Linear friction welding of Ti-6Al-4V: processing, microstructure, and mechanical-property inter-relationships", *Metallurgical and Materials Transactions*, Vol 36A, (2005), pp. 2149-2164.

9 G. Lütjering, "Influence of processing on microstructure and mechanical properties of $(\alpha + \beta)$ titanium alloys", *Materials Science and Engineering* A243 (1998) pp. 32-45.

10 U.F. Kocks, C.N. Tomé and H.-R. Wenk, "Texture and anisotropy: preferred orientations in polycrystals and their effect on materials properties". Cambridge University Press, UK, 1998.

FRICTION STIR WELDING AND PROCESSING V

Session II

Session Chair

Murray W. Mahoney

Friction Stir Welding and Processing V
Edited by: Rajiv S. Mishra, Murray W. Mahoney, and Thomas J. Lienert
TMS (The Minerals, Metals & Materials Society), 2009

MICROSTRUCTURE AND PROPERTIES OF FRICTION STIR WELDED 1.3WT% N CONTAINING STEEL

Yutaka S. Sato[1], Kei Nakamura[1], Hiroyuki Kokawa[1], Shuji Narita[2], Tetsuya Shimizu[2]

[1]Tohoku University; 6-6-02 Aramaki-aza-Aoba, Aoba-ku; Sendai, 980-8579, Japan
[2]Daido Steel; 2-30 Daido-cho, Minami-ku; Nagoya, 457-8545, Japan

Keywords: Friction stir welding, High nitrogen steel, Microstructure, Property

Abstract

In this study, FSW was applied to a high nitrogen steel (HNS) containing 1.3wt% nitrogen using a PCBN tool, and feasibility of FSW for HNS, and microstructure and properties of the weld were examined. FSW produced defect-free welds in the HNS at several welding parameters. The stir zone had roughly the same nitrogen content as the base material, which suggested that both the PCBN tool wear and the nitrogen desorption hardly occurred during FSW. FSW refined the grain structure in the stir zone, which resulted in the higher hardness than the base material. Simultaneously, FSW resulted in rapid formation of Cr_2N precipitates on the grain boundaries in the stir zone, which caused reduction of the corrosion resistance. This study showed that FSW is an effective method to produce a defect-free weld with high hardness in the HNS, although the corrosion resistance of the stir zone is reduced.

Introduction

High nitrogen steel (HNS) is advanced steel with nitrogen partly or completely replacing Ni to obtain the austenitic structure. Austenitic steels can benefit from high nitrogen on some aspects: 1) nitrogen in solid solution is a beneficial alloying element to raise the strength without significant loss of the ductility and toughness; 2) nitrogen is a strong austenite stabilizer, thereby reducing the amount of Ni required for austenite stabilization; and 3) nitrogen remarkably improves resistance to intergranular, pitting and crevice corrosion and stress corrosion cracking [1].

When the HNS is used as a structural material, welding and joining processes are required. However, fusion welding of the HNS results in nitrogen desorption from the weld pool because the solubility of nitrogen in the solid metal is much lower than that in the liquid [2]. The nitrogen desorption often produces large porosity in the fusion zone. Moreover, the HNS often experiences precipitation of cellular and intergranular Cr nitrides (Cr_2N) in the heat-affected zone (HAZ) [3]. The nitride precipitation reduces seriously the local mechanical and corrosion resistance. To alleviate these problems, careful control of shielding gas, filler metal composition, and thermal cycle are usually required during welding.

Friction stir welding (FSW) is expected to be an effective joining process to prevent the nitrogen desorption from the HNS, due to the solid-state process accompanying no melting and solidification. Additionally, since this is a low heat-input process, there is a high possibility that this process can alleviate the Cr nitride precipitation in the HAZ. Recently, much progress has been made in FSW of high temperature materials by numerous investigators [4-13], and a previous study on FSW of HNS has also been reported. Park et al. [14] attempted FSW of an HNS containing about 0.5wt% nitrogen using polycrystalline cubic boron nitride (PCBN) tool, and then examined the microstructural feature of the weld. FSW produced a weld with some tunnel-type defects in the stir zone of the HNS. Precise microstructural analyses revealed a remarkable increase in nitrogen content and formation of Cr borides in the stir zone, which were caused by severe tool wear of PCBN tool during FSW. They suggests that FSW can be one of the effective processes to avoid the nitrogen desorption and formation of porosity in the HNS, but there have been few studies examining the defect-free weld with no contamination of the welding tool in the HNS. Moreover, the Ni-free HNS containing nitrogen higher than 1.0wt% has been recently developed, but feasibility of FSW for the Ni-free HNS with more than 1.0wt% nitrogen has not been examined yet.

The present study applied FSW to an HNS with 1.3wt%N and few Ni using PCBN tool, and then examined microstructure, and mechanical and corrosion properties of the weld. The objective of this study is to clarify effect of FSW on microstructure and properties of the 1.3wt%N HNS.

Experimental procedures

The base material used in this study is an HNS obtained by pressurized melting process, whose chemical composition is 20.84Cr-0.20Ni-8.02Mn-3.91Mo-0.04C-1.31N (wt%). Bead-on-plate FSW was applied to the 4mm-thick material using a convex-scrolled-shoulder-step-spiral (CS4) pin tool made of PCBN. The welding speed of 1.0 mm/s, tool tilt angle of 0 deg and axial load to the spindle of 30 kN were constantly used, and the rotational speed of the tool varied between 400 and 800 rpm.

Nitrogen contents of the base material and stir zone were measured in an inert gas atmosphere by fusion gas chromatographic analyzer produced by LECO (model TC-436DR oxygen and nitrogen analyzer). The samples of about 0.5 g for nitrogen content analysis were carefully cut from only the stir zone using an electrical-discharge machine. The samples were then cleaned by filing away the surface and washed with acetone for 300 s by an ultrasonic cleaner. At least three samples were analyzed for both the stir zone and base material.

Microstructures of the welds were examined by optical microscopy, scanning electron microscopy (SEM), and electron backscatter diffraction (EBSD) and transmission electron microscopy (TEM). Samples for optical microscopy and SEM were electrolytically etched in a 10 vol.% oxalic acid aqueous solution at 30 V for 5 s to observe the microstructure distinctly. Morphology and size of the grains were evaluated by EBSD. Cross sections for EBSD analysis were cut perpendicular to the welding direction and then electrolytically polished in 10 vol.% $HClO_4$ ethanol solution at 223 K (-50 °C). Crystallographic data were obtained from center of the stir zone. EBSD data were collected in a Hitachi H4300SE SEM, operating at 20 kV. Thin-foil disk specimens, 3 mm in diameter, for TEM were cut from the welds using an electrical discharge machine, and then were prepared by jet eletropolishing in a 20 vol.% $HClO_4$ ethanol solution at 223 K (-50 °C). These thin foils were observed at 200kV in a JEOL JEM-2100 TEM.

Local mechanical properties in the welds were evaluated by Vickers hardness test. Vickers hardness profiles across the stir zone were measured at the mid-thickness of the weld on the cross section perpendicular to the welding direction using a load of 9.8 N for 15 s. Corrosion properties are examined by ferric sulfate-sulfuric acid test [15]. The specimen having dimension of 10 x 50 x 6 mm was soaked in the test solution for 345.6 ks (72 h) during the test. Tested specimens were observed by SEM.

Results and discussion

Cross sections of the welds produced at rotational speeds of 400, 600 and 800 rpm are shown in Fig. 1. FSW could produce defect-free welds in the HNS at the welding parameters used in this study. The contrast of the stir zone in this figure depends on the rotational speed, i.e. the lower rotational speed results in the lighter contrast in the stir zone. The intergranular corrosion was observed in a part of the heat-affected zone (HAZ) of all welds after etching in the oxalic acid aqueous solution. It does not occur in the HAZ proximal to the stir zone, and the location of corrosion is farther from the border between stir zone and TMAZ at the higher rotational speed.

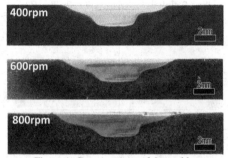

Figure 1. Cross sections of the welds.

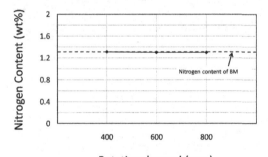

Rotational speed (rpm)

Figure 2. Effect of rotational speed on nitrogen content of the stir zone.

Effect of rotational speed on nitrogen content of the stir zone is given in Fig. 2. Nitrogen contents of the stir zones are about 1.3 wt%, which is roughly the same as that of the base material. This result shows that the nitrogen desorption hardly occurred during FSW. On the other hand, Park et al. [16] reported a drastic increase in nitrogen content in the stir zone of an

HNS after FSW using PCBN tool, which was attributed to wear of the PCBN tool. Therefore, this result strongly suggests that the tool was scarcely worn during FSW in this study.

Optical micrographs of the base material and centers of the stir zones are shown in Fig. 3. The base material has an annealed microstructure with a high density of twin boundaries. Average grain size of the base material was about 24 μm. In the stir zones, on the other hand, the grain boundaries are hardly observed due to presence of the pits, which are significant in the stir zone produced at 400 rpm. The number of the pits increases with decreasing rotational speed. EBSD was employed to clarify size and morphology of the grain structure in the stir zone, and quantified to be 0.5, 1.0 and 1.5 μm as average grain sizes of the stir zones produced at 400, 600 and 800 rpm, respectively. This grain refinement would be due to recrystallization during FSW.

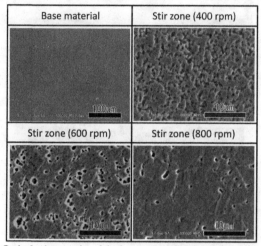

Figure 3. Optical micrographs of the base material and centers of the stir zones.

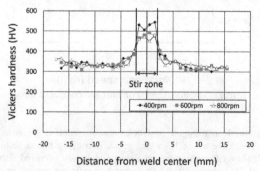

Distance from weld center (mm)

Figure 4. Hardness profiles of the welds.

Hardness profiles across the stir zones in the welds are presented in Fig. 4. Hardness of the base material was scattered between 320 and 350 Hv. The stir zone exhibits higher hardness than the base material. Hardness of the stir zone increases with decreasing rotational speed. This result suggests that the hardness mainly depends on the grain size in the stir zone.

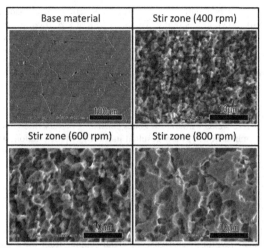

Figure 5. SEM images of the base material and centers of the stir zones after ferric sulfate-sulfuric acid test

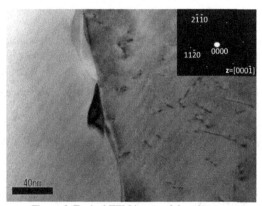

Figure 6. Typical TEM image of the stir zone.

SEM images of the base material and centers of the stir zones subjected to ferric sulfate-sulfuric acid test are shown in Fig. 5. A small number of pits can be seen in the base material, while the stir zones are deeply corroded. Degree of the corrosion is more significant in the stir zone produced at the lower rotational speed.

To identify the microstructural feature leading to deterioration of corrosion resistance in the stir zone, TEM observation was conducted. A typical TEM image of the stir zone is shown in Fig. 6. Any precipitates were hardly observed in the base material, while the stir zones have tiny precipitates along the grain boundaries. Selected area diffraction (SAD) pattern taken from the tiny precipitates is also shown in this figure. The SAD pattern identified this precipitate as Cr_2N with hexagonal structure. The number of Cr_2N precipitates increased with decreasing rotational speed, which may contribute to an increase in hardness in the stir zone with the grain size. Since formation of Cr_2N precipitates can result in Cr depleted zone around the precipitate and the grain boundaries, it would cause the reduction of the corrosion resistance of the stir zone.

Summary

FSW produced defect-free welds with the stir zones consisting of very fine grain structure, which raised the hardness. Simultaneously, FSW resulted in the Cr_2N precipitation, which deteriorated the corrosion resistance. The present study shows that FSW is an effectively welding method to produce relatively high-quality weld without any defects in the HNS compared to fusion welding processes.

Acknowledgements

The authors are grateful to Mr. A. Honda for technical assistance and thank Prof. Z.J. Wang and Dr. S. Mironov for their helpful discussions. Financial support from the Japanese Ministry of Education, Science, Sports and Culture with a Grant-in-Aid from the Global COE program in Materials Integration International Center of Education and Research at Tohoku University are gratefully acknowledged.

References

1. L. Zhao, Z. Tian, and Y. Peng, "Microstructure and Mechanical Properties of Heat-affected Zone of High Nitrogen Steel Simulated for Laser Welding Conditions," *ISIJ International*, 47 (2007), 1351-1356.

2. W. Dong, H. Kokawa, S. Tsukamoto, and Y.S. Sato, "Nitrogen Desorption by High-Nitrogen Steel Weld Metal during CO2 Laser Welding," *Metallurgical and Materials Transactions B*, 36B (2005), 677-681.

3. I. Woo, M. Aritoshi, and Y. Kikuchi, "Metallurgical and Mechanical Properties of High Nitrogen Austenitic Stainless Steel Friction Welds," *ISIJ International*, 42 (2002), 401-406.

4. T.J. Lienert, W.L. Stellwag, Jr., B.B. Grimmett, and R.W. Warke, "Friction stir welding studies on mild steel – Process results, microstructures, and mechanical properties are reported," *Welding Journal*, 82 (2003), 1s-9s.

5. A.P. Reynolds, W. Tang, T. Gnaupel-Herold, and H. Prask, "Structure, properties, and residual stress of 304L stainless steel friction stir welds," *Scripta Materialia*, 48 (2003), 1289-1294.

6. S.H.C. Park, Y.S. Sato, H. Kokawa, K. Okamoto, S. Hirano, and M. Inagaki, "Rapid formation of the sigma phase in 304 stainless steel during friction stir welding," *Scripta Materialia*, 49 (2003), 1175-1180.

7. S.H.C. Park, Y.S. Sato, H. Kokawa, K. Okamoto, S. Hirano, and M. Inagaki "Corrosion resistance of friction stir welded 304 stainless steel," *Scripta Materialia*, 51 (2004), 101-105.

8. Y.S. Sato, T.W. Nelson, and C.J. Sterling, "Recrystallization in type 304L stainless steel during friction stirring," *Acta Materialia*, 53 (2005), 637-645.

9. H. Fujii, L. Cui, N. Tsuji, M. Maeda, K. Nakata, and K. Nogi, "Friction stir welding of carbon steels," *Materials Science and Engineering A*, 429 (2006), 50-57.

10. L. Cui, H. Fujii, N. Tsuji, K. Nakata, K. Nogi, R. Ikeda, and M. Matsushita, "Transformation in Stir Zone of Friction Stir Welded Carbon Steels with Different Carbon Contents," *ISIJ International*, 47 (2007), 299-306.
11. A.P. Reynolds, W. Tang, M. Posada, and J. DeLoach, "Friction stir welding of DH36 steel," *Science and Technology of Welding and Joining*, 8 (2003), 455-460.

12. Y.S. Sato, P. Arkom, H. Kokawa, T.W. Nelson, and R.J. Steel, "Effect of microstructure on properties of friction stir welded Inconel Alloy 600," *Materials Science and Engineering A*, 477 (2008), 250-258.

13. S. Mironov, Y.S. Sato, and H. Kokawa, "Microstructural evolution during friction stir-processing of pure iron," *Acta Materialia*, 56 (2008), 2602-2614.

14. S.H.C. Park, Y.S. Sato, H. Kokawa, K. Okamoto, S. Hirano, and M. Inagaki "Microstructure of Friction-Stir-Welded High-Nitrogen Stainless Steel," *Materials Science Forum*, 539-543 (2007), 3757-3762.

15. R. Bandy and D. van Rooyen "Effect of Thermal Stabilization on Low Temperature Stress Corrosion Cracking of Inconel 600" *Corrosion*, 40 (1984), 281-289.

16. S.H.C. Park, Y.S. Sato, H. Kokawa, K. Okamoto, S. Hirano, and M. Inagaki "Boride formation induced by pcBN tool wear in friction-stir-welded stainless steels," *Metallurgical and Materials Transactions A*, (2008), in press.

Friction Stir Welding and Processing V
Edited by: Rajiv S. Mishra, Murray W. Mahoney, and Thomas J. Lienert
TMS (The Minerals, Metals & Materials Society), 2009

FRICTION TAPER STUD WELDING OF CREEP RESISTANT 10CrMo910

D.G. Hattingh[1], M Newby[2], A. Steuwer[1,3], I.N. Wedderburn[1], P Doubell[2], M.N. James[4]

[1]NMMU, Gardham Avenue, PO Box 77000, Port Elizabeth, 6031, South Africa
[2]Eskom Holdings Ltd, Lower Germiston Road, Rosherville, Johannesburg, 2022, South Africa
[3]ESS Scandinvia, Stora Algatan 4, 22350 Lund, Sweden
[4]University of Plymouth, Drake Circus, Plymouth PL4 8AA, UK

Keywords: friction taper stud welding, heat treatment, residual stress, creep resistant steel, neutron diffraction

Abstract

Friction Taper Stud Welding (FTSW) is a novel welding technique that involves forcing a rotating consumable tool into a tapered (conical) cavity of nearly matching shape. The resultant generated heat causes a plasticized layer which bonds to the bottom of the hole and radially to the adjacent hole side. This is similar to other friction welding techniques such as linear and inertia friction welding, but involves a conical interface. Possible applications are repair welds in steel pipes. However, detailed knowledge of the residual stress distributions is essential for structural integrity interactions. This manuscript introduces the main concepts of FTSW and discusses the effects of pre and post weld heat treatment on the triaxial residual stress field (measured by neutron diffraction) generated by FTSW in a creep resistant steel manufactured from 10CrMo910 steel.

Introduction

Since Bevington filed the first patent for the use of friction welding and processing materials in the late 19th century (1) around 17 techniques of using friction constructively to join, cut, clad and process materials have evolved. Friction stir welding (FSW) invented at The Welding Institute (TWI) and patented in 1991 provided the means to perform butt joints on long seams. Friction Taper Stud Welding (FTSW) a variant of the original Friction Hydro Pillar Processing (FHPP) is a comparatively recent solid-phase welding technique. This technique is the focus of considerable R&D interest because of its potential in fabrication and manufacturing where it offers a number of novel production routes (2, 3). While development of the FTSW technique is ongoing, it already shows promise for joining and repairing even thick plate in ferrous and non-ferrous materials. Good quality FHPP welds have been produced, using a parallel hole geometry, in steel and certain non-ferrous materials, and these have been characterized by good impact, tensile, and bend results (4, 5, 6). Good mechanical integrity can be achieved and Impact tests have demonstrated that a significant improvement in toughness properties can also be achieved. Metallographic examination has shown that both FHPP and FTSW deposit material which is hot-worked with a very fine-grained microstructure.

Developments in the field have reached the stage where benefits from the process can be exploited in the power generation milieu where arc welding traditionally represented the vast majority of welding processes. Friction weldments exhibit inherent good properties. Microstructure phase transformation, that is ferrite to austenite and back, occurs to a lesser extent. This limits the occurrence of short range stresses due to volume changes and the complicated negative effects of gas solubility changes which are particularly pertinent to negative effects of the diffusion of dissolved hydrogen from different areas around the weldment. The lower temperatures involved also results in a very narrow heat affected zone (HAZ). This in turn limits the volume of potentially hard and brittle microstructure with associated problems such as hydrogen embrittlement often experienced on high strength low alloy steels. For stainless steels the occurrence of sensitization can also be significantly reduced. The completed weldment exhibits a very fine homogeneous microstructure, a feature that is conducive to good impact toughness properties. The process is ideally suited for automation. Friction welding involves a solid state welding process, no volume change during solid to liquid state occurs and thermal stresses are thus significantly less if compared to a welding process that relies on a molten weld pool. This implicates much reduced effort for the weld set-up especially if post weld heat treatment (PWHT) can be omitted. PWHT for thermal stress relieving purposes requires significant time to affect and often requires pipe blocking. The physical properties of the friction welding process lend itself to special niche applications that render previously scrapped components candidates for repairs. Inherent good mechanical properties promote the repairs of high integrity weldments such as turbine shafts, discs and blade roots. From a process point of view virtually no fumes are generated, minimal distortion experienced and no spatter has to be removed afterwards. Limitations of the process are mainly as a result of the process still being developed to full potential and many potential areas of application poses technical challenges with regards to the application with suitable equipment.

Development of FTSW process for 10CrMo910

A comprehensive literature survey was performed to gather all the information required to compile a suitable test matrix. To avoid incurring unnecessary expenses prematurely, the chances for success was first gauged through a low budget feasibility study performed at the Nelson Mandela Metropolitan University (NMMU) which indicated that good weldments can be achieved through further development of the basic weld test matrix. A limited number of FHPP weldments were performed on low carbon steel test plates, yielding areas of good fusion on some of the samples. The principal objective of this investigation was to test the viability of the Friction Hydro Pillar Process (FHPP) welding technique as a process for plugging holes in steel sections. Due to the lack of literature concerning the FHPP technique there was some difficulty in selecting start parameters for this viability study. Parameters that required consideration before testing could commence were:

- parent material and consumable tool material
- hole and consumable tool geometry: diameter, depth and profile
- process parameters such as rotational speed, feed rate and downward force (Z force)

Typical test coupons prepared for the feasibility study is depicted in Figure 1.

Figure 1: Test Samples showing consumable tools and test blocks with prepared holes.

A modified Nicholas-Corea milling machine was used to carry out the test welds. The machine has been modified to carry out Friction Stir Welding research making it suitable for this investigation. The mill head has a custom fitted telemetry device used to monitor spindle speed (rpm), down force (N) and torque (Nm). Motor modifications allow for a max spindle speed of 800 rpm. Figure 2 shows a weld in progress.

Figure 2: Weld in progress showing tool upset in plasticized condition.

Initial tests did not successfully achieve the Friction Hydro Pillar Process, however, it did show enough evidence that the process is possible. Although sectioning revealed large cavities on most of the welds some areas displayed adequate bonding. It was obvious that obtaining the correct relationship between the rotary speed of the tool and the down force on the tool will be critical in perfecting the FHPP. This will in turn be determined by the material characteristics of the weld material. Optimum FTSW relies on correct hole/stud geometries of the taper angle, rotation speed and downward pressure. Despite the relative narrow band of parameter ranges, the FTSW study yielded good repeatability with consistent good side-wall fusion as revealed in Figure 3. Resultant samples were sectioned, etched and polished for metallographic evaluation and reported with macro and microphotographs while hardness surveys were conducted on selected samples. For both processes hardness checks yielded elevated hardness, however initial PWHT studies showed temper response to be good for the weldments.

Figure 3: Cross-sectional view of the FTSW.

Portable FTSW platform
A key factor for effective application of the technology within Industrial plant was the development and manufacture a portable device to machine and subsequently to plug blind holes on selected components. These would mainly be thick-walled components such as boiler tube headers mostly manufactured from creep resistant Cr-Mo materials. Another important application will be rotating turbine components such as blades, discs and shafts manufactured from martensitic stainless steels and high strength low alloy steels.

A Portable Friction Welding Platform (PFWP), capable of carrying out hole repairs on thick walled steel pipe sections, was designed, registered and manufactured by the friction stir processing research group at the Nelson Mandela Metropolitan University (NMMU), Port Elizabeth, South Africa. Important specifications for the device were accommodating a core drill tip that can produce a maximum 12 mm diameter profiled (inclined to approximately 1 in 5 taper) hole up to a maximum depth of 25 mm. The platform was based on a

77

combination of electric/pneumatic/hydraulic actuating, clamping and rotating components to apply the necessary clamping force and rotational torque to perform a FTSW. The platform must be capable to operate in close proximity of a component weld preheat up to 200°C. The modular design facilitates easy transport and positioning, which can be remotely operated at a distance of 30 meters. Figure 4 shows the basic assembled modular arrangement of the machine without the power and control systems and the machine during commissioning with control, power and hydraulic lines attached.

Figure 4: Arrangement of Portable Friction Welding Machine (PFWP).

The welding cycle is fully automated and pre-programmed with set points for rotational speed, welding force, tool travel, forging force and forging time. Once these parameters are set and the weld cycle started the machine automatically completes the weld. This results in extremely consistent control of the process parameters, which are also monitored and recorded for verification purposes. Maximum operating conditions for the equipment are a spindle speed of 5000rpm and a welding/forging force of 40kN. Vertical head travel is limited to 100mm.

The equipment consistently produced a fully bonded interface on a 25mm deep friction taper plug weld in a 40mm thick rounded creep resistant steel section, with a typical weld time of 20 to 25 seconds.

Characterization of pre and post weld heat treatment effects on the residual stress field of FTSW in creep resistance steel (10crmo910)

The repair of holes in thick walled steam pipes used in power generation, manufactured from creep resistant 10CrMo910 steel predictably require heat treatment. This work aimed to characterise the effects of pre and post weld heat treatment on the triaxial residual stress field generated by FTSW. In order to optimise this repair process there is a requirement for research work to be performed to identify optimum heat treatment parameters in order to reduce detrimental residual stresses (10). We therefore require an understanding of the effect of the heat treatment processes on the residual stress field of the weld. This information will be used to optimise the process parameters necessary to achieve the required combination of mechanical properties and residual stresses. The main variables governing the FTSW process are the rotary speed of the tool and the magnitude of the loading force applied to the tool during rotation, which intensifies the frictional contact. A third parameter namely 'tool off-set' can also be controlled. This represents the amount of 'shortening' the consumable tool undergoes during welding, in other words the vertical downwards feed distance of the consumable tool after the point that the tool comes into contact with the hole bottom. Table 1 shows the typical ranges for these parameters applicable to a specific hole and tool geometry.

Table 1 – FTSW Variable Process Parameters

TOOL SPEED (N)	TOOL LOADING	TOOL OFF-SET

4200 - 5000rpm	10 - 20kN	10 – 20mm

In this experiment, welds were produced with fixed weld parameters, namely the rotary tool speed, tool loading (vertical force) and tool off-set. The heat treatment processes used was aligned with the currently employed processes used in the power generation industry for weld heat treatments on 10CrMo910 steels which is a heat resistant Martensitic steel manufactured by *Hot Isostatic Pressing*. The steel has a fine grained microstructure ideally suited for diffraction investigation. For this material it is specified that pre-heating is required for thick walled sections and post weld annealing is mandatory. The temperature ranges and processing parameters used for producing the samples are given in Table 2.

Table 2 – Process data and Heat Treatment Temperature Ranges

Weld No.	Spindle Speed (RPM)	Tool Pressure (Bar)	Tool Force (kN)	Burn-off (mm)	Holding Time (s)	Pre-weld Heat Treatment	Post Weld Heat Treatment	Comments
LL08_W01	5000	68	17.775	12.0	18.0	-	-	-
ILL08_W02	5000	68	17.775	12.0	18.0	200 -300°C	-	20 minutes soak
LL08_W03	5000	68	17.775	12.0	18.0	-	650 - 750°C	Soak for 2 minutes per mm thickness. Slow still air cooled

Figure 5: FTSW test coupon for SALSA Neutron Scanning Experiment.

The dimensions of the final friction-processed coupon as shown in Figure 5 were approximately 12.5cm x 12.5cm and 6cm thick in the centre. The high intensity neutron source at the Institut Laue Langevin, (ILL), in Grenoble, France, coupled with an automated, precise and fast diffraction instrument such as SALSA allow the non-destructive determination of the triaxial residual stress field in these samples. The gauge volume was approximately 2-3mm^3 with typical counting times of the order of 15min, using a 110 and 211 reflections, wave length of 1.74Å was used. The measurements were carried out on one half of a symmetry plane bisecting the sample using neutron count rather than scan time as a measurement control parameter. Figure 6 shows a typical set-up on the Hexapod at SALSA for the scanning of the longitudinal (Hoop Strain) direction.

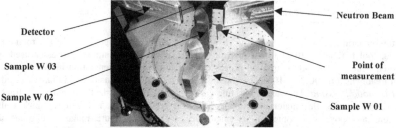

Detector

Sample W 03

Sample W 02

Neutron Beam

Point of measurement

Sample W 01

Figure 6. Set-Up for Longitudinal (Hoop Strain) Scanning.

The matrix map in Figure 7 indicates all the points considered for measurement during the different scan directions.

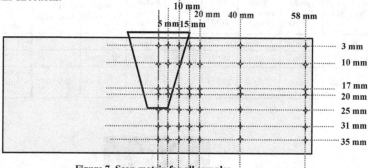

Figure 7. Scan matrix for all samples.

For the unstrained reference (D0) sample (Figure 8), an identical sample was produced at identical parameters and machined into a comb shape on the same half-symmetry plane as shown in Figure 8 . D0 scans 3 orientations was performed on both as welded as well as in post weld heat treated (720 °C for 4 minutes) condition.

Figure 8. EDM - Do sample prior to heat treatment.

Residual Stress Summary

This neutron diffraction experiment was performed at the ILL on the SALSA beamline with a wavelength of 1.74 Å. Scans were performed across one half from the centre of three FTSW samples to obtain representative hoop (longitudinal scan), Axial (transverse scan) and radial (reflection scan) strain data which will be correlated to the different heat treatment conditions. The hoop and axial strain scanning was performed along the 110 reflection plane while for the reflection strain scanning this was changed to the 211 reflection plane in order to accommodate the larger volume material to be penetrated. Extensive D0 measurements

80

were also performed on both as welded and heat treated sample. The data will allow predictions to be made for particular combinations of heat treatment parameters for a set of FTSW conditions.

Figure 9, 10 and 11 shows the strain plot of the hoop residual strain distribution across the as-welded, pro heat treatment and post heat treatment transverse section of welds made with constant weld parameters but different heat treatment conditions.

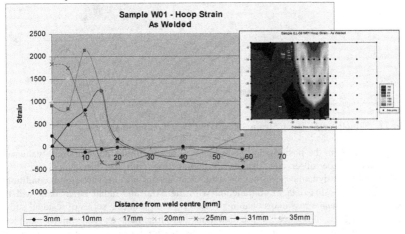

Figure 9: Sample W01 – Hoop Strain – As Welded sample.

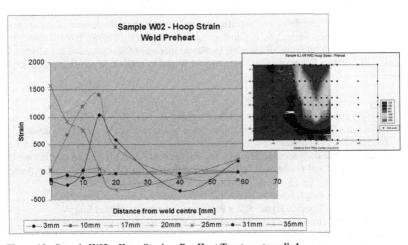

Figure 10: Sample W02 – Hoop Strain – Pre-Heat Treatment applied.

81

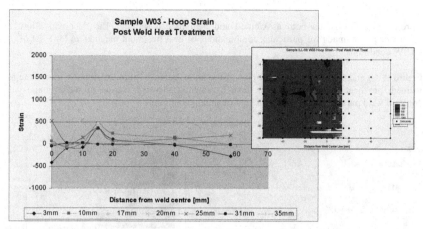

Figure 11: Sample W03 – Hoop Strain – Post Weld Heat Treatment applied.

A clear difference in the position and appearance of the maximum and minimum values of hoop strains can be observed. Taken in conjunction with the mechanical property, metallographic and dynamic performance data, clear insight into heat treatment conditions can be extracted. This data in conjunction with the axial and radial strain data will be processes into residual stress maps enabling researchers to produce a series of residual stress maps describing the influence of pre and post heat treatment on the welded samples. Detailed knowledge of the residual stress and strain distributions is indispensable in linking the FTSW parameters with mechanical properties and fatigue performance.

This data is of high technological significance to advancing performance in many high technology applications limited by residual stress and structural integrity interactions. The residual stress data will help to validate FE model predictions of this process, and ultimately the work will lead to a better understanding of FSTW and to solid-state welding processes in general.

References
1.   Friction processes - 100 years on; TWI Connect, no.106, May-June 2000, p.3
2.   Jefferson's Welding Encyclopedia; Compiled by the AWS ; Published by Global Engineering Documents
3.   Emerging friction joining technology for stainless steel and aluminium applications; Thomas W M and Nicholas E D; Presented at 'Productivity beyond 2000': IIW Asian Pacific Welding Congress, Auckland, New Zealand, February 1996
4.   TWI, Leading Edge, Friction Hydro Pillar Processing, Connect, June 1992
5.   Nicholas E.D; Friction Hydro Pillar Processing, Advances in Welding Technology,11th Annual North American Welding Research Conference, 7-9 November 1995
6.   Nicholas E.D; Fabrication by Friction, Fabr. Eng., April 1985, p254-255
7   NICHOLAS, E.D., Leading Edge - Friction Hydro Pillar Processing, TWI, Connect, June 1992
8   THOMAS, W.M. & NICHOLAS, E.D., Friction Takes the Plunge, TWI, Connect, September 1993
9   BLAKESTONE, G.R., Friction Welding on Live Pipelines, Pipeline Technology, Volume 1, 1995

10   M.N. James, et al., "Residual stresses and fatigue performance", Engineering Failure Analysis, Vol 14 (2), 384-395 (2007)

Friction Stir Welding and Processing V
Edited by: Rajiv S. Mishra, Murray W. Mahoney, and Thomas J. Lienert
TMS (The Minerals, Metals & Materials Society), 2009

Exploring Geometry Effects for Convex Scrolled Shoulder, Step Spiral Probe FSW Tools

Carl Sorensen, Bryce Nielsen

Brigham Young University, 435 CTB, Provo, UT 84602, USA

Keywords: Friction Stir Welding, Tool Geometry, CS4 Tools

Abstract

A new tool design for FSW is the convex scrolled shoulder, step spiral probe (CS4) tool. Compared with traditional FSW tools, the CS4 tool has been demonstrated to offer larger process windows, lower operating forces, and the possibility of operating at zero tilt angle. This paper presents a parametric geometric description of the CS4 tool. Based on this description, a series of experiments has been performed to determine the effects of tool geometry on operating forces and weld surface finish. The effects of tool geometry on operating forces is explained.

Introduction

The Convex Scrolled Shoulder, Step Spiral Probe (CS4) tool has recently been developed for use in friction stir processing. This tool is able to accommodate variable plate thickness, run at zero tilt, and reduce process forces in many cases. However, the initial tools were designed largely by intuition; no thorough study of the tool parameters has been performed. There is only anecdotal understanding of the effects of each of the different geometric parameters of the CS4 tool.

The objective of this research project is to develop a fundamental understanding of the effects of CS4 geometric parameters on the forces, spindle torque, tool temperature, and quality of the weld surface. This is accomplished through statistical experimentation, with the goal of producing an empirical model of the system behavior, rather than trying to drive the process outputs to a particular point.

This report covers two experiments. The first is a preliminary or pilot study whose goal is to determine a safe region for completing a statistical experiment. The second is a statistical screening experiment to identify the tool design parameters with the most significant effect on tool performance.

CS4 Tool Parameters

To characterize each of the different parameters of the CS4 geometry, a standard CS4 parameterization has been developed as described by Sorensen and Nielsen (2007). The same variables are used in this report to describe the tool geometry. The variables circled in Figure 1 are the nine parameters that were varied throughout this study. All remaining parameters remain constant or their values vary as a function of other parameters and their respective values can be found in Table 1.

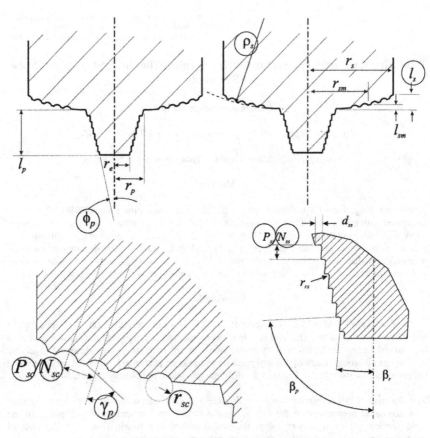

Figure 1: Tool Geometry Variables

Table 1: Constant Tool Parameters

Parameter	Value	Parameter	Value
Probe length l_p	5.1 mm (.200 in)	Mean shoulder length l_{sm}	$F(\rho_s, r_s, l_s, r_p)$
Probe end radius r_e	.75 mm (.030 in)	SS Relief Angle β_r	0 deg
Probe base radius r_p	$F(l_p, r_e, \varphi_p)$	SS Pressure Angle β_p	90 deg
Outside shoulder radius r_s	13.2 mm (.498 in)	Root radius r_{rs}	Small as possible
Mean shoulder radius r_{sm}	$F(r_s, r_p)$	SS Depth d_{ss}	$F(P_{ss}, N_{ss}, \varphi_p)$

Pilot Study

The CS4 tool designs used in this study were slightly different from those used in previous work. The shoulder concave radius, ρ_s, was 25 mm (1 in), instead of 89 mm (3.5 in) as in previous tools. In addition, the cross sectional geometry of the scroll was a circular arc, rather than a complex geometry developed by using an angled grinding wheel to shape the scroll. Further, one of the designs had a concave region of the shoulder near the probe, a configuration that had never before been tested. Because these design differences might lead to process failure or tool failure, it was determined that a pilot study should be performed to explore the performance of these tools and give guidance for the statistical screening design.

Accordingly, a pilot study was developed using five tools, each of which had one parameter that varied significantly from the nominal design which was created. The parameters that were held constant in the pilot study are listed in Table 2. The values for each of the variable parameters in the pilot study are listed in Table 3

Table 2: Constant Tool Parameters for the Pilot Study

Parameter	Value	Parameter	Value
Tread height P_{ss}/N_{ss}	.63 mm (.025 in)	Number of scrolls N_{sc}	$F(\rho_s, r_s, l_s, r_p)$
Scroll pressure angle γ_p	60 deg	Scroll pitch P_{sc}	$3N_{sc}r_{sc}\sin(\gamma_p)$

Table 3: Variable parameters for the Pilot Study

Tool	φ_p, deg.	l_s, mm(in)	N_{ss}	r_{sc}, mm (in)	, mm (in)
1	40°	1.6 (0.063)	2	12.7 (0.5)	25.4 (1)
2	20°	1.6 (0.063)	2	12.7 (0.5)	25.4 (1)
3	30°	1.6 (0.063)	2	12.7 (0.5)	11.3 (0.444)
4	30°	3.2 (0.126)	2	12.7 (0.5)	25.4 (1)
5	30°	1.6 (0.063)	4	19.1 (0.75)	25.4 (1)

The tools for the pilot study were run at a range of parameters that had been successful with the previous CS4 tools, in order to make comparisons with earlier work. Pew et al.[1] (2007) had previously run CS4 tools at feed rates from 126-280 mm/min (5-11 ipm) and spindle speeds of 200-800 rpm. To test the tools at a variety of combinations of these parameters each tool was used to make three different welds at fixed feed rates of 126 mm/min (5 ipm), 203 mm/min (8 ipm), and 280 mm/min (11 ipm). In each of these welds the spindle speed would start at 200 rpm. When process loads maintained a steady state for a period of about 30 seconds, the spindle speed was increased to 500 rpm. Once steady state was reached, the spindle speed was increased to 800 rpm. Axial, longitudinal, and transverse forces and spindle torques were measured during each of the steady-state periods.

Welds in the pilot study were run in position control mode. Depths were chosen to give a constant engaged shoulder diameter for all welds. Because the shoulder geometries were significantly different, different tools required different depths to achieve the same width. In order to find the appropriate depths, a few practice welds were run to find the depth that produced a weld width approximately 75-80% of the diameter of the tool. Tools 1, 2 and 5 were run at 6.0 mm (.235 in) plunge; Tool 3 was run at 5.5 mm (.215 in) plunge and Tool 4 was run at 7.0 mm (.275 in) plunge.

One of the primary purposes of the pilot study was to establish that changing the tool geometry would make significant changes in the process forces and overall weld quality. The tools were able to produce welds at most combinations of parameters. Some of the parameters left considerable amounts of flash and there was a wide range of surface finish quality for each of the welds. As expected, most of the forces for the welds increased as the feed rate increased and decreased as the rotation speed increased. However, there were a few exceptions. For example; Tool 1 showed a decrease in axial forces at the high feed rate and some of the tools showed minimums for longitudinal and lateral forces at intermediate spindle speeds. The complete results and analysis for the Pilot Study are described by Sorensen and Nielsen[2] (2007).

The other purpose of the pilot study was to determine if any of the tool features would lead to process failure or tool failure. There was no process failure for any of the weld parameters. There was one tool failure. The probe on Tool 2 sheared off at the high feed rate. Therefore the probe diameter was insufficient for the process forces.

Screening Experiment

The pilot study demonstrated that a wide variety of CS4 tools would produce sound welds over a range of parameters and increased the confidence that a bold experimental plan could be successful. The next step is to determine which tool design parameters have the greatest influence on process forces, torque and tool temperature and quantify those relationships. The pilot study also showed that it would be important to consider the surface finish of the welds and the amount of flash produced as dependent variables in the experiment. A modified Plackett-Burman screening experiment was chosen, because it will allow the exploration of up to 11 factors with only 12 runs.

The constant tool parameters from Table 1 are used again in the screening design with the exception of the probe end radius. Instead of keeping the probe end radius constant, the midpoint of the probe between the base and the tip is kept at a constant radius to prevent the probe from having such a small diameter at the base. This new midpoint radius r_{mp} is held constant at 2.5 mm (.100 in). The values in Table 2 that were held constant for the pilot study are varied for the screening experiment.

The remaining parameters of the tool vary in the Plackett-Burman study, either directly or as the result of a constraint equation. The variable parameters include the cone angle of the pin (φ_p), the radius of curvature on the shoulder (ρ_s), the shoulder length (l_s), the number of step spirals (N_{ss}), the pitch of the step spirals (P_{ss}), the radius of the scroll cut (r_{sc}), the pressure angle of the scroll cut (γ_p), the number of scroll starts (N_{sc}) and also the distance between the scrolls. To help define the distance between the scrolls we introduce a new variable (L_a) which describes the fraction of the original shoulder that has not been removed by the scroll cuts. This shoulder land fraction allows us to define the pitch of each scroll (P_{sc}) with the following equation:

$$\frac{P_{sc}}{N_{sc}} = 2(1+\frac{L_a}{1-L_a})r_{sc}\sin(\gamma_p)$$

(1)

Table 4 shows the variable parameters used in the screening design, along with the high, low and center values of each of these parameters. Table 5 lists the variable settings for each of the factors for each run of the experimental design, which consists of a 12-run Plackett-Burman design plus an additional center point, for a total of 13 runs. The high, low and center values are represented with a 1, -1 or 0 respectively.

Table 5: Factor Levels for each variable in Screening Experiment

Level	X1 φ_p, deg	X2 $1/r_{sc}$ mm^{-1} (in^{-1})	X3 l_s mm (in)	X4 N_{ss}	X5 N_{ss}/P_{ss} mm^{-1} (in^{-1})	X6 γ_p, deg	X7 N_{sc}	X8 r_{sc} mm (in)	X9 L_a
Low	15	0.020 (0.5)	1.5 (0.06)	2	0.79 (20)	45	2	0.5 (0.02)	0.25
High	35	0.060 (1.5)	3.0 (0.12)	4	1.97 (50)	75	4	0.75 (0.03)	0.5
Center	25	0.040 (1)	2.3 (0.09)	3	1.37 (35)	60	3	0.63 (0.025)	0.375

Table 6: Plackett-Burman Experimental Design

Tool	X1	X2	X3	X4	X5	X6	X7	X8	X9
1	1	1	-1	1	1	1	-1	-1	-1
2	1	-1	1	1	1	-1	-1	-1	1
3	-1	1	1	1	-1	-1	-1	1	-1
4	1	1	1	-1	-1	-1	1	-1	1
5	1	1	-1	-1	-1	1	-1	1	1
6	1	-1	-1	-1	1	-1	1	1	-1
7	-1	-1	-1	1	-1	1	1	-1	1
8	-1	-1	1	-1	1	1	-1	1	1
9	-1	1	-1	1	1	-1	1	1	1
10	1	-1	1	1	-1	1	1	1	-1
11	-1	1	1	-1	1	1	1	-1	-1
12	-1	-1	-1	-1	-1	-1	-1	-1	-1
13	0	0	0	0	0	0	0	0	0

One observation from the pilot study was that the worst welds (poor surface finish or high forces) were the welds run at high spindle speed and low travel speed or the welds run at low spindle speed and high travel speed. To reduce this effect it was determined that each tool would be run at two fixed feed rates; at 152 mm/min (6 ipm) and at 254 mm/min (10 ipm). As before in the pilot study, each weld would consist of three different spindle speeds with enough time to reach a steady state at each spindle speed. The difference between the screening experiment would be that a variation of a Pseudo Heat Index (PHI) proposed by Chimbli et al.[3] (2007) would be used to vary the spindle speeds:

$$PHI = \frac{(RPM)^2}{IPM} \qquad (2)$$

Instead of running the welds at the same combination of spindle speeds for each feed rate, the PHI would stay the same so that the welds run at higher travel speeds would be run at higher spindle speeds as shown in Table 7.

Table 7: Spindle Speeds for Screening Experiment

Travel Speed	Spindle Speeds (RPM)		
mm/min (ipm)	Low	Medium	High
152 (6)	259	433	606
254 (10)	335	559	782

In the pilot study, the welds were run at different plunge depths to achieve the same width of weld. For the screening experiment it was determined that plunge depth would be one of the factors investigated. It was determined that the welds would be run at a plunge depth that varied as a function of the probe length (constant for all tools) and as a fraction of the shoulder length (l_s).

$$plunge\ depth = .2 + p * l_s \qquad (3)$$

Three different plunge depths were desired for each tool. The plunge fractions and their corresponding plunge depths for the different shoulder lengths are shown in Table 8. Therefore, one weld was run at each plunge depth for the two travel speeds for a total of six welds with each tool. Since each weld has three different spindle speeds, there is a total of 18 different weld parameters for each of the 13 tools created.

Table 8: Plunge Depths for Screening Experiment

Plunge fraction (p)	.4 (Low)	.55 (Medium)	.7 (High)
l_s = 1.5 mm (.06 in)	5.7 mm (.224 in)	5.9 mm (.233 in)	6.1 mm (.242 in)
l_s = 2.3 mm (.09 in)	6.0 mm (.236 in)	6.3 mm (.250 in)	6.7 mm (.263 in)
l_s = 3.0 mm (.12 in)	6.3 mm (.248 in)	6.7 mm (.266 in)	7.2 mm (.284 in)

The tools each contained a thermocouple hole, where a Type K Thermocouple was inserted axially to the center of the base of the probe. After each of the welds had achieved a steady state, the average values of the forces, torques and temperatures were measured and used as the yield for the screening experiment analysis.

Screening Experiment Results

After all of the welds were completed and the average forces, torques and temperature had been recorded, five different regressions were performed. The variables in the regression were the factor levels for each of the nine geometric features chosen previously, the weld parameters (feed rate, plunge depth and PHI) and the two level interactions between the geometric features and each of the weld parameters. Interactions between geometric features could not be investigated because of the constraints of the screening design. The yields for the five different regressions were the axial force, longitudinal force, lateral force, spindle torque and tool temperature.

The purpose of the regression analysis was to determine which of the nine geometric features were significant for inclusion in a more comprehensive response surface analysis. The significance level was chosen to be an alpha value of .05. Any factors or interactions that did not meet this significance were dropped from the regression. Using the stepwise regression function in Minitab, all of the insignificant variables were eliminated from the five regressions.

The significant variables were then ranked according to the magnitude of their coefficients and the coefficients of their interactions with the weld parameters. These rankings can be seen in Table 9. A rank of 1 means that the variable had the highest significant coefficients, 2 is the second highest, and so forth. Some of the variables were not significant for that particular response variable or their coefficients were very small when compared to other coefficients and their corresponding slots in the table have been left blank. For almost all of the response variables, the weld parameters were highly significant.

Table 9: Rank of Coefficients for Screening Experiment

	Axial Force	Longitudinal Force	Lateral Force	Spindle Torque	Tool Temperature
X1		3	1	7	
X2	1	2		1	1
X3	2	1	7	8	
X4	6			5	
X5		4	3		
X6	3		2	2	
X7	4		4	4	
X8	5		6	3	
X9			5	6	

Discussion of Results

After completion of the five regression analyses, all nine of the geometric features were significant in one or more of the regressions. A composite design involving all nine of the geometric features would require over 100 tools for a central composite design and close to 100 for a small composite design or Box-Behnken design, either of which don't give as much information. After inspecting the coefficients of each of the regressions, it was observed that four of the geometric features had much higher effect on the response surface of each of the yields than the other five features. These four were the cone angle of the pin (φ_p), the radius of curvature on the shoulder (ρ_s), the shoulder length (l_s) and the pressure angle of the scroll cut (γ_p). These four variables are the first and second most important variables for each of the response variables. A central composite design involving four variables only requires 25 tools. It was also noted that including one more variable to the design only added an additional 2 tools because the design changes from a full to fractional design. After looking at the data again, it was noted that the pitch of the step spirals (P_{ss}) divided by the number of step spirals (N_{ss}) was very significant in increasing or decreasing the variation in the longitudinal forces (a possible indicator of poor weld quality) of the welds.

Future Work

With the most important geometric features identified, it is important to quantify their influece on each of the response surfaces. The next step in this study is to create a Central Composite Design with the five selected variables and determine a response surface based on the results of that experimental design. After the Central Composite experiment is carried out and the results analyzed, a response surface will be created that will predict each of the forces, spindle torque and tool temperature for each set of variables and process parameters.

References

1. Pew, J.W., T.W. Nelson and C.D. Sorensen, "Torque based weld power model of friction stir welding", *Science and Technology of Welding and Joining*, 12:4 (July 2007), pp. 341-347.

2. Sorensen, C.D. and B. Nielsen, "Design parameters for convex scrolled shoulder, step spiral pin FSP tools", Interim report for the Center for Friction Stir Processing, May 2007.

3. S.K. Chimbli. et al., "Minimizing lack of consolidation defects in friction stir welds", Friction Stir Welding and Processing IV, TMS (The Minerals, Metals & Materials Society), 2007.

Friction Stir Welding and Processing V
Edited by: Rajiv S. Mishra, Murray W. Mahoney, and Thomas J. Lienert
TMS (The Minerals, Metals & Materials Society), 2009

Friction Stir Welding of "T" Joints in HSLA-65 Steel

Murray Mahoney[1], Russell Steel[2], Tracy Nelson[1], Scott Packer[3], and Carl Sorensen[1]
1) Brigham Young University, 2) MegaDiamond, 3) Advanced Metal Products

Key Words: HSLA-65 Steel, PCBN Tooling, FSW

Abstract

Our objective is to demonstrate a practical approach for friction stir welding (FSW) of "T" joints in long lengths of HSLA-65 steel. Friction stir welding of HSLA-65 steel offers challenges but achieving a defect-free weld in a "T" joint geometry in HSLA-65 steel is even more challenging. In addition to producing a sound weld nugget, the "T" fillets require additional considerations. An excessive fillet volume can create thinning of the top sheet and an unbonded lap adjacent to the leg of the "T" where metal simply extrudes into the fillet cavity. Conversely, if the fillet volume is small and the FSW tool designed to mix metal within this volume, as opposed to simply extruding metal into the fillet volume, there is a practical risk of the FSW tool contacting the support tooling, especially over long weld lengths. From a practical perspective, this is also not acceptable. Our FSW studies use both different PCBN tool designs and "T" joint geometries in attempts to circumvent these concerns and create a weld approach that is practical for the required material and geometry. Metallographic results, tool designs, and different "T" joint geometries are presented.

Background

FSW was initially developed for aluminum alloys. However, many practitioners of FSW are beginning to realize the potential to expand FSW to higher temperature materials such as ferrous alloys.[1-14] Currently, the U.S. Navy is considering HSLA-65 steel as a higher-strength alternative to DH-36 steel as HSLA-65 offers potential for weight reduction using thinner plates. Most FSW research, especially on ferrous alloys, has focused on demonstrating the ability to friction stir weld butt joints. However, in ship fabrication, "T" stiffeners welded to plate are commonplace. Due to limited design opportunities associated with material selection, a "T" weld geometry effectively creates a lap interface. Further, tooling designs to prevent weld drop-through offer additional challenges for FSW. That is, extrusion of the weld metal into a fillet cavity creates an un-bonded lap whereas mixing of metal between the fillet and the leg of the "T" creates a fully involved weld. This fillet geometry issue is better described and visualized in the results section of this manuscript.

Experimental Approach

Beginning in the early 1990's, several major HSLA-65 projects were conducted leading to the certification of HSLA-65 steel for navy ship non-primary structures. The higher strength HSLA-65 plate allows for the use of thinner plates and structural weight reductions were estimated to be 1,500 long tons per aircraft carrier, the largest weight reduction measure available to the design.[15] Thus, HSLA-65 can be an cost-effective weight saver. In this study, we used 6 mm thick HSLA-65 steel of max composition 0.1C -1.1 to 1.65 Mn - 0.025P - 0.01S - 0.1 to 0.5 Si - 0.35 Cu - 0.4Ni - 0.2Cr - 0.08Mo - 0.1V - 0.05Nb - 0.08Al. The combinations of FSW procedures, FSW tools, and support tooling (anvil designs) are summarized in table 1 and described in detail throughout the manuscript.

Table 1. Summary of different weld boundary conditions.

Anvil Design	FSW Tool	Weld Passes	Joint Design
Chamfer – fig. 1a	CS4 – fig. 2a	2	lap
Small radii – fig. 1b	CS4 + tip – fig. 2b	2	lap
Rect. reservoir – fig. 1c	CS4 large pin – figs. 3a & 3b	1	lap
Chamfer – fig. 1a	CS4 – fig. 2a	2	butt

Results

Friction stir welding was performed to evaluate the ability to create a defect free "T" joint in HSLA-56 steel. A "T" joint configuration requires additional considerations for both FSW tool design and design of the support tooling. The support tooling needs to create a smooth fillet with concurrent mixing of metal from both sections of the "T" rather than simply accommodating metal flow into the fillet without mixing. Without mixing, a lap or crack-like feature can result. We evaluated four support tooling (anvil) approaches to join a "T" stiffener to a plate, fig. 1. In the first three approaches, figs. 1a, 1b and 1c, the long leg of the "T" is butted to the top plate resulting in a lap joint when welded from the top surface. The difference between the three anvils is the design of the fillet. In fig 1a, the fillet is a modest chamfer, in fig. 1b the fillet is a small radius, and in fig. 1c a shallow rectangular reservoir is used. In each case, the volume and shape of the post-weld fillet is changed.

The FSW tool used for the weld geometry shown in fig. 1a is shown in fig. 2a, i.e., a convex step spiral scroll shoulder (CS4) tool and all welds were completed in two passes. The FSW tool used for the geometry shown in fig. 1b is shown in fig. 2b. This is a modified CS4 tool design. For a lap joint, there are potential benefits if the FSW tool uses features to deform the horizontal interface (lap-joint) without vertical flow. The friction stir weld tool shown in fig. 2b was designed with a small feature that protruded beyond the step spiral pin and extended into the lap joint interface. This added feature used three flats for the purpose of minimizing vertical flow at the horizontal faying surfaces and producing sufficient deformation to create a defect free weld. Further, this new lap tool was designed to conform more closely to the small radii support tooling. That is, this new tool is held in close proximity to the support tooling and produces metal mixing in the gap between the support tooling and the structure rather than extrusion of metal into the cavity.

The FSW tool used to weld the joint geometry shown in fig. 1c is shown in fig. 3a showing a large pin diameter designed to complete the lap weld in one pass. In initial weld parameter trails on lap joints, a defect free weld could be achieved at 300 rpm and 0.42 mm/s (1 ipm) using the fine spiral pin, fig. 3a. At higher weld travel speeds, a tunnel defect was evident. From a practical perspective, this travel speed is too slow and a second tool design using this same tool but with a coarse spiral on the pin was evaluated, fig. 3b. First, lap weld studies were performed to evaluate increasing the weld travel speed using this large single-pass tool with the coarse step-spiral tool illustrated in fig 3b. Two 6mm thick HSLA-65 plates were stacked to make a lap geometry. This was a lap weld not a "T" joint performed at 250 rpm and 0.42, 0.85, and 1.26 mm/s (1, 2, and 3 ipm). Figure 4 illustrates typical metallographic results for all three weld travel speeds showing a defect free weld except for a lack of metal fill on the surface of the advancing side. The defect free weld nugget is a significant accomplishment for this large pin diameter tool. Since this lack of surface fill was created for both "T" and lap welds (shown later), it cannot be attributed to the support tooling and metal flow into the filet cavity. Thus, we believe the inability to flow metal to the advancing side on the surface may be attributable to inadequate argon flow increasing the flow behavior of the surface material especially since the weld interior is defect free. These results illustrate the ability of the coarse step-spiral tool to more efficiently move material from the front of the tool to the tool's wake. Loads for the lap welds are illustrated in fig. 5. As expected, all loads increased with increasing tool travel speed exceeding 10,000 lbs. at 1.26 mm/s (3 ipm). This may indicate an upper limit for tool travel speed for large PCBN tools due to the high loads albeit the upper load limit is not known.

In the fourth "T" weld approach, fig. 1d, the long leg of the "T" is inserted between two top plates and stands above the surface to produce a "proud" weld nugget resulting in two butt joints. This joint design can be used if the top plate and leg of the "T" are the same alloy. Welds with this geometry were also completed in two passes. All two-pass welds were evaluated with the advancing side on the exterior in one case and with the advancing side on the interior for a second weld trial.

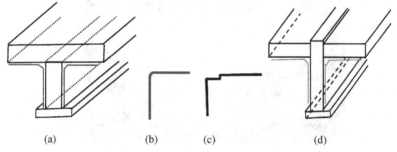

(a)   (b)   (c)   (d)

Figure 1. Illustration of different weld joint designs and support tooling to weld "T" joints. a) chamfered anvil, b) small radii anvil, c) shallow rectangular reservoir, and d) butt joints.

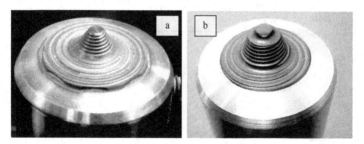

Figure 2. a) Concave scroll shoulder, step spiral tool, CS4, used to weld "T" joints shown in fig. 1a. b) A modified CS4 tool with a feature to mix the lap interface and improve metal flow in the weld fillet.

Figure 3. a) A large diameter, fine step-spiral pin designed to weld the joint shown in fig. 1c in 1-pass, and b) a coarse step-spiral pin used to increase the weld travel speed.

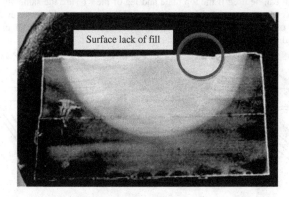

Figure 4. Lap weld using the tool shown in fig. 3b, 250 rpm @ 0.85 mm/s.

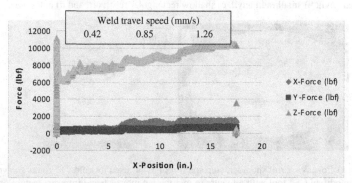

Figure 5. Loads associated with a lap weld at travel speeds of 0.42, 0.85, and 1.26 mm/s and 250 rpm. The FSW tool illustrated in fig. 3b was used.

Results from this study illustrate the potential to successfully weld "T" joints in HSLA-65 steel. Using the joint design in fig. 1a and the FSW tool in fig. 2a, and welding in two passes, the best approach was to maintain the advancing side on the weld interior (300 rpm, 1.7 mm/s, 2° tilt). Figure 6 illustrates a metallographic cross-section of this weld showing a small tunnel defect in the weld center. This type defect can be eliminated with either a change in the weld parameters, an increase in the applied Z-load, and/or a change in the features on the FSW tool pin. Metal flow is complete in the fillet region on both sides of the weld. However, on closer inspection, a lap is evident between the fillet and leg of the "T", fig. 7. Tests are not complete to determine how this lap might influence mechanical or fatigue properties. Using the butt joint design shown in fig. 1d and the same FSW tool, fig. 2a, metallographic results are comparable to those shown in figs. 6 and 7. That is, a very small tunnel defect (~800μm) was present as well as the unbonded lap. Thus, joint designs shown in figs. 1a and 1d are both acceptable when the correct weld parameters are selected. In both cases, the advancing side positioned to the interior produced the superior weld microstructure.

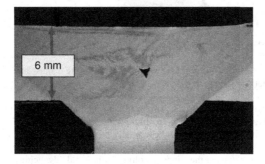

Figure 6. "T" joint weld in HSLA-65 steel completed using two passes with the advancing side on the interior.

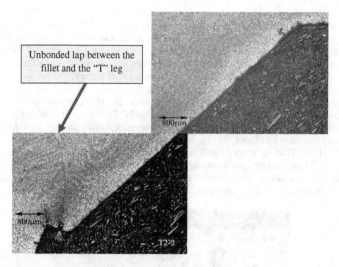

Figure 7. Illustration of the fillet from the weld shown in figure 4 illustrating complete fill but a lap between the fillet and the leg of the "T".

Figure 8 illustrates a single-pass friction stir weld using the tool shown in fig. 2b and the joint design and support tooling shown in fig. 1b. In this case, the weld is defect free (300 rpm at 1.27 mm/s) and the fillet has a small radius and essentially no cold lap. This result is ideal but it must be understood that this result was obtained under laboratory conditions. Even with laboratory control, the FSW tool contacted the support tooling on the weld advancing side. This highlights the difficulty of tool alignment over a long weld length especially when considering tool deflection and FSW system elasticity. However, for well controlled conditions, this result illustrates the ability to create an excellent weld fillet topography and a defect free weld nugget.

Figure 8. Macrographs illustrating the weld fillet using the FSW tool shown in fig. 2b and the joint design shown in fig. 1b (single pass FSW, 300 rpm, 1.7 mm/s).

"T" joint welds were completed using the anvil illustrated in fig. 1c and the coarse step-spiral tool shown in fig. 3b; 250 rpm @ 0.42 mm/s. Since the coarse step-spiral tool appeared to perform better in initial lap weld trails, this tool was used for all "T" welds. Figure 9 shows the interior of the weld nugget to be defect free and considerable penetration into the leg of the "T". Unfortunately, the surface appearance was inconsistent showing intermittently a good surface and at times a surface with a flat planar defect on the weld advancing side, fig. 10. Metal build-up was observed on the surface of the retreating side that did not flow to the advancing side leaving a surface cavity on the advancing side. Increasing the tool depth and increasing the tool rotation rate to 400 rpm did not eliminate the surface defect. Although based on the heat tint width this weld would be considered a hot weld, almost no flash was created for the entire weld length. The surface defect appears to be created by partial metal transfer across the top surface with much of the metal remaining on the weld retreating side resulting in metal build-up rather than filling the crown surface on the advancing side. Subsequent to these weld studies, it was discovered that the argon flow was inappropriate. It is possible that the flow stress was sufficiently altered to restrict flow on the top surface. The exit hole did not show a defect and the rectangular weld fillets were well formed. However, the weld fillet and anvil were occasionally joined on the weld advancing side. Separating the anvil and welded plate required some moderate force. No sticking occurred on the weld retreating side.

Loads for this "T" weld are illustrated in fig. 11 for a weld travel speed of 0.42 mm/s (1 ipm). For this large pin, the z-axis load is reasonably low (8000 lbs. max) and the translational loads are very low (<500 lbs.). It was difficult to align the "T" leg height with the support tooling. That is, at the weld start, the "T" leg was positioned ~0.58 mm above the tooling, at the weld middle, ~0.25 mm above the tooling, and at the weld finish, the "T" leg and tooling were at the same height. This may have accounted for the changing loads shown in fig. 11. In any case, the loads are reasonable for such a large tool. This change in alignment of the "T" leg in relation to the support tooling allowed evaluation of the relative positions. That is, for an optimum weld should the "T" leg be above the support tooling or flush with the tooling surface? In fig. 9, the "T" leg was 0.58 mm above the support tooling. Figure 12 shows the advancing and retreating side fillets at higher magnification. The retreating side fillet is completely formed with the weld interface well mixed into the weld nugget. On the advancing side, the fillet is well formed but evidence of the prior interface is evident. Clearly, interface mixing was not efficient. It remains to determine if this is a moderately dispersed remnant oxide or a defect that could impact post-weld mechanical properties or fatigue strength. Metallographic results for the different "T" height positions were similar. That is, the weld nugget is defect free, penetration is well into the "T" leg, and the surface cavity is present. The similarity in the macrostructures even with a difference of 0.58 mm in position of the plate with respect to the "T" leg is encouraging. This provides flexibility in manufacturing. However, more effort will be required to better deform the lap interface and to better mix the fillet with the "T" leg on the advancing side.

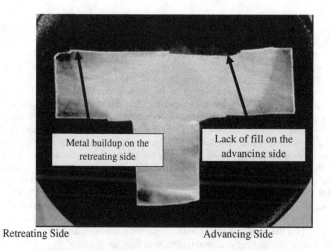

Metal buildup on the
retreating side

Lack of fill on the
advancing side

Retreating Side   Advancing Side

Figure 9. "T" joint weld completed using the tooling shown in fig. 1c and the tool in fig. 3b.

Figure 10. Crown surface of the weld illustrated in fig. 9 showing the inability of metal to flow to the advancing side.

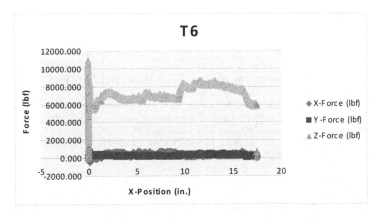

Figure 11. Loads associated with a single pass "T" weld; 250 rpm @ 0.42 mm/s.

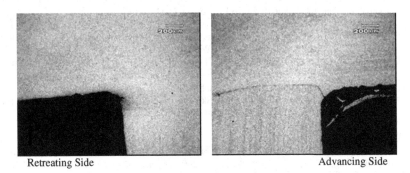

Retreating Side   Advancing Side

Figure 12. Microstructure of the retreating & advancing side weld fillets for the "T" weld shown in fig. 7.

Summary

Results have demonstrated the ability to friction stir weld "T" joints in 6 mm thick HSLA-65 steel and create a defect free weld nugget. Defect free welds were completed at travel speeds up to 1.26 mm/s (3ipm) using either one or two weld passes depending on the tool geometry. Different support tools (anvils) were designed to tailor the fillet geometry. Weld fillets were well formed with significant mixing between the surface and leg of the "T". With the correct combination of FSW tool and anvil, the weld fillets appeared well bonded to the structure. However, caution must be used to prevent contact of the FSW tool with the anvil.

Acknowledgements
The authors would like to acknowledge the financial support of the Office of Naval Research, Mr. Johnnie Deloach program manager.

References
1. Feng, Z. (2005), "Friction Stir Welding of API Grade X-65 Steel Pipe," *Annual AWS Conference*, Dallas, TX.
2. Johnson, R, dos Santos, J, and Magnasco, M (2003), "Mechanical Properties of Friction Stir Welded S355 C-Mn Steel Plates", *Proceedings of the Fourth International Symposium on Friction Stir Welding,* TWI, Park City, Utah, May 14-16, 2003, paper on CD.
3. Konkol, PJ, Mathers, JA, Johnson, R, and Pickens, JR (2003), "Friction Stir Welding of HSLA-65 Steel for Shipbuilding", *Journal of Ship Production*, vol. 19 no. 3, August 2003, pp. 159-164.
4. Konkol, P.J. and Mruczek, M.F., "Comparison of Friction Stir Weldments and Submerged Arc Weldments in HSLA-65 Steel", *Welding Journal*, July 2007, pp.187-s to 195-s.
5. Leinert, TJ, Stellwag, WL Jr., Grimmett, BB, and Warke, RW (2003). "Friction Stir Welding Studies on Mild Steel", *Welding Journal,* January 2003, pp 1-s to 9-s.
6. Li, Tim, Gau, Wei, and Khurama, Shuchi, "Friction Stir Welding of L80 and X70 Steels", *Proceedings of the Sixth International Symposium on Friction Stir Welding,* TWI, Saint-Sauveur, Canada, October 10-13, 2006, paper on CD.
7. Packer, Scott M., Steel, Russell J., Nelson, Tracy, W. Sorensen, Carl D., and Mahoney, Murray W., "Tool Geometries and Process parameters Required to Friction Stir Weld High Melting Temperature Materials", 2005-AWS-20.
8. Posada, M, DeLoach, J, Reynolds, AP, Fonda, R, and Halpin, J (2003), "Evaluation of Friction Stir Welded HSLA-65", *Proceedings of the Fourth International Symposium on Friction Stir Welding,* TWI, Park City, Utah, May 14-16, 2003, paper on CD.
9. Posada, M, DeLoach, J, Reynolds, AP, Skinner, M, and Halpin, JP (2001), "Friction Stir Weld Evaluation of DH-36 and Stainless Steel Weldments", *Friction Stir Welding and Processing*, TMS, pp. 159-171.
10. Reynolds, AP, Posada, M, DeLoach, J, Skinner, MJ, and Lienert, TJ (2001), "FSW of Austenitic Stainless Steels", *Proceedings of the Third International Symposium on Friction Stir Welding*, TWI, Kobe, Japan, September 2001, paper on CD.
11. Steel, RJ, Pettersson, CO, Sorensen, CD, Sato, Y, Sterling, CJ and Packer, SM (2003), "Friction Stir Welding of SAF 2507 (UNS S32750) Super Duplex Stainless Steel", *Proceedings of Stainless Steel World 2003*, Paper PO346.
12. Sterling, CJ, Nelson, TW, Sorensen, CD, Steel, RJ, and Packer, SM (2003), "Friction Stir Welding of Quenched and Tempered C-Mn Steel", *Friction Stir Welding and Processing II*, TMS, pp 165-171.
13. Thomas, W. (1999), "Friction Stir Welding of Ferrous Materials; A Feasibility Study", *Proceedings of the First International Symposium on Friction Stir Welding,* TWI, Thousand Oaks, California, June 14-16, 1999, paper on CD.
14. Thomas, WM, Threadgill, PL, and Nicolas, ED (1999). "Feasibility of Friction Stir Welding Steel", *Science and Technology of Welding and Joining*, 4:6, pp 365-372.
15. Czyryca, Errnest, Kihl, David, and DeNale, David, "Meeting the Challenge of Higher Strength, Lighter Warships", *The AMPTIAC Quarterly*, Vol. 7, No. 3, 2003, pp. 63-70.

Friction Stir Welding and Processing V
Edited by: Rajiv S. Mishra, Murray W. Mahoney, and Thomas J. Lienert
TMS (The Minerals, Metals & Materials Society), 2009

Quantifying Post-Weld Microstructures in FSW HSLA-65

Tracy W. Nelson, Lingyun Wei, and Majid Abassi

Mechanical Engineering
Brigham Young University
435 CTB
Provo, UT 84602
USA

Key Words: Friction Stir Welding, High Strength Low Alloy Steels, Microstructure

Abstract

A comprehensive microstructural investigation of friction stir welds in HSLA-65 steel has been undertaken. Friction stir welds were made in 6.4 mm HSLA-65 steel over a range of process parameters using a polycrystalline cubic boron nitride (PCBN) convex scroll-shoulder step-spiral (CS4) tool. The post weld microstructure was investigated by optical microscopy (OM) and Orientation Imaging Microscopy (OIM). OM revealed primarily lath upper bainite microstructures in the stir zone. OIM was used to establish quantitative measures of the prior austenite grain size, bainite packet size, and lath size. Prior austenite grain sizes in the stir zone were as large as 50μm. This new approach to acquiring quantitative microstructural data is presented.

Introduction

High strength low alloy (HSLA) steels are a class of low carbon steels exhibiting higher yield and ultimate strength in comparison with their regular carbon steel counterparts due to the addition of small percentages of alloying elements.[1] Their high strength, toughness, weldability, formability, and corrosion resistance render them suitable for use in a wide variety of structural applications, e.g., automotive components, pressure vessels, etc.

HSLA 65 steel is a relatively new structural steel for use as a hull structure, offshore structure, and line pipes. It is the uniformly refined microstructures in HSLA steels that provide the superior combination of high-strength and excellent toughness.[2] These microstructures are achieved by thermal mechanical controlled processing (TMCP) producing the desired grain size and phase fractions. Arc welding of TMCP steels generally degrades the TMCP microstructure, especially in the heat-affected zone (HAZ).

Research has demonstrated the feasibility of friction stir welding (FSW) [3] for joining HSLA type steels [4-7]. Posada *et al.* reported excellent post-weld mechanical properties in FSW of DH-36 relative to arc welds.[4] Konkol *et al.* successfully demonstrated single-pass weldments in 6.4mm and a double-sided weld in 12.7mm thick HSLA-65 using FSW.[5] Weldments were evaluated by performing metallographic evaluations, hardness, Charpy V-notch toughness, salt spray corrosion, and transverse tensile and bend tests. Friction stir welded weldments exhibited satisfactory properties relative to arc welds. Miles et al. has demonstrated success in joining some of the higher strength steels for automotive applications.[6] They showed mechanical properties in dual phase steels are highly dependent on FSW parameters.

To date, much of the work in FSW of HSLA steels lacks detailed characterization of the post-weld microstructures: an essential component to understand the process as-well-as the performance of welds. Ayer *et al.* reported on the microstructure of friction stir welded API grade X80 and L80 steels.[7] This study presented a qualitative comparison of microstructural features between friction stir welds and typical fusion welds. Similarly, Possada [4], Konkol [5] and Miles [6] have reported only limited microstructural details in DH-36 and HSLA-65.

More detailed microstructural characterization has been reported in aluminum and other non-ferrous alloys. Friggard [8], Fonda [9], Nelson [10], and others reported microtextures associated with FSW in various aluminum alloys. Sato et al. utilized microtexture in his studies of recrystallization phenomenon of FSW type 304L stainless steel.[11] Oh-ishi et al. used both electron backscatter diffraction (EBSD) and transmission electron microscopy (TEM) in their investigations of FSP of NiAl bronze.[12] The author acknowledges that there are many others who have used the various assessment tools to investigate microstructures associated with FSW in a variety of materials.

In spite of these efforts, much of the microstructural information reported in the literature in the FSW community is very qualitative. Oh-ishi *et al.* are one of the few investigators who reported a quantitative assessment of phases present in post-FSP NiAl bronze for several parameters.[13] The majority of microstructural data available in the literature report data for an isolated weld, or trends in grain and lath size, or strength of texture as a function of a very limited number of welds.

Quantitative microstructural data is essential to develop a more fundamental understanding of FSW. Quantitative microstructure data can be used in regression analyses to develop models between process parameters, post-weld microstructure, and mechanical properties. These models would be of great use to both researchers and end users to understand the influence of weld process variables on post weld performance.

The absence of quantitative FSW microstructural data in the literature is not for lack of interest or effort. It is likely the consequence of the difficulty and magnitude of the effort required to acquire such data. This is especially true in materials that undergo an allotropic transformation, or those in which a significant portion of their strength is governed by precipitation.

This paper presents an approach to acquire quantitative microstructural data in FSW HSLA-65 steel. The objective of this approach is to obtain quantitative microstructural data, and specifically, prior austenite grain size (PAGS), bainite or martensite lath size, bainite packet size, and ferrite grain size are the microstructural data of interest.

Approach

The composition of the HSLA-65 used in this investigation is given in Table 1. Test plates were sheared from a larger plate to a dimension of 762 mm (30 inch) in length along the rolling direction and 203.2 mm (8 inch) wide. The thickness of the plates was 9.525 mm (0.375 inch). The plate thickness was deliberately kept greater than the pin length to avoid interaction between the tool and the anvil under the higher Z-force. Each plate was lightly ground on both sides to remove the oxide and surface scale prior to welding. Bead on plate welds were used in this investigation.

Welding was performed using a Convex-Scrolled-Shoulder-Step-Spiral (CS4) tool as shown in Figure 1. The tool consists of a shoulder diameter of 36.83 mm (1.45 inch) with the pin being 4.47 mm (0.176 inch) in length. The shoulder and pin section of the tool were manufactured from solid PCBN. No tool tilt was required for the CS4 tool.

Before welding, plates were degreased with a methanol solvent and tightly clamped to the welding anvil. An argon atmosphere, at a flow rate of 2.8×10^5 mm^3/s, was introduced as shielding gas to protect both the tool and the weld area from surface oxidation. The welding direction was parallel to the plate rolling direction. FSW was performed under position control at tool rotation speeds of 400 and 600 RPM (revolutions/min), and feed rates of 125mm/min (5ipm), and 200mm/min (8ipm).

Table 1. Measured Chemical Composition of HSLA-65 Steel

Element	wt.%	Element	wt.%
C	0.081	Cu	0.26
Mn	1.43	N	0.009
Si	0.2	V	0.055
S	0.003	Ti	0.013
P	0.022	Nb	0.021
Ni	0.35	Al	0.018
Mo	0.063	Fe	Bal.
Cr	0.15		

Transverse metallographic samples were removed from each weld with a waterjet cutter. Samples were mounted in Bakelite then ground and polished successively through 1μm diamond. Samples for optical microscopy were etched with 2% Nital. OIM analysis was performed on an FEI XL-30 SFEG at 25Kv.[14]

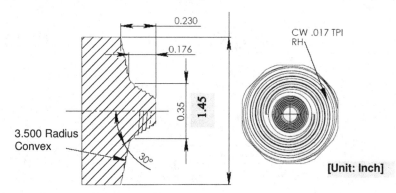

Figure 1. Geometry of CS4 tool used in the present study.

Results and Discussion

In order to develop relationships between FSW parameters and post weld microstructures, quantitative microstructural data is needed. Prior austenite grain size (PAGS), bainite or martensite lath size, bainite packet size, and ferrite grain size are all essential microstructural characteristics that are affected by both dependent and independent FSW parameters. Similarly, post-weld mechanical properties will be determined by these microstructural features. Traditional approaches to metallographic examination of FSW in HSLA type steels have proven inadequate in quantifying these microstructural features.

The optical micrographs in Figure 2 are from friction stir welds in HSLA-65 for four different heat inputs. The tool rotation speed (rpm) and travel speed (mm/min) are noted in the right hand corner of each micrograph. Heat inputs are noted in the bottom left-hand corners. From these, several microstructural characteristics can be observed. First, the lath size continuously increases from the lowest (Figure 2a) to highest heat input (Figure 2d) weld. However, it is difficult to quantify the lath size even at higher magnifications.

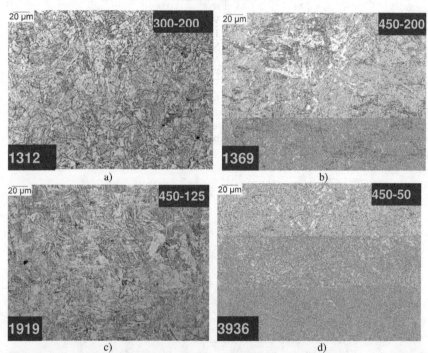

Figure 2. Optical micrograph of advancing side hard zone in FSW HSLA-65.

Also from these figures, the PAGS appears to increase with increasing heat input. The PAGS in Figure 2a is greater than 20μm, while in Figure 2d the PAGS is greater than 30μm, with some as large as 50μm. However, any assessment of PAGS from these figures would be strictly qualitative. For the most part, prior austenite grain boundaries are absent. Those grain boundaries that are visible are discontinuous at best. Even though considerable effort was expended to find an etchant to enhance the prior austenite grain boundaries, little benefit was gained. The best result of this effort is shown in Figure 2a. One notable microstructural feature from the micrographs is the very large PAGS. Grain sizes appear to range from 20 to 50 μm. This is counter intuitive given the fact that FSW is generally known for dramatic reductions in grain size. Similar results were reported previously by Ayer et al. [7]

Orientation Imaging Microscopy (OIM™) has been used widely to characterize microstructure.[14] In the FSW community, OIM has primarily been used to evaluate texture while some have used it to investigate transformation characteristics. Inverse pole figures (similar to those shown in Figure 3) are sufficient for texture studies, but insufficient to obtain any quantitative microstructural data.

The authors believe OIM is underutilized in obtaining quantitative microstructural data. By increasing the resolution of OIM, more detail information regarding the microstructure can be observed and subsequently analyzed. Figure 3 shows results from three OIM scans at difference step sizes. At a step size of 1μm (Figure 3a and b) there are no observable microstructural features. At a 0.5μm step size, some lath structure and faint grain boundaries are visible in the IQ map (Figure 3d). Even at these resolutions it would be difficult to acquire any quantitative measurement of these.

IPF and IQ maps of an OIM Scan at 0.2mm step size are shown in Figures 3e and 3f. In this figure, bainite lath and packets are clearly evident. There is evidence of PAG's and prior austenite grain boundaries (indicated by the arrows). The combination of step size and magnification make a significant difference in the observable microstructural features.

By using a combination of IPF, IQ, grain boundary maps, and OIM data, it is possible to acquire quantitative microstructural data. Bainite lath size can be measure from an IQ map (Figure 3f). This can be obtained by plotting image quality, confidence index, misorientation and grey scale along a line across bainite packets (illustrated by the color graduated line in Figure 3f). Lath spacing is the distance between peaks of these quantities as shown in Figure 4c.

Bainite packet size and PAGS are acquired by reconstruction. This approach has been used by Cayron et al. [15] for reconstruction of prior γ grains in steels, and Wynne et al. [16] for reconstruction of prior β grains in titanium alloys. Since there is little or no retained austenite in the HSLA, PAGS must be reconstructed from the room temperature ferrite microstructure. First, the misorientation between each ferrite variant of the Kudjumov-Sachs and Nishiyama-Wasserman orientation relationships was determined. Second, redundant variants were eliminated. Third, the variants were coded into the OIM software for analysis. Forth, the data set was analyzed, matching pairs of variants to a parent ferrite grain. Ferrite grains nucleating from the same austenite grain should be related by one of these variants, within some tolerance. Grains that meet the tolerance are assigned to the parent ferrite grain.

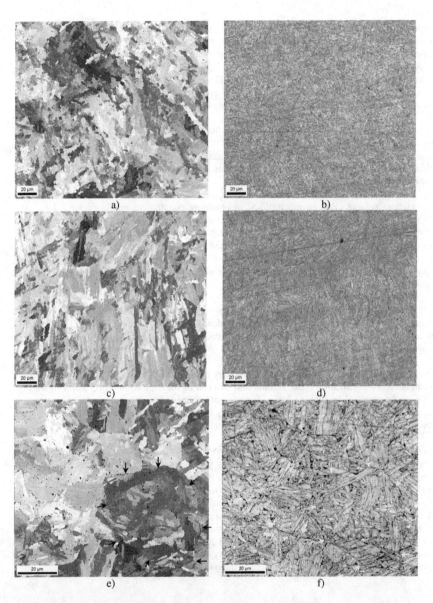

Figure 3. Comparison of OIM results from scans taken at different step sizes: a and b were taken at a step size of 1µm, c and d were taken at a step size of 0.5µm, and e and f were taken at a step size of 0.2µm.

108

Preliminary results of PAGS reconstruct from the data shown in Figure 3e are shown in Figure 4b. This figure was reconstructed using the K-S variants only. Future work will incorporate the N-W ORs. Comparing the grain map in Figure 4a, the IPF map in Figure 3e, and the PAG map in Figure 4b, the results are encouraging. As can be observed, the PAGS is non-uniform. Statistical averages can be calculated from an area fraction of the reconstructed PAGS maps. Similarly, the bainite packet size can also be extracted from these data.

The authors are currently working to streamline the analysis and automate the process. Presently, the analysis requires some manual intervention. This is useful for validation purposes. Future work will include: addition of the N-W ORs, validation of the analysis, incorporation of grain shape and/or misorientation tolerances criteria, incorporation of lath size measurements, and coding.

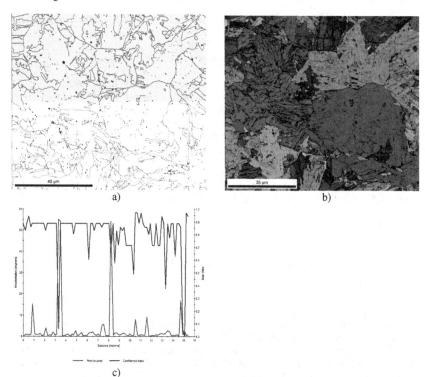

Figure 4. OIM grain map (a) and preliminary reconstruct of PAG (b) from OIM data in Figure 3e and f, and c) plot of misorientation and confidence index measured across and bainite packet.

Summary

In summary, good quantitative microstructural data has been one of the limiting factors in developing a greater fundamental understanding of FSW in HSLA steels. Regression analyses are required to develop models between dependent process variables and microstructure, as well as microstructure and properties. In order to perform such analyses, statistically based process, post weld mechanical property, and microstructural data are needed. Quantitative post weld mechanical property data are readily available. Over the past decade the evolution of FSW equipment has facilitated the acquisition of quantitative process data. To date, microstructural analysis in FSW continues to be, more often than not, qualitative.

The approach discussed herein will enable the acquisition of more quantitative microstructural data. These data will be used to develop models between FSW parameters and microstructure, and post weld microstructure and mechanical properties.

Acknowledgment

This work was supported by the Office of Naval Research, contract No. N00014-08-1-0025, program manager Dr. Julie Christodoulou.

References

1. Suresh S. "Fatigue of materials", Cambridge: Cambridge University Press; 1991.
2. K. Sampath, "An understanding of HSLA-65 Plate Steels," *Journal of Materials Engineering and Performance,* Volume 15(1) February 2006, pp. 32-40.
3. W.K. Thomas, E.D. Nicholas, J.C. Needham. M.G. Murch, P. Templesmith, and C.J. Dawes, "Friction Stir Butt Welding", G.B. Patent Application No. 9125978.8, Dec. 1991; U.S. Patent No. 5460317, Oct. 1995.
4. M. Posada, J. DeLoach, A.P. Reynolds, R. Fonda, and J. Halpin, "Evaluation of Friction Stir Welded HSLA-65," *4th International Friction Stir Welding Symposium*, Park City UT, USA.
5. P.J. Konkol, J.A. Mathers, R. Johnson, and J.R. Pickens, "Friction Stir Welding of HSLA-65 Steel for Shipbuilding", *Journal of Ship Production*, Vol. 19, No. 3, August 2003, pp. 159-164.
6. M.P Miles, J. Pew, T.W. Nelson, and M. Li, (2006) "Comparison of Formability of Friction Stir Welded and Laser Welded Dual Phase 590 Steel Sheets", *Science and Technology of Welding and Joining*, 11(4), pp 384-388.
7. R. Ayer, D. P. Fairchild, S. J. Ford, N. E. Nissley, H.W. Jin and A. Ozekcin, "Friction Stir Welding Study of Linepipe Steels," Seventh International Symposium on Friction Stir Welidng, Awaji, Japan, May 2008.
8. O. Friggard, O. Grong, and O.T. Midling, "A Process Model for Friction Stir Welding of Age Hardening of Aluminum Alloys", *Met Trans A* 32A(5) 2001, pp. 1189-2000.
9. R.W. Fonda, J.F. Bingert, and K.J. Colligan, "Development of Grain Structure during Friction Stir Welding", *Scripta Met* 51 (2004) pp. 243-248.
10. T.W. Nelson, B. Hunsaker, and D.P. Field, "Local Texture Characterization of Friction Stir Welds in 1100 Aluminum", Proceedings, First International Symposium on Friction Stir Welding, Thousand Oak, CA, May 2001.
11. S. Sato, T.W. Nelson, and C.J. Sterling, (2005), "Recrystallization in type 304L stainless steel during friction stirring", *Acta Materialia*, 53(3), pp. 637-645.
12. Keiichiro Oh-Ishi, Alexander P Zhilyaev, and Terry R McNelley. "A Microtexture Investigation of Recrystallization during Friction Stir Processing of As-Cast NiAl Bronze", *Met. Trans. A* 37A(7) 2006, pp. 2239-2251.

13. Keiichiro Oh-Ishi, Alexander P Zhilyaev, and Terry R McNelley. "The Influence of Friction Stir Processing Parameters on Microstructure of As-Cast NiAl Bronze", Met. Trans. A 36A(6) 2005, pp. 1575-1585.

14. B.L. Adams, S.I. Wright, and K. Kunze, "Orientation Imaging: The Emergence of a New Microscopy," *Metall. Trans.*, Vol. 24A, 1993, pp. 819-833.

15. C. Cayron, "ARPGE: A Computer Program to Automatically Reconstruct the Parent Grain from Electron Backscatter Diffraction Data", *J. Applied Crystallography*, (2007) 40, pp. 1183-1188.

16. B.P. Wynne, P.L. Threadgil, P.S. Davies, M.J. Thomas, B.S. Ng, "Microstructure and Texture in Static Shoulder Friction Stir Welds of Ti-6AL-4V", Seventh International Symposium on Friction Stir Welding, Awaji, Japan, May 2008.

Friction Stir Welding and Processing V
Edited by: Rajiv S. Mishra, Murray W. Mahoney, and Thomas J. Lienert
TMS (The Minerals, Metals & Materials Society), 2009

FRICTION STIR WELDING OF DUAL PHASE STEEL

Wei Yuan, Jeff M. Rodelas, and Rajiv S. Mishra

Center for Friction Stir Processing, Department of Materials Science and Engineering,
Missouri University of Science and Technology, Rolla, MO 65409, USA

Keywords: Friction stir welding, dual phase steel, mechanical properties, phase transformation

Abstract

Friction stir welding (FSW) of a dual phase (DP590) steel was evaluated with a cemented carbide tool. Different tool traverse speeds at 1000 rpm tool rotation rate were employed to compare the microstructural changes (phase transformation, grain morphology, and grain size) and mechanical properties in the nugget region. Mechanical properties of the parent material and nugget regions were characterized by mini-tensile and microhardness tests. The yield strength and ductility increased in the nugget region after FSW. The weld efficiency was about 83% for a 1 inch per minute (ipm) run and over 90% for 2 ipm and 4 ipm runs. Microhardness was higher in the nugget than the base material, and the heat affected zone was the softest region. Detailed microstructural evolution and corresponding thermal history of various regions are compared and discussed.

Introduction

Weight reduction is the method the automotive industry relies on most to produce more fuel efficient vehicles. Steel is the most commonly used material in the automotive industry. It has gained wide acceptance because of its optimal mechanical properties, ease of processing, availability, and recyclability. The emphasis on reducing vehicle weight and improving safety has resulted in increased use of thin sheet advanced high strength steels (AHSS) and a recent focus on the weldability of these alloys.

Fusion welding is the most commonly used technique to join steels. However, AHSS present a challenge. That is, AHSS have a high alloying element content leading to property degradation of the weld. In addition, the heat affected zone (HAZ) shows a large grain structure when fusion welding is used [1-3]. Friction stir welding (FSW), a solid-state welding technique, has been applied successfully to join metals with moderate melting temperatures, such as aluminum, magnesium, and copper. The most significant challenge in applying FSW to high strength and high melting point metals is tool life. Tools made from tungsten alloys, polycrystalline diamond (PCD), and polycrystalline cubic boron nitride (PCBN) have been used and reported in the literature [4].

The consequences of phase transformations accompanying thermomechanical processing during FSW are complex and have not been well documented in the literature. The current study examined FSW of advance high strength dual phase (DP) steel 590 using a WC-12wt%Co tool. The mechanical properties and microstructural evolution of these friction stir welded steel sheets were examined.

Experimental Approach

This work focused on friction stir lap welding of 1 mm thick uncoated DP590 steel. Steel sheets were sheared to 5-inch long and 1.5-inch wide coupons. Before welding, the surface of these coupons was cleaned with acetone to remove oil and grease. A WC-12wt%Co welding tool was used. The tool had a 10 mm concave shoulder diameter and a pin length of 1.1 mm. The pin tapered from 4.5 mm at the root to 3 mm at the pin tip. The tool used in this study showed limited shoulder deformation, i.e., an increase of approximately 0.1 mm in shoulder diameter after the initial runs. The shoulder diameter did not change further after the first nine welds.

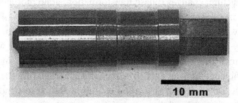

Figure 1. Photograph of a WC-Co tool.

Three sets of FSW DP590 runs were made using a conventional FSW machine operated in the position control mode. Tool traverse speeds of 1 inch per minute (ipm), 2 ipm and 4 ipm were applied while tool rotation speed was kept constant at 1000 rpm, plunge depth at 0.05 inch, and tool tilt at 3 degrees. To reduce oxidation, an argon shielding atmosphere was introduced through a rectangular perforated copper diffuser surrounding the tool.

Properties of welds were evaluated based on mini-tensile tests of the nugget region and microhardness measurements. Schematics of the friction stir welding process and mini-tensile sample are illustrated in Figure 2. Mini-tensile samples of nuggets for various welding conditions were produced by milling the weld region 0.3 mm deep into the upper sheet steel. The samples were then cut from the center of the nugget using a mini CNC machine. Sample gauge length was perpendicular to the tool traverse direction. Samples were ground and polished to a thickness of approximately 0.35 mm and tested at room temperature using a mini-tensile machine at a strain rate of 1×10^{-3} s^{-1}. The tensile properties of each weld were evaluated using at least three samples. Microhardness tests were performed on the cross-section of each weld. Vickers microhardness measurements were taken at 0.5 mm below the top sheet surface with a 0.5 mm indent spacing, 0.5 kgf load, and 10 s dwell time.

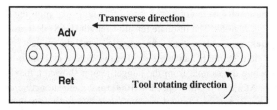

(a). Upper view of friction stir welding process.

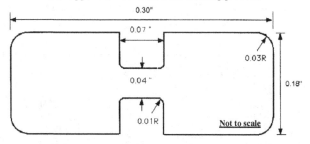

(b) Dimensions of mini-tensile sample.
Figure 2. Schematics of (a) friction stir welded specimen and (b) mini-tensile sample.

After welding, each weld was cross-sectioned perpendicularly to tool traverse direction, mounted, polished, and etched with 4% Nital. The microstructure of the weld was observed by optical microscopy.

Results and Discussion

FSW successfully produced defect-free DP590 welds. A cross-section of the weld made at a traverse speed of 2 ipm is shown in Figure 3. Unlike FSW of aluminum, there is no clear distinction among nugget, thermomechanically affected zone (TMAZ), heat-affected zone (HAZ), and parent material (PM) region. The unique microstructural regions in friction stir welded DP590 are not clear and it is difficult to delineate their boundaries. Figure 3 shows a large welded region with a relatively light color. The solid-state phase transformations during cooling make it difficult to observe details of the mixing of upper and lower sheets.

Figure 3. Cross-section of FSWed DP590 steel at 1000 rpm and 2ipm.

To evaluate the effects of the welding thermal cycle on the properties of welds, mini-tensile tests of the nugget region were performed using various welding process parameter combinations and the results were compared. Figure 4 shows the mini-tensile properties of the PM and the nugget regions. The ultimate tensile strength (UTS) decreased after FSW for all tested weld parameters. However, the sample from the nugget exhibited greater elongation and the yield strength increased in all runs except at 1 ipm. The PM showed elongation of 35% while the sample from the nugget region showed a maximum elongation of 40% at 2 ipm run. At a tool rotation speed of 1000 rpm, yield strength and UTS increased as the tool traverse speed increased from 1 ipm to 2 ipm. No significant change in properties was observed when the traverse speed was increased from 2 ipm to 4 ipm. The welding efficiency, as defined by the ratio of UTS of the nugget to that of the PM, was about 83% for 1 ipm tool traverse speed run and over 90% for 2 ipm and 4 ipm runs.

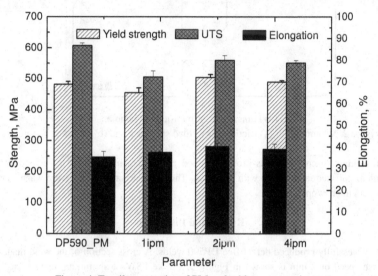

Figure 4. Tensile properties of PM and mid. nugget regions.

Figure 5 shows the microhardness profile of the upper sheet 0.5 mm below the top surface at three different tool traverse speeds. The location of the shoulder circumference is indicated by a dashed line. Hardness results show a softening region with a minimum hardness of 140 HV located about 5 mm from the weld center, corresponding to the outer edge of the shoulder. The hardness in the stir zone, however, was higher than that of the PM for 1 ipm and 2 ipm welds. The 4 ipm weld nugget showed almost the same hardness as the PM. Maximum observed hardness of the nugget was about 190 HV, whereas that of the PM was about 170 HV. Sarwar and Priestner have reported increasing the martensite content and its aspect ratio raised tensile strength and ductility for dual-phase steels [5]. However, the hardness of processed steel will vary when a phase transformation is involved. Significantly, nugget hardness for the 4 ipm run was less than that for the 1 ipm run. However, the tensile strength for the 4 ipm run was higher than that for the 1 ipm run. The location of sample gauge may explain this variation. As indicated, the gauge length of the mini-tensile sample was only 1.27 mm. Microhardness measurements of material in this small gauge may not indicate the stress of the material with two testing indents. Another possible explanation is that although the hard phase in the nugget affects the hardness, it may not determine final failure strength.

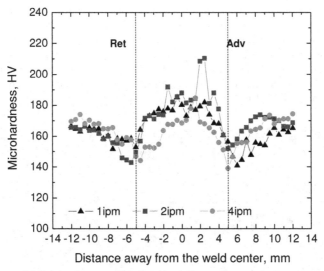

Figure 5. Microhardness profile of welds made at various tool traverse speeds.

To better understand both property changes after FSW and correlation between properties and microstructure, microstructures were characterized at various zones and with various welding parameters and then compared to the PM. Figure 6 shows microstructures of the DP590 PM and various zones for a 4 ipm run. The PM features a mixture of ferrite matrix and martensite islands decorating grain boundaries with an average ferrite gain size of about 12 μm. The details of the martensite cannot be resolved under an optical microscope.

In the HAZ, the material was heated to a peak temperature between A1 and A3. During cooling, the austenite appeared to transform to finer martensite or bainite, depending on the cooling rate. Outside of the HAZ, there should have been a HAZ II, with a peak temperature below A1, where martensite would have decomposed to a more stable structure with a mixture of ferrite and cementite. No such structure could be observed under an optical microscope. In the TMAZ, a finer grain structure was displayed with an average grain size of about 5μm. Ozekcin et al., reported grain refinement with a similar grain size in a similar region for friction stir welded carbon steels [6]. The microstructure of the TMAZ suggests the material in this region is heated above A3 and remains above that level briefly. The result is a fully austenitized structure. At the same time, severe plastic deformation introduced by friction and stirring, along with high strain rates, results in dynamic recrystallization of austenite. Fine grained ferrite and hard phase (martensite and/or bainite) form during cooling.

117

This work observed the microstructures for the HAZ and TMAZ on the retreating side of the weld. However, there were no obvious variations in microstructure between the advancing and retreating sides.

Figure 6 (d) shows the microstructure of the nugget region about 0.5 mm below the upper surface. A high volume fraction of acicular ferrite with increased hard phases formed, resulting in increased hardness in the nugget. A similar observation has been reported by others. However, the hard phase identification in the welded region is not consistent. Feng et al., reported the structure as bainite/acicular ferrite during friction stir spot welding (FSSW) of DP600 [1]. Khan et al., also analyzed the FSSW of DP600 and reported it as a mixture of lath martensite, fine particles or rods of matensite, bainite, and fine acicular ferrite [7]. The hard phase could not be identified by optical microscopy.

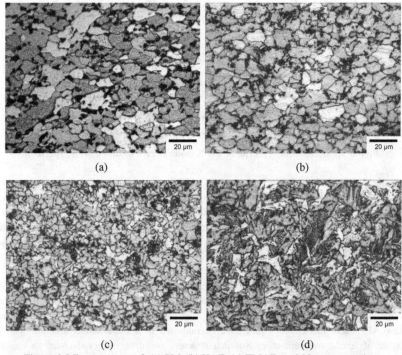

Figure 6. Microstructure of: (a) PM, (b) HAZ, (c) TMAZ, and (d) nugget region.

Figure 7 shows microstructures of the weld nugget at three different tool traverse speeds. The acicular ferrite structure in the nugget was formed in all three weld conditions, but most notably in 1 and 2 ipm runs. Higher traverse speeds appear to generate finer grains and a lower volume fraction of acicular ferrite. Acicular ferrite is widely recognized to be a desirable microstructure for its superior mechanical properties, especially its toughness.

Acicular ferrite nucleates intragranularly and heterogeneously on small nonmetallic inclusions, such as oxides or other compounds, and radiates in many different directions from nucleation sites. The prior austenite grain size affects the formation of acicular ferrite. A microstructure with large austenite grains has a better chance to form acicular ferrite due to the reduced number of grain boundary nucleation sites [8].

The highest reported temperature of steel during FSW is in the range of 1000~1200 °C and that a fully austenite structure forms in the nugget [6,9]. The peak temperature decreases and cooling rate increases with increasing tool traverse speed [10]. Additionally, austenite grains become coarser with higher temperatures and longer durations above A3. Therefore, the largest austenite grains are expected at a 1 ipm run and the smallest at a 4 ipm run. This may explain why the largest grains and highest volume fraction of acicular ferrite form at a 1 ipm run, whereas the smallest grains and lowest volume fraction form at a 4 ipm run. Since FSW is a thermomechanical process, however, the shear stress during FSW and the residual stress in the welds may have consequences for the formation and development of acicular ferrite. The deformation of austenite prior to its transformation will also increase the nucleation sites at the austenite grain boundaries. These sites are favorable to acicular ferrite formation [8].

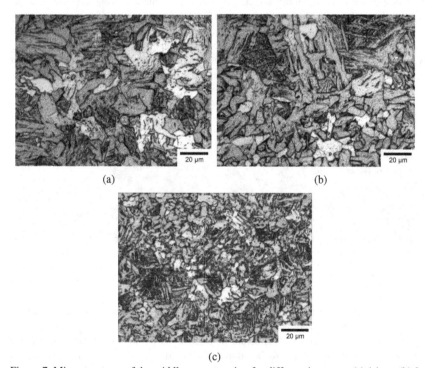

(a)   (b)

(c)

Figure 7. Microstructures of the middle nugget region for different ipm runs: (a) 1 ipm, (b) 2 ipm, and (c) 4 ipm.

As shown in Figure 3, a narrow gray region is observed on the advancing side of the nugget. A high magnification view of this gray region is shown in Figure 8. The microstructure indicates an acicular martensite structure in this region for all three runs. This likely explains the sudden increase in hardness on the advancing site of the weld for 2 ipm and 4 ipm runs whereby the martensite region moves close to the center of the nugget. No variation in hardness was observed on the advancing side for a 1 ipm run because the martensite region is deep inside the nugget and below the line where hardness was measured. Ozekcin et al., also reported acicular phase formation such as lath martensite, degenerate upper bainite, and lower bainite in the advancing side of the weld and indicated that increased hardenability in this region resulted in the formation of martensite [6]. However, the origin of increased hardenability is not clear. Ozekcin et al., suggested increased hardenability might result from austenite grain coarsening and boron alloying caused by the dissolution of BN particles from the tool [6]. These phenomena, however, cannot explain why the martensite region was present only on the advancing side.

By varying the weld process parameters, the location of the martensite region can shift horizontally and vertically in the nugget, but it remains on the advancing side. These results might indicate that formation of the martensite is related to the process parameters affecting not only the material flow, but also the shear stress during FSW and possibly the residual stress as well. The material flow may affect cooling rates promoting martensite formation. Shear stress and residual stress can also induce martensite formation, however, it is not clear how these factors affect phase transformation and further research is required.

Fig. 8. Microstructures of martensite region at the advancing site of the nugget.

Conclusions

Dual phase (DP590) steel was successfully friction stir welded using a cemented carbide tool. Microstructural changes and mechanical properties in the nugget region were evaluated for different welding parameters. A structure of acicular ferrite with some hard phases was observed in the nugget region for all three welding parameters. The volume fraction of acicular ferrite decreased as the tool traverse speed increased from 1 ipm to 4 ipm. The grains

120

in the nugget became finer with higher traverse speed. Ductility increased in the nugget region, as did yield strength, except for the 1 ipm run. The weld efficiency was about 83% for the 1 ipm run and over 90% for the 2 and 4 ipm runs. Hardness was higher in the nugget than in the base material and the heat affected zone was the softest region.

References

[1] Z. Feng, M.L. Santella, and S.A. David, "Friction Stir Spot Welding of Advanced High Strength Steels-A Feasibility Study," *Society of Automotive Engineers Technical Papers.* 2005-01-1248; 2005.

[2] Y. Hovanski, M.L. Santellab, and G.J. Grant, "Friction stir spot welding of hot-stamped boron steel," *Scripta Materialia*, 57 (9) (2007), pp. 873-876.

[3] R. Nandan, T. DebRoy, and H.K.D.H. Bhadeshia, "Recent advances in friction-stir welding – Process, weldment structure and properties," *Materials Science*, 53 (2008), pp. 980–1023.

[4] C.D. Sorensen and T.W. Nelson, Microstructure and Mechanical Properties of Friction Stir Welded Titanium Alloys. In: R.S. Mishra, M.W. Mahoney, editors. Friction Stir Welding and Processing, Materials Park: ASM International; 2007, pp. 123-154.

[5] M. Sarwar and R. Priester, "Influence of ferrite-martensite microstructural morphology on tensile properties of dual-phase steel," *Journal of Materials Science*, 31 (1996) 2091-2095.

[6] A. Ozekcin et al., "A Microstructural Study of Friction Stir Welded Joints of Carbon Steels," *International Journal of Offshore and Polar Engineering*, 14 (4) (2004), pp. 284-288.

[7] M.L. Khan et al., "Resistance and friction stir spot welding of DP600: a comparative study," Science and Technology of Welding and Joining, 12 (12) (2007), pp. 175-182.

[8] H. K. D. H. Bhadeshia, *Bainite in Steels* (Institute of Materials, 2001), pp. 237-276.

[9] T. J. Lienert et al., "Friction Stir Welding Studies on Mild Steel," *Welding journal*, 82 (2003), 1.S-9.S.

[10] L. Cui et al., "Friction stir welding of a high carbon steel," *Scripta Materialia*, 56 (7) (2007), pp. 637-640.

FRICTION STIR WELDING AND PROCESSING V

Session III

Session Chair

Thomas J. Lienert

Friction Stir Welding and Processing V
Edited by: Rajiv S. Mishra, Murray W. Mahoney, and Thomas J. Lienert
TMS (The Minerals, Metals & Materials Society), 2009

EFFECTS OF ROTATION SPEED AND WELDING SPEED ON MATERIAL FLOW AND SITR ZONE FORMATION DURING FSW/P

S. Cui and Z.W. Chen

Department of Mechanical and Manufacturing Engineering, AUT University
34 St Paul Street, Auckland, 1005, New Zealand

Keywords: tool induced flows, stir zone size, torque and forces

Abstract

Friction stir (FS) experiments have been conducted to investigate the effects of rotation speed and welding speed on the dominance of each mode of material flow forming the stir zone. The size of the total stir zone has been found to be only affected slightly by rotation speed. In the low rotation speed range the shoulder induced flow dominated requiring a high torque value and in the high rotation speed range the pin induced flow dominated which lowered the torque. Welding speed has been found to strongly affect the total stir size. The dominant effect of lowering welding speed was lowering the flow resistance of the workpiece material thus increasing substantially the volume of rotational shear material (RSM). At a low welding speed a high volume of RSM resulted in its occupation of the whole and large nugget zone. Increasing welding speed reduced RSM flow and thus the size of pin induced flow zone.

Introduction

The significance of understanding the material flow during FSW has clearly been described and general features of the different flow paths forming the weld zones have recently been summarized by Reynolds [1]. The detailed mechanisms particularly with regard to the rounding flow and periodic banding clearly need further research to clarify. Material flow and the subsequent microstructural evolution are particularly important to be understood for materials processing using friction stir principle and are currently an important aspect of friction stir research [2].

We have recently conducted friction stir (FS) experiments and identified the mechanism [3,4] governing the formation of banding (ring structure) in FS 5083 aluminium alloy using parameters that are common [3,4]. We have also further identified non-ring nugget structure with severe segregation of deformed material in the nugget formed during FS experiments using a common cast aluminium alloy and a typical FS condition [5]. We also conducted detailed metallography analysis providing explanation on how the segregation formed [6].

For cast aluminium alloys, there has been a strong focus in exploring FSP (processing), first developed by Mishra et al. [7] which is based on the same thermomechanical principle of FSW, to refine the coarse as-cast microconstituents and to eliminate porosity [8-14]. It is clear that the understanding of the detailed flow phenomena during FSP for attaining the best refined microstructure, as reflected in a recent review of FSP technology [15], is insufficient and further research is needed.

In the present work, a series of FS experiments were conducted using a common cast aluminium alloy and a range of rotation speed and welding speed in order to examine how these parameters affect the flow mode and FS zone formation. During FS experiments, a dynamometer was used to monitor the forces and torque. Features of the weld zones together with the force and torque values have been examined to understand how rotation speed and welding speed affect shoulder and pin induced flows leading to the formation of stir zones with various microstructure features.

Experimental Procedures

FS experiments were conducted using a milling machine. Workpiece material was A356 (Al-7Si-0.3Mg) cast alloy (machined to 6.35 mm from an ingot). FS tools (lefthand-thread pins) were made using H13 tool steel. Shoulder diameter was 18 mm, outer pin diameter was 6 mm and pin length was 5.7 mm. FS parameters were tilt angle at 2.5°, welding speed (v) at 28, 224 and 450 mm/min (machine step settings) and rotation speed (ω) at 250, 500, 710, 1,000, and 1,400 rpm (machine step settings). The sizes of all workpiece plates were approximately the same and the length from the plunging location to the exit hole was kept to ~ 120 mm. After welding, metallography samples were prepared for examination, following the normal metallorgraphy procedure.

In the present work, torque (T), down force (F_Z) and horizontal force (F_H) were measured using a LowStir^{TM} unit [16]. Figure 1 shows the data obtained using a sampling rate of 5/s during a FS experiment. The torque and down force have the usual meanings. The horizontal force is however an aggregate of forces in all directions. The values of torque and forces before ~ 42 s (Figure 1) correspond to the manual plunging operation and the torque and forces increased to reach more stable values after the tool started moving forward (at ~ 42 s). The peculiar increase in torque and decreases in forces at the time of ~ 60 s was due to the tool travelled across the location where two machined pieces joined together, bearing no meaning to the evaluation. The experiment completed at ~ 73 s.

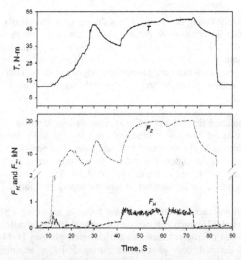

Figure 1. Data of torque, down force and horizontal force measured using a LowStir^{TM} unit during a FS experiment (machine setting: ω = 710 rpm and v = 224 mm/min).

Results

Stir zone

Images taken using a steromicroscope of cross sections are presented Figure 2 for examining the effect of rotation speed on stir zone formation. In Figure 2b, the combined total stir (thermomechanically affected) zone has been outlined showing the generally well established bell (upside down) shape with the upper portion affected dominantly by the shoulder and mid to lower portion by the pin. Outside the total stir zone, as-cast α(Al) dendrites appear dark and interdendritic Al-Si eutectic appears light grey. Inside the total stir zone, if the material has been slightly deformed, the original dendritic feature can be recognized and if it is heavily deformed no original as-cast feature can be identified.

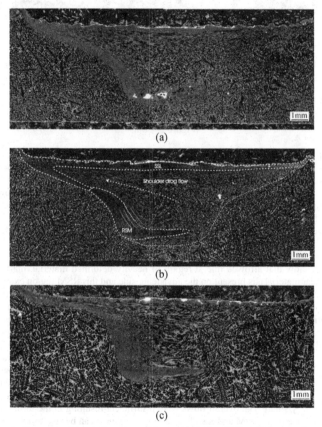

(a)

(b)

(c)

Figure 2. Cross sections (advancing side on the left) of FS A356 cast alloy made using machine settings $v = 224$ mm/min and (a) $\omega = 250$ rpm, (b) $\omega = 500$ rpm, (c) $\omega = 710$ rpm, (d) $\omega = 1,000$ rpm and (b) $\omega = 1,400$ rpm.

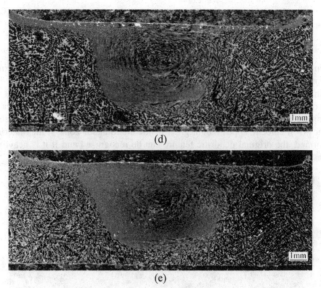

(d)

(e)

Figure 2. Cont.

As shown in Figure 2b, inside the "bell", there is a top layer denoted as SSL (shoulder shear layer) and it hooks back and down to meet RSM (rotational shear material) on the border region of the advancing site. Beneath the top SSL, the deformed dendrite trunks can be identified lining towards the advancing side, as indicated by the arrow which has been drawn in the lowest part of such flow region. Thus, in the region above that arrowed line, the workpiece material appears to have been dragged across (named as shoulder drag flow) flowing from retreating side to advancing side, as described by Reynolds [1]. This flow has also a strong forward component dragged by the forward motion of the shoulder [17].

In a distance below that arrowed line, a region of heavily deformed material, marked as RSM (rotational shear material), is present on the advancing side and this RSM region extends but tails off towards the retreating side near but not the bottom of the "bell". The rest is the material that has been deformed but much less extensively, particularly more towards the retreating side, and thus the originally dendritic feature is still apparent. The less deformed material was largely the result of the flow dragged into the region by RSM and the pin crests during FS, as has been explained [6]. We name this flow as RSM/pin drag flow.

The shape of the "bell" and features of zones inside the "bell" of the (total) stir zone gradually change as rotation speed increases. For a constant welding speed (Figure 2 for $v = 224$ mm/min), at 250 rpm, there is a relatively thick SSL, the shoulder drag flow is relatively large and RSM region is relatively small. The overall flows were insufficient resulting in the formation of small voids between the bottom RSM and RSM/pin drag flow. Thus it can be suggested that at this low rotation speed, shoulder shear flow (to form SSL) and shoulder drag flow are dominant. Increasing rotation speed to 500 rpm, although the shoulder flow is still dominant, RSM flow and RSM/pin drag flow have improved (in volume).

When rotation speed increased to 710 rpm (Figure 2c), RSM flow increased further resulting in a larger, in particular the side (advancing) portion of, RSM region. This increase in rotation speed resulted in a clearer circular pattern in the RSM/pin induced flow region. At the same time, the relative effect of shoulder has decreased, meaning a thinner SSL and smaller shoulder drag flow zone. These resulting features of increasing rotation speed continued, as can be seen in Figure 2d for 1,000 rpm. By 1,400 rpm, RSM flow, extended upward from below and on the retreating side, has become very dominant that it occupied most of the whole pin induced flow zone. The shoulder drag flow zone has on the other hand decreased considerably. The upper boundary of the RSM region on advancing site, extended close to the surface of the stir zone, can now be clearly identified.

The area of the total stir zone for each cross section in Figure 2 has been estimated and data are given in Figure 3. This stir area does not seem to be a very strong function of rotation speed although from these limited data there appears to be a minimum value at rotation speed ~ 710 rpm. It appears that as the dominance of shoulder flow reduced after 500 rpm, which has been described above, the total stir zone area decreased quite significantly and then increased more gradually as the rotation speed increased after ~ 710 rpm.

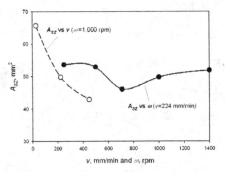

Figure 3. Total stir zone area plotted against welding speed and rotation speed.

Cross sectional images for examining the effect of welding speed on stir zone formation are shown in Figure 4. At a constant rotation speed (1,000 rpm), the major effect of welding speed is on the volume of RSM flow. At a very low welding speed (28 mm/min), RSM formed the entire pin induced flow zone, appearing as a commonly observed and distinctive FS nugget. In this case, there is a clear boundary, even on the retreating side, dividing the heavily deformed material (no dendritic remnants) inside and dendritic as-cast structure outside. Accompanying with this dominant RSM flow is a large nugget size (Figure 3), both wider and more penetrating, and a smaller shoulder drag flow, although SSL is quite thick (Figure 4).

Increasing welding speed (Figure 4) from 28 to 224 mm/min clearly decreased the amount of RSM flow and the overall pin induced flow zone reduced significantly. Thus in this pin induced flow dominant condition, the reduction of the pin induced flow resulted in a considerable decrease in total stir zone size, as shown in Figure 3. Further increase of welding speed to 450 mm/min resulted in a further decrease in RSM flow (Figure 4) and thus the stir zone size (Figure 3). SSL also became thinner at the higher feed rates. The circular flow feature in the pin induced flow region remained strong for the three different welding speeds at the same rotation speed.

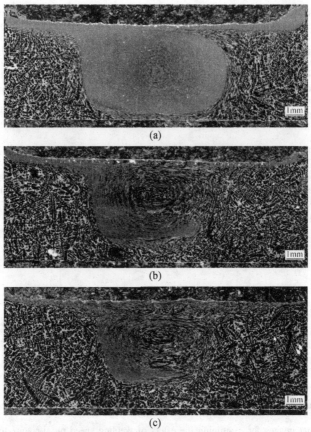

(a)

(b)

(c)

Figure 4. Cross sections (advancing side on the left) of FS A356 cast alloy made using ω = 1,000 rpm and (a) v = 28 mm/min, (b) v = 224 mm/min and (c) v = 450 mm/min. An example of total thermomechancially affected zone is illustrated in (b).

Torque and forces

Data of torque and forces for the ranges of rotation speed and feed rate are presented in Figure 5. As has been generally observed [18], increasing rotation speed in the lower speed range decreased rapidly the torque, as can be seen in Figure 5a. The decrease in down force with the increase in rotation speed was more gradual. The overall horizontal force was low and the downward trend at low rotation rate was significantly more notable. As for the effect of welding speed at the same rotation speed of 1,000 rpm, at a very low welding speed (28 mm/min), the torque and down force were low and the horizontal force was close to zero. When welding speed increased from 28 to 224 mm/min, torque and forces increased significantly. Further increase of

130

welding speed to 450 mm/min resulted in further increases in torque and down force. The rate of further increase in horizontal force was however significantly higher.

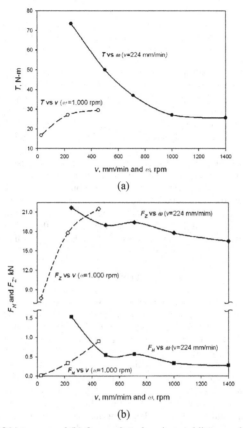

(a)

(b)

Figure 5. Values of (a) torque and (b) forces plotted against welding speed and rotation speed.

Discussion

Rotation speed affects the shearing of workpiece material by the shoulder and by the pin quite differently. At a low speed (Figure 2a), the amount of pin shearing material (RSM into thread spaces and delivering in the trialing end of the pin during FS [5,6]) was low. This in general is reasonable as the volume of transportation through thread spaces should be low at a low rotation speed. As the RSM flow is low, the RSM/pin drag flow is also low, thus the general pin induced flow is low. The shoulder induced flow was on the other hand large, both the SSL flow and the drag flow. This suggests that at low rotation speed the tool shoulder shears the workpice effectively.

131

Increasing rotation speed thus increases the rate of RSM volume into thread spaces and a higher amount of it forming and occupying the stir zone. This trend is clearly presented in Figure 2, for the series of FS experiments conducted at a constant welding speed and increasing rotation rate from 250 gradually increased to 1,400 rpm. These metallography observation and measured values of stir zone size in Figure 3 suggest strongly that the effectiveness of shoulder shearing and dominance of shoulder flow decreased quite considerably after 500 rpm while the influence of increasing rotation rate on stir zone size is not very strong.

The trend of torque (Figure 5a) increasing rapidly as rotation speed decreases in low rotation speed range does not correlate well with the overall trend of stir zone size as a function of rotation speed (Figure 3). It may be suggested that at low rotation speed frictional heat is low and the dominant shoulder flow with an overall low temperature stir zone and thus higher flow stresses resulted in higher forces and particularly a very high torque value. The flow resistance at the low rotation speed FS experiment thus resulted in a lack of joint (Figure 2a).

Welding speed on the other hand affects the stir zone size strongly, particularly in the lower range of the speed (Figs. 3 and 4). The dominant influence of lowering welding speed is increasing rapidly the flow rate of RSM. This increase not only resulted in RSM having occupied the whole pin induced flow zone but also a substantial increase in the volume of the zone. The explanation on this is that with the substantial decrease in welding speed the heat generated during FS is relatively high and thus a bigger volume of lower flow stresses is obtained. This allows the high volume (per feed distance) RSM flow both coming in the thread space from a larger volume and flowing out in a large volume.

Conclusions

Changing either rotation speed or welding speed in a wide speed range used in the present study has not change the individual flow mode but affected the individual flow rate substantially. Further, rotation speed and welding speed affected the shoulder induced flow and pin induced flow quite differently. Rotation speed affected the relative portions of flows significantly but affected the overall stir flow volume only slightly. At low rotation speeds, shoulder induced flow dominated but such flow reduced substantially in volume when the speed was higher than 500 rpm. Pin induced flow became dominant at 710 rpm and the rate of this flow increased gradually as rotation speed increased in the high rotation speed range. On the other hand, the stir zone volume is a strong function of welding speed, particularly for the low speed range. At a low welding speed, a high volume of heavily deforming material (named RSM) flowed into thread spaces and delivered to the trailing site. This resulted in a large nugget zone and the occupation of RSM in the whole zone. As welding speed increased, RSM flow reduced resulting in the increase in the portion of RSM/pin drag flow forming the nugget and in reduction of the nugget and thus the overall stir zone size.

References

1. A.P. Reynolds, *Scripta Materialia*, 58 (2008), 338-342.
2. R.S. Mishra, *Scripta Materialia*, 58 (2008), 325-325.
3. Z.W. Chen, T. Pasang and Y. Qi, *Materials Science and Engineering A*, 474 (2008), 312-316.
4. Z.W. Chen and S. Cui, *Scripta Materialia*, 58 (2008), 417-420.
5. Z.W. Chen and S. Cui, (Paper presented at the 7th International Symposium on Friction Stir

Welding, Japan, May 2008).

6. Z.W. Chen and S. Cui, submitted (2008).
7. R.S. Mishra, M.W. Mahoney, S.X. McFadden, N.A. Mara and A.K. Mukherjee, *Scripta Materialia*, 42 (2000), 163-168.
8. S.R. Sharma, Z.Y. Ma and R.S. Mishra, *Scripta Materialia*, 51 (2004), 237-241.
9. Z.Y. Ma, R.S. Mishra and M.W. Mahoney, *Scripta Materialia*, 50 (2004), 931-935.
10. Z.Y. Ma, S.R. Sharma and R.S. Mishra, *Materials Science and Engineering A*, 433 (2006), 269-278.
11. Z.Y. Ma, S.R. Sharma and R.S. Mishra, *Scripta Materialia*, 54 (2006), 1623-1626.
12. K. Nakata, Y.G. Kim, H. Fujii, T. Tsumura and T. Komazaki, *Materials Science and Engineering A*, 437 (2006), 274-280.
13. Z.Y. Ma, S.R. Sharma and R.S. Mishra, *Metallurgical and Materials Transactions A*, 37 (2006), 3323-3336.
14. S.R. Sharma and R.S. Mishra, *Scripta Materialia*, 59 (2008), 395-398.
15. Z.Y. Ma, *Metallurgical and Materials Transactions A*, 39 (2008), 642-658.
16. *LOWSTIR User Manual* (The Lowstir Consortium, January 2007).
17. Z.W. Chen, R. Peris, R. Maginness and Z. Xu, *International Journal of the Society of Materials Engineering for Resources*, 20 (2006), 4081-4086.
18. T. Long, W. Tang and A.P. Reynolds, *Science and Technology of Welding and Joining*, 12 (2007), 311-317.

Friction Stir Welding and Processing V
Edited by: Rajiv S. Mishra, Murray W. Mahoney, and Thomas J. Lienert
TMS (The Minerals, Metals & Materials Society), 2009

Bending Limits in Friction Stir Processed 5083 Aluminum Plate

M.P. Miles[1], C. Smith[2], J. Opichka[2], L. Cerveny[2], M. Mahoney[3], A. Mohan[4], R.S. Mishra[4]

[1]Brigham Young University, Provo, UT 84602
[2]Friction Stir Link, Waukesha, WI 53186
[3]Consultant, Midway, UT 84049
[4]Missouri University of Science and Technology, Rolla Missouri, 65409

Keywords: AA 5083, Friction Stir Processing, Plate Bending

Abstract

Bending performance of aluminum plates at room temperature can be enhanced by friction stir processing (FSP), which can locally anneal and refine grain size at the pre-tensile side of the plate. Plates with thicknesses from 8 - 25 mm of AA 5083 have been friction stir processed and then bent into a v-die to investigate the increase in ductility that results from FSP. A finite element model was also developed to predict bending limits of the friction stir processed plate, as well as an unprocessed plate. The material property gradient in the friction stir processed plate was obtained by machining tensile specimens at various locations through the thickness of the plate and then testing the specimens to generate flow stresses for the model calculations. This approach allowed for good agreement between experiments and model prediction of plate bending limits.

Introduction

In marine applications where weight is important, aluminum plate is often used for structural components. However, due to extrusion and bending limitations, few structural shapes, such as c-channels, I-beams, or angles are employed. Alternatively, flat plate is often gas metal arc welded into shapes or, if formed, bent with very large radii. The former is expensive and creates excessive distortion, while the latter is not desirable. Significant or useful bending of aluminum metal plates at room temperature is difficult, because of its high strength and relatively low ductility as measured by total elongation in a tensile test. For example, AA 5083 has a total elongation of about 20%, which limits bending performance. It is desirable to bend thick plates, rather than cut and weld, to achieve better structural properties compared to welded structures and to reduce fabrication costs. Plates bent to the proper shape avoid a loss of strength that accompanies welding.

Friction stir processing (FSP) can be used to locally increase the ductility of aluminum plates subsequently bent into a useful shape. FSP is a variation of traditional friction stir welding (FSW) and can be used to locally modify the properties in a part [1-9]. The effect of FSP on the plate is to locally anneal and refine the grains in the material. This increases the surface ductility to a depth that usually goes beyond the FSP tool penetration depth and significantly enhances the plate bending limit.

Experimental Methods

The alloy used for this study was AA 5083 ranging in thickness from 8 mm to 25 mm. Since the objective was to compare the effect of FSP on the mechanical behavior of the plate, tensile specimens were taken from the FSP zone as well as the surrounding material. Tensile specimens of gauge length 1.3 mm, gauge width 1.0 mm and thickness 0.5 mm were machined from slices taken through the thickness using a mini-CNC machine. The tensile axis was transverse to the FSP path. Tensile tests were carried out at an initial strain rate of $1x10^{-3}$ s^{-1} on a custom-built, computer-controlled tensile tester at constant cross-head speed.

Initial friction stir processing trials were performed with a range of rotation speeds and travel speeds, as well as several FSP tool designs. The FSP tool designs allowed for a range of processing depths and widths. Initially, the overlap distance and direction were set to be constant in all of these initial trials. The purpose of these initial trials was to investigate the stability of the friction stir processing from a manufacturing perspective and also to understand the performance of various processing parameters and tool designs.

To initially gauge the capability of the FSP process at the various processing conditions, small coupons were bent at each of the processing conditions. This was performed in a wrap-around bend testing machine, with the FSP side in tension. A range of bend diameters was tested. This was a simple test that could be performed, before spending significant effort developing detailed mechanical property maps.

Results and Discussion

One set of feasibility trials was performed on 25 mm AA 5083 plate. The 25mm AA 5083 plate was bent in a v-die in the as-received condition, resulting in a failure at approximately 37° included bend angle, where the onset of cracking on the tensile surface of the plate probably started to occur at closer to 35°. The punch used to perform the bending was a cylinder of 75 mm diameter. A photo of the plate after failure is shown in Figure 1.

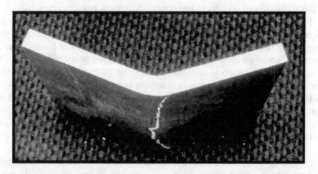

Figure 1. As-received AA 5083 plate failure in bending at about 37° included bend angle.

Tensile testing of the base material resulted in a total elongation of about 20%. When the AA5083 plate was friction stir processed, the bending performance improved dramatically, allowing for a full 90° bend without visible failure on the tensile surface of the plate. A photograph of the plate after bending is shown in Figure 2.

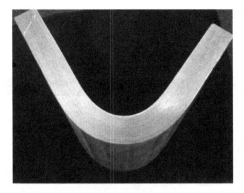

Figure 2. A greater than 90° bend of 25mm thick AA 5083-H151 plate. FSP created a ductile layer of material on the side of the plate undergoing tension, resulting in better ductility than the base material plate.

Tensile specimens were machined from the friction stir processed plate, allowing measurement of flow stresses used for the simulation. The first specimen was taken from the layer on the side of the plate processed in the first 7.5 mm of the plate thickness. Subsequent specimens were taken from greater depths within the plate, and the resulting tensile curves are shown in Figure 3.

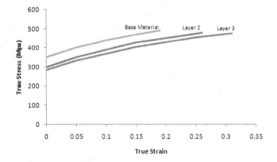

Figure 3. Tensile curves for two layers of material in the friction stir processed plate, as well as the base material. Layer 3 represents the tensile behavior of the first 7.5mm (depth in the thickness direction) of material on the processed side of the plate. Layer 2 represents the tensile behavior for the next 7.5mm of material, while the rest of the plate was modeled with base material properties.

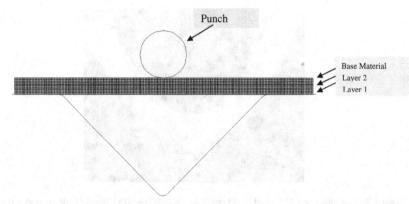

Figure 5. Plate bending model for friction stir processed plate. Three layers of material behavior were used to represent the gradient of properties created by FSP. Layer 1 is the first 7.5mm of thickness on the processed side of the plate, Layer 2 is the next 7.5mm of thickness, while the last 10mm of thickness are modeled as base material properties.

A model of plate bending was developed in order to understand how FSP improves bending performance and also to provide a tool for predicting bending performance under for different plate thicknesses and bending angles and radii.

The model was used first to predict the bending limit of the base material. For plane-strain conditions, which is the assumption of the model based on the bending of a wide plate (width to thickness ratio of about 8), the limit strain is calculated as $\frac{\sqrt{3}}{2}$ multiplied by the limit strain from a uniaxial tension test [10, 11]. In the case of AA5083, the limit strain in uniaxial tension is 0.18-0.19. Therefore, in plane strain tension, the condition on the outside surface of the plate during bending, the limit strain is 0.165. Using this as a failure criterion for the model, we obtain a predicted maximum bending angle of about 38°, as shown in Figure 6.

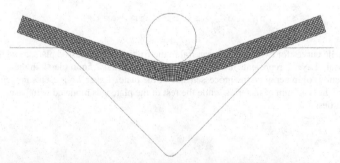

Figure 6. As received AA 5083 plate at a bending limit of 38° (included angle).

The comparison of the model result with the experimental plate from Figure 1 is good, although the model predicts the onset of cracking to occur in the experiment at closer to 38°, where the plate probably started to crack at around 35°.

The model for the friction stir processed plate predicted onset of cracking at a bend angle of about 90°, as seen in Figure 7, while the experiment produced a bend angle greater than 90° without cracking.

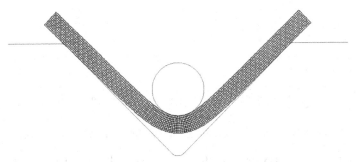

Figure 7. Bending model of friction stir processed plate, with predicted bend angle of about 90° prior to failure.

Thus, the model underpredicts the bend limit with respect to the experiment by about 10°, but this is reasonable as a first result and can be refined in future efforts where more careful measurement of the flow stress gradient through the plate thickness should improve the model prediction.

Conclusions

Bending limits of AA5083 plate were increased dramatically by FSP. For a 25 mm thick plate the bending limit was improved from 38°, for the as-received plate, to greater than 90° for the friction stir processed plate. This demonstrates that FSW can be used to significantly enhance the formability of aluminum without degrading bulk properties. This can allow for significant cost reductions or performance improvements in structural marine applications, and can potentially expand the ability to use aluminum for complex shaped structures without needing to compensate for reduced properties. A finite element model was developed in order to predict bending limits in friction stir processed plates. Good agreement between the experiments and the predictions were observed, where the stress strain behavior in the processed plate was taken into account by integrating three different stress-strain curves into the model to account for the ductility gradient caused by FSP. The model results will be refined and used in future efforts to predict bending performance under different conditions of FSP, plate thickness, and alloy composition, which will speed development of bending applications for Navy structures.

Acknowledgments

This work was funded by the Office of Naval Research under contract N00014-08-C-0089 an STTR Program, Topic N06-T038, Friction Stir Processing to Aid Forming. Friction Stir Link, Inc. is the lead organization and the Missouri University of Science and Technology is the partnering research organization.

References

1. S.P. Lynch, D.P. Edwards, A. Majumdar, S. Moutsos and M.W. Mahoney: *Mater. Sci. Forum*, 2003, vols. 426-432, pp. 2903-2908.

2. M. Mahoney, A.J. Barnes, W.H. Bingel and C. Fuller: *Mater. Sci. Forum*, 2004, vol. 447-448, pp. 505-512.

3. M.W. Mahoney, W.H. Bingel, S.R. Sharma, and R.S. Mishra: *Mater. Sci. Forum*, 2003, 426-432: pp. 2843-2848.

4. Z.Y. Ma, R.S. Mishra, M. Mahoney: *Scripta Mater.*, 2004, vol. 50/7, pp. 931-935.

5. C. Fuller, M. Mahoney, and W. Bingel: Proceedings 4th International Symposium Friction Stir Welding, May 2003, Park City, Utah, TWI Ltd. Cambridge, UK.

6. M. Mahoney, R.S. Mishra, T. Nelson, J. Flintoff, R. Islamgaliev, and Y Hovansky: Friction Stir Welding and Processing, 2001, TMS, Warrendale, PA, pp. 183-194.

7. Z. Y. Ma, R.S. Mishra, M. W. Mahoney, and R. Grimes: *Mat. Sci. Eng.*, 2003, vol A351, p. 148.

8. C. Fuller, M. Mahoney, and W. Bingel: Proceedings 5th International Symposium on Friction Stir Welding, Sept. 2004, Metz, France, TWI Ltd., Cambridge, UK.

9. Z.Y Ma, S.R. Sharma, R.S. Mishra, and M.W. Mahoney: *Mater. Sci. Forum*, 2003, vol. 426-432, pp. 2891-2896.

10. D.A. Barlow: *J. Mech. Phys. Solids*, 1954, vol. 2, pp. 259-264.

11. M.P. Miles, M. Mahoney, C. Fuller: *Metallurgical and Materials Transactions*, 37A, 2006, pp. 399-404.

Friction Stir Welding and Processing V
Edited by: Rajiv S. Mishra, Murray W. Mahoney, and Thomas J. Lienert
TMS (The Minerals, Metals & Materials Society), 2009

Microstructure and Mechanical Properties of an Al-Mo in situ Nanocomposite Produced by Friction Stir Processing

I. S. Lee*, P. W. Kao and N. J. Ho

Department of Materials and Optoelectronics Science; Center for Nanoscience and Nanotechnology, National Sun Yat-Sen University, Kaoshiung 804, Taiwan.
* Correspondence author.

Keywords: friction stir processing (FSP), Al-Mo intermetallic, $Al_{12}Mo$

Abstract

In this work, friction stir processing (FSP) was applied to produce aluminum based nanocomposites from powder mixtures of Al-Mo. This technique combines the hot working nature of FSP and the exothermic reaction between Al and Mo. Fully dense Al-matrix composites with large amounts of nanometer sized reinforcement particles, formed in-situ, can be fabricated by FSP without further consolidation. The microstructure was characterized using TEM, SEM and XRD. The Al-Mo intermetallic particles were formed in situ during FSP and were identified as $Al_{12}Mo$. These particles have an average size of ~200nm. Due to the fine dispersion of $Al_{12}Mo$ particles, the aluminum matrix has an ultrafine-grained structure (~1μm). In addition, the reaction mechanism, and microstructural evolution during FSP, as well as the mechanical properties of the Al-Mo in situ composites, will be presented.

Introduction

The mechanical properties of particulate-reinforced metal–matrix composites (MMCs) are controlled by the size and volume fraction of the reinforcements as well as the nature of the matrix-reinforcement interface [1]. Superior mechanical properties can be achieved when fine and stable reinforcements with good interfacial bonding are dispersed uniformly in the matrix. In conventionally processed power metallurgy composites, the reinforcing particles are formed prior to being added to the matrix metal. Therefore, the scale of the reinforcing phase is limited by the starting powder size; typically on the order of microns to tens of microns and rarely below 1 μm. However, it is possible to synthesis fine reinforcing particles in-situ in the matrices of MMCs [2]. The advantages of in-situ MMCs include a more homogeneous microstructure and strong interfacial bonding between the reinforcement particles and the matrices [2]. Recently, other investigators fabricated in-situ metal matrix composites by friction stir processing (FSP) [3-7].

Friction stir processing (FSP) was developed by Mishra et al. [8] for microstructural modification based on the principle of friction stir welding (FSW); a solid state welding technique invented in 1991 at The Welding Institute [9]. In FSP, a rotating tool, with pin and shoulder, is inserted and moved along the material providing localized heating and plastic deformation. The material experiences very large plastic strain ($\varepsilon_{eff} > 40$) by being extruded around the rotating tool [10]. A comprehensive literature review on friction stir welding has been given by Mishra and Ma [11], and a review on the recent developments in FSP was provided by Ma [12].

Aluminum reinforced with large amounts (up to ~50 vol. %) of nanometer-sized Al_3Ti particles was fabricated from Al–Ti elemental powder mixtures via friction stir processing (FSP) [5]. These authors suggested that the rapid Al-Ti reaction in FSP is a combined result of the large plastic strain and high temperature introduced during FSP and the exothermic reaction between Al and Ti. In MMCs, high reinforcement volume fractions are typical, giving rise to excellent properties. Furthermore, Al–TM (TM: transition metals) alloys are expected to show superior elevated temperature mechanical property because of the microstructure stability originating from the low diffusivity of TM in Al. In this work, the Al-Mo binary system was investigated because Mo has very low diffusivity and solubility in Al [13]. The objective of the present study is to produce ultrafine-grained aluminum reinforced by large amounts of nanometer-sized Al–Mo intermetallic particles formed in-situ and dispersed in the aluminum matrix during FSP.

Experimental

In this work, a powder mixture of Al-5 at. % Mo (denoted as Al-5Mo) was prepared from pure aluminum powder (99.7% purity, -325 mesh), and pure molybdenum powder (99.99% purity, 4μm). After proper mixing, the Al–5Mo powder was cold compacted to a billet of $12 \times 12 \times 88$ mm^3 in a steel die set by using a pressure of 225 MPa. To improve the billet strength for easier handling during FSP, the green compact was sintered in Ar at 823°K for 20 min. The tool pin used during FSP was the standard M1.2*6 (diameter of 6 mm and pitch height of 1.2 mm). The tool spindle angle (angle between spindle and workpiece normal) was 3°. The rotating tool was traversed along the long axis of the billet. Based on trial studies, a tool-rotation speed of 1400 rpm, traverse speed of 30-45 mm/min, and four FSP passes were applied to give defect free specimens consistently. For multiple FSP passes, the rotating tool was moved along the same line for each pass and was applied after the workpiece had cooled to room temperature from the previous pass.

X-ray diffraction (XRD, CuK$_\alpha$, 40 kV, 30 mA) was used to identify the phases present in the specimens andcanning electron microscopy (SEM, JSM-6330) used to study the distribution of second phase particles. Thin foils prepared by ion-milling were examined using transmission electron microscope (Tecnai F20 G2 Field-Emission TEM) operated at 200 kV. The Vickers microhardness was measured with a 300 gm load for 15 sec. Mechanical properties of specimens machined from the stirred zone (SZ) were measured on an Instron 5582 universal testing machine with an initial strain rate of 1×10^{-3} s^{-1}. The dimensions of the gauge section of the tensile specimens were 3 mm in diameter and 14 mm in gauge length. For tension tests, the loading axis was parallel to the FSP tool traversing direction.

Results and Discussion

The X-ray diffraction pattern indicates no reaction between the Al and Mo during sintering, figure 1, line C. After four FSP passes, diffraction peaks corresponding to the $Al_{12}Mo$ intermetallic phase appear in the XRD pattern (Fig. 1, lines A and B). In addition, the lower tool traveling speed (30mm/min) produces a larger amount of $Al_{12}Mo$ intermetallic phase. However, the diffraction peaks corresponding to Mo still exist in the sample produced by 4 FSP passes. This indicates that the Al-Mo reaction was not completed after 4 FSP passes.

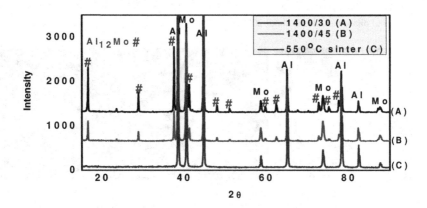

Figure 1. XRD patterns for (a) specimen processed by 4 FSP passes (1400rpm-30mm/min), (b) specimen processed by 4 FSP passes (1400rpm-45mm/min), and (c) as-sintered specimen.

The cross-section of the stirred zone in specimens produced by FSP was examined using SEM. The microstructure of the as-sintered condition is shown in Fig. 2a, and can be characterized as a dispersion of clusters of fine Mo particles in an aluminum matrix. The typical microstructure in the stir zone after 4 FSP passes is shown in Fig. 2b, and shows a uniform dispersion of second phase particles of submicrometer size. Three phases, each differing in contrast can be characterized in the BEI images (Figs. 2b and 2c). The brightest phase is Mo, the gray phase is $Al_{12}Mo$, and the dark gray matrix is Al. The reaction between Al and Mo is not complete even after 4 FSP passes because Mo particles still exist in the stirred zone.

143

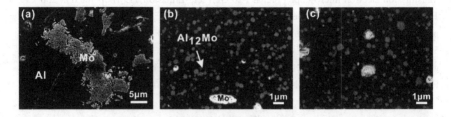

Figure 2. SEM backscattered electron image (BEI) showing the microstructure of (a) as-sintered condition, and in the stirred zone of specimens produced by 4 FSP passes (b) 1400 rpm - 45 mm/min, and (c) 1400 rpm – 30 mm/min.

The microstructure of the SZ in specimens produced by FSP was also examined by the use of TEM. Typical microstructures are shown in Fig. 3. As shown in Fig. 3a, large amounts of fine second phase particles are uniformly distributed in the aluminum matrix. The average size of the aluminum grains, as revealed by the dark field image, is about $2\mu m$ (Fig. 3b). In general, the second phase particles have an equiaxed shape as shown in Fig. 3(c), and the average size of second phase particles is about 200 nm.

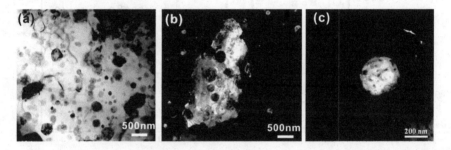

Figure 3 The microstructure of FS processed specimens revealed by TEM.(1400 rpm – 30 mm/min, 4 FSP passes) (a) Bright field image showing nanometer size particles distributed in the aluminum matrix, (b) dark field image showing fine $Al_{12}Mo$ particles dispersed in an aluminum grain, and (c) $Al_{12}Mo$ particle revealed by dark field image.

During FSP, the time that the material is subjected to the thermomechanical action is very short, i.e., only on the order of tens of seconds. The reaction between the Al and Mo must proceed very fast during FSP. The large plastic strain in FSP can shear the metal powders and break the oxide film surrounding Al and Mo powders causing intimate contact between the Al and Mo accelerating the reaction. The reaction product can be effectively removed from the interface by the large plastic strain realized during FSP such that direct contact between the Al and Mo can be maintained, and consequently, the reaction can proceed rapidly at the interface.

In addition, as suggested by Hsu et al.,, the heat release associated with the Al-TM reaction at the interface might cause local melting of Al, possibly further enhancing the reaction between Al-TM [3,4].

Decreasing the FSP traveling speed will enhance the Al–Mo reaction, resulting in more reinforcement particles, and higher hardness. The average hardness in the stirred zone of specimens produced by 4 FSP passes can be raised from 62 Hv to 74 Hv by decreasing the tool traveling speed from 45 to 30 mm/min. The tensile stress-strain curve of the specimen produced by 4 FSP passes with a tool travel speed of 30 mm/min is shown in Fig. 4, where the results of Al-10Ti [4] and Al-10 Fe [6] are also shown for comparison. The tensile properties of Al-5Mo, Al-10Ti [4] and Al-10 Fe [6] are summarized in Table 1. All the Al-TM composites shown in Table 1 have enhanced Young's modulus as compared to that of aluminum (70 GPa). Such improvement in the bulk modulus can be attributed to the presence of large amounts of the intermetallic phases in the composite, each having high moduli, i.e., E = 216, 130, and 116 GPa for Al_3Ti, $Al_{13}Fe_4$, and $Al_{12}Mo$, respectively. Due to the presence of large amounts of nanometer sized reinforcing particles and the UFG structure of the aluminum matrix, the Al-5Mo composite exhibits high strength. As compared with Al-10Ti [4] and Al-10 Fe [6], Al-5Mo also has good tensile ductility (~16.5%). However, the strength of Al-5Mo is inferior to that of Al-10Ti. This may be attributed to the coarser particles (~200 nm) in the Al-5Mo as compared to 83 nm in Al-10Ti [4]. The lower strength in the Al-Mo system also may be affected by the incomplete reaction in the Al-5Mo. If Mo in the Al–5Mo composite is fully reacted, the volume fraction of $Al_{12}Mo$ can reach 0.69. This should provide a considerable strengthening effect.

Based upon this work and previous studies [3-6], it is suggested that more research work is needed in the following two aspects: (1) to enhance the rate of the in-situ reaction during FSP, and (2) to reduce the size of the reinforcement particles formed by the in-situ reaction. Besides the processing parameters of FSP, several factors may be considered such as: (a) thermodynamic characteristics of the reactants such as heat release in the reaction, (b) particle size of the reactants, and (c) the stability of the reinforcement phase at elevated temperature, which may be affected by the melting point, and the diffusivity and solubility of the constituent elements in the aluminum matrix.

Table 1. Tensile properties of Al–TM composites produced by FSP

Sample	E (GPa)	σ_y (MPa)	UTS (MPa)	Elongation (%)
Al-5Mo	78	160	177	16.5
Al-10Fe [6]	91	177	217	3.7
Al-10Ti [4]	86	316	366	7.2

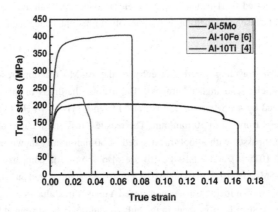

Figure 4. Tensile stress-strain curves of specimens produced by FSP.

Summary

In this work, an Al-Mo in-situ composite was successfully produced by using a technique combining friction stir processing (FSP) and the exothermic reaction between Al and Mo. The Al-Mo intermetallic particles, identified as $Al_{12}Mo$, were formed in-situ during FSP. The microstructure of the composite can be characterized as a fine dispersion of nanometer sized $Al_{12}Mo$ particles (~200nm) in an aluminum matrix with an ultrafine-grained (UFG) structure (~2μm). Due to the presence of large amounts of nanometer sized reinforcing particles and the UFG structure of aluminum matrix, the composite exhibits both high strength and good ductility.

References

1. T.W. Clyne, P.J. Withers, An Introduction to Metal Matrix Composites, Cambridge, Cambridge University Press; 1993.

2. S.C. Tjong, Z.Y. Ma, "Microstructural and mechanical characteristics of in situ metal matrix composites", *Mater. Sci. Eng. A* 29 (2000) 49–113.

3. C.J. Hsu, et al., Scripta Mater 53 (2005) 341.

4. C.J. Hsu, et al., "Al-Al3Ti Nanocomposites Produced in situ by Friction Stir Processing", *Acta Materialia* 54(2006) 5241-5249.

5. C.J. Hsu, et al., "Intermetallic-reinforced aluminum matrix composites produced in situ by friction stir processing", *Materials Letters* 61(2007) 1315-1318.

6. I.S. Lee, et al., "Microstructure and mechanical properties of Al–Fe in situ nanocomposite produced by friction stir processing", *Intermetallics* 16 (2008) 1104–1108.

7. Z.Y. Ma, J.H. Li, M. Luo, X.G. Ning, Y.X. Lu, J. Bi and Y.Z. Zhang, "In-situ formed

Al$_2$O$_3$ and TiB$_2$ particulates mixture-reinforced aluminum composite", *Scripta Metall. Mater.* 31 (1994) 635-639

8. R.S. Mishra et al., "High strain rate superplasticity in a friction stir processed 7075 Al alloy", *Scripta Mater*, 42 (2000), 163-168.

9. W.M. Thomas et al., "Friction Stir Butt Welding", G.B. Patent Application No. 9125978.8, December 1991; US Patent No. 5460317, October 1995.

10. P. Heurtier et al., "A thermomechanical analysis of the friction stir welding process", *Mater Sci Forum*, 396-402 (2002), 1573-1542.

11. R.S. Mishra, Z.Y. Ma, Mater Sci Eng R 2005;50:1.

12. Z.Y. Ma, "Friction Stir Processing Technology: A Review", *Met and Mater A*, 39 (2008) 642-658.

13. W.F. Gale, T.C. Totemeier. Smithells Metals Reference Book, 8th ed. Elsevier; 2004.

Friction Stir Welding and Processing V
Edited by: Rajiv S. Mishra, Murray W. Mahoney, and Thomas J. Lienert
TMS (The Minerals, Metals & Materials Society), 2009

CONTROL OF STRUCTURE IN CONVENTIONAL FRICTION STIR WELDS THROUGH A KINEMATIC THEORY OF METAL FLOW

H.A. Rubisoff[1], J.A. Schneider[1], and A.C. Nunes, Jr.[2]

[1]Mississippi State University;
P.O. Box ME; Mississippi State, MS 39762
[2]NASA Marshall Space Flight Center
MSFC, Huntsville, AL 35812

Keywords: friction stir welding, AA2219, material flow

Abstract

In friction stir welding (FSW), a rotating pin is translated along a weld seam so as to stir the sides of the seam together. Metal is prevented from flowing up the pin, which would result in plowing/cutting instead of welding, by a shoulder on the pin. In conventional FSW, the weld metal rests on an "anvil", which supports the heavy "plunge" load on the tool. In this study, both embedded tungsten wires along and copper plating on the faying surfaces were used to trace the flow of AA2219 weld metal around the C-FSW tool. The effect of tool rotational speed, travel speed, plunge load, and pin thread pitch on the resulting weld metal flow was evaluated. Plan, longitudinal, and transverse section x-ray radiographs were examined to trace the metal flow paths. The results are interpreted in terms of a kinematic theory of metal flow in FSW.

Introduction

FSW is a solid state joining process developed by The Welding Institute [1]. Since the development of FSW, researchers have worked to model weld metal flow in the vicinity of the tool and its relation to the weld structure. Early theories suggested a "chaotic-dynamic mixing" in the material [2]. Later tracer studies, using steel shot [3], aluminum shims [4], copper foil [5], bi-metallic welds [6-7], and tungsten wire [8], revealed defined streamline flow paths of the tracers interpretable in terms of an orderly flow of metal around the pin-tool. However, the effect of process parameters on the resulting metal flow is still not physically understood [9, 10]. This becomes increasingly important in designing robust tooling and process schedules to avoid defects such as wormholes.

The rotating weld tool used in FSW consists of a shoulder which rides along the workpiece surface, and a threaded cylindrical pin, which extends through the material thickness. As this weld tool rotates, the rotation motion causes deformation of the material adjacent to the surface of the shoulder and the intruding pin. This study considers the effects of process parameters and tool design on FSW microstructures and relates the microstructures to flow field components in a kinematic model of FSW metal flow.

Figure 1 illustrates the kinematic model of the FSW flow field [9]. The model decomposes the flow field around the pin into three incompressible flow field components, any combination of which yield an incompressible FSW flow field. These components are a rigid body rotation around the pin, a uniform translation as the tool transverses the weld seam, and a ring vortex flow field around the tool established by threaded pin features. The flow components combine to create two currents in the flow field: a "straight-through" current of which the flow elements remain within the rotating flow for less than a complete rotation of the tool and a "maelstrom" current of which the elements remain within the rotating flow for multiple rotations [9].

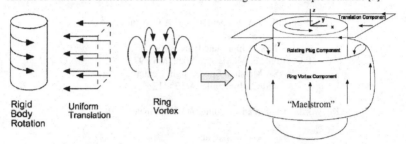

Rigid Body Rotation   Uniform Translation   Ring Vortex   "Maelstrom"

Figure 1. The flow around a FSW pin tool can be decomposed into three components: 1) a rigid body rotating plug component, 2) a translation component, and 3) a ring vortex component. Apparently, complex FSW structural features can be understood in terms of these three components and their interactions. These components can be related to welding parameters and tool design. The components present a conceptual bridge between weld process, which can be controlled, and weld structure (and weld properties).

The rigid body rotation field component comprises structural metal attached to the tool and rotating with the tool. It is bounded by a surface attached to the tool and a shearing surface, observed to be very thin, with the rotating plug of metal attached to the tool on one side and the body of weld metal moving at a relatively slow weld speed with respect to the tool on the other side. Metal crossing this boundary is subjected to extreme shear rates (typically 10^3 to 10^5 sec^{-1}) comparable to those of metals crossing a similar shear plane in metal cutting operations. As the rotating field moves through the metal, it entrains elements of metal, rotates them, and abandons them in the wake of the weld tool.

The ring vortex field component superposes a radial velocity component and an axial velocity component on the flow field at the shearing surface. A negative radial velocity component retains metal elements longer in the rotating flow and tends to shift their exit into the weld cross section toward the advancing side (AS) of the tool. Some of the flow may be retained for many revolutions flowing axially and exiting only where the radial flow velocity component shifts from inward to outward. This effect has been reported in other studies where markers have traced out metal paths rotating multiple times around the weld tool [11]. The ring vortex flow field outside the shear surface adds distortions of its own to the weld structure.

It is generally possible, as will be demonstrated here, to attribute changes in tracer patterns to changes in the flow field components of the kinematic model, and to relate the flow field component changes to changes in weld parameters and tool geometry. Thus, through the kinematic model it is possible to relate weld parameters and tool geometry to weld structure and to control weld structure.

Experimental Procedure

AA2219-T87 panels 0.25″ thick were used in this study. The weld tool was machined from tool steel with a left hand pitch of 20 threads per inch. Pure copper (98.7%) was thermally sprayed to a thickness of 0.006″ along the faying surface of one panel to mark the seam. Tungsten wires of 0.001″ diameter were also placed longitudinally along the weld seam at depths of 20%, 50%, and 80% of the panel thickness or 0.05″, 0.13″, and 0.20″ respectively from the shoulder surface. One panel was used for each wire depth providing 3 repetitions of processing parameters for each copper plated faying surface.

The weld parameters in this study included rotation speed, travel speed, and force. The weld schedule for each panel is described in Table I. Each panel was welded using a systematic variation of one parameter as illustrated in Figure 2 to yield 3 weld specimens per panel. After the tungsten wire was positioned in a grove at the required depth, run on tabs were tack welded in place to hold the panels together.

Table 1. Weld schedule for conventional and self-reacting friction stir weld panels
a) FSW schedule

Panel	Rotational Speed (rpm)	Travel Speed (ipm)	Downward Force (lbs)	Wire position (in)	speed/rotation (in/rev)
C7	200	4.5	6500, 7000, 8000	0.05	0.023
C8	200	4.5	6500, 7000, 8000	0.13	0.023
C9	200	4.5	6500, 7000, 8000	0.20	0.023
C22	150, 200, 300	4.5	7000	0.05	0.03- 0.015
C23	150, 200, 300	4.5	7000	0.13	0.03-0.015
C24	150, 200, 300	4.5	7000	0.20	0.03-0.015
C37	200	3, 4.5, 6	7000	0.05	0.015- 0.030
C38	200	3, 4.5, 6	7000	0.13	0.015- 0.030
C39	200	3, 4.5, 6	7000	0.20	0.015- 0.030

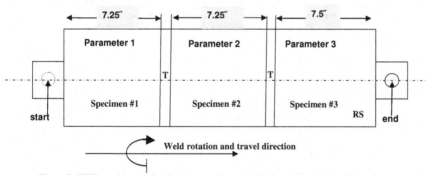

Figure 2. FSW panel layout showing run-on tabs on the 2 ends. A 1″ transition (T) section separates the parameter variations. One panel was used for each embedded wire placement depth.

Once all the panels were welded, x-ray radiographs of the welds were taken in 3 views, Figure 3. Plan x-rays recorded the as-welded panels. To radiograph in the longitudinal direction, the width of the weld panels was trimmed to isolate the weld region. Next, the samples were sectioned to an approximate thickness of 0.25″ and radiographed in the transverse direction. The 0.25″ transverse thickness captured 8-16 revolutions of the weld tool.

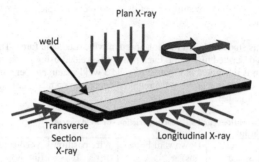

Figure 3. Layout directions of x-ray radiographs were taken of the weld region.

The x-ray radiographs were then examined to locate and highlight the post weld position of the tungsten wire and copper plating to determine what parameters affect material flow and how the process parameters move material through the weld nugget. Corresponding metallographic samples were prepared of the plan and transverse views to compare marker placement with variations observed in macro images.

Results and Discussion

Metal flow around the weld tool can be considered as a bundle of stream lines, Figure 4a. As the metal is wiped around with the weld tool, a shear zone exists in the workpiece separating the region of fine metal grains from the relatively coarse grains of the parent material microstructure. Figure 4b is a plan view of a friction stir weld metallographically polished to remove the deformation induced by the shoulder. The fine grained region surrounding the cavity where the weld tool was removed is circular and displaced toward the RS so that a thicker region is noted on the RS than on the AS.

Weld markers introduced from the weld stream into the rotating flow are whisked around and deposited in the wake of the weld. Ignoring the effect of the ring vortex flow component, i.e., lateral and axial shifts, a wire marker exits along the same line as it enters the rotating flow. Figure 5 show plan views of tungsten wire markers introduced close to and distant from the shoulder. Two features are noteworthy. The wire is fragmented in the rotating flow. The fragments exhibit appreciable lateral scatter close to the shoulder.

Fragmentation occurs when the shear forces of the metal in the rotating flow on the segment of wire introduced into the rotating flow produce a sufficient load at the anchor point of the wire to tear the wire apart. When this happens, the wire segment is swept away in the rotational flow around the pin and out into the weld structure in the wake of the tool. Copper marker deposits on weld seam surfaces are torn apart into fragments in a similar fashion. Hence, in subsequent radiographs, streamlines are marked by discontinuous fragments of copper. Bundles of streamlines averaged in radiographs may look like a continuous line or, where broadly distributed, like a field of separate fragments. Metallographic images, which exhibit a very thin surface plane only, do not show continuous traces, but only discontinuous fragments from which drawing conclusions regarding streamlines may not be feasible.

Scatter occurs due to oscillation of the diameter of the shear surface. This varies the conditions of exit from the rotating field in a complex way and produces a complex series of lateral displacements in the tracer [12]. The shear surface is anchored to the tool shoulder at the edge of a no-slip surface for which the friction force is greater than the metal flow stress. Periodically, metal is emitted from under the shoulder. This periodic emission process forms the peaks and valleys of the "tool marks" on the weld surface and the band structures that appear as "onion rings" in weld transverse sections. As metal escapes from under the shoulder edge, local normal pressure may drop making slip against the shoulder edge easier than shear within the metal. The anchor point of the shear surface follows the onset of slip on the shoulder and oscillates. The oscillation and the scatter induced by it are thus greatest at the shoulder and diminish away from the shoulder. The scatter effect tends to produce broader distributions of marker fragments toward a tool shoulder.

For present purposes, marker fragmentation and scatter are artifacts that need to be recognized to avoid confusion but that do not significantly obscure the interpretation of the data in terms of the basic kinematic model

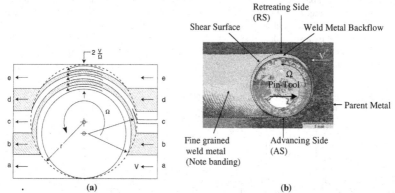

(a)   (b)

Figure 4. (a) Expected stream line flow of weld metal due to rotational motion of the weld tool. (b) Metallographic mount of the plan view of a friction stir weld termination showing differences in the refined metal region on the AS versus RS of the weld pin.

153

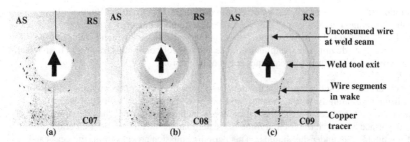

Figure 5. Plan inverted x-ray of friction stir welds processed at 200 RPM, 4.5 ipm, and 8000 lbs plunge force. Tungsten wire segments have been digitally enhanced to document post weld position of (a) 0.05″, (b) 0.13″ and (b) 0.20″ below shoulder surface.

Inverted x-ray radiographs of the transverse surface of a 0.25″ thick friction stir weld are shown in Figure 6 for the range of travel and plunge force investigated in this study. The dark region corresponds to the deposit of copper originally located on the faying surface of the weld seam. Within the parameter range investigated for travel and plunge load, little variation in the resulting copper tracer placement is observed. If the copper on the faying surface were simply whisked around the pin tool, a vertical band of copper at the position of the initial weld seam location would be seen. Evidence of the trace from the former shoulder and pin surfaces can be observed on the AS of the transverse section. The copper bands in Figure 7 show distortions from a purely vertical band with the heaviest concentration of copper located on the RS. In all the weld sections of Figure 7, the expected trace loop is observed extending from the shoulder AS to the pin RS and back around to the original seam location. But this "primary loop" is somewhat obscured by a secondary distortion. This slight distortion is observed as a slight dip about mid thickness of the weld panel bringing the copper trace back to the position of the former seam location.

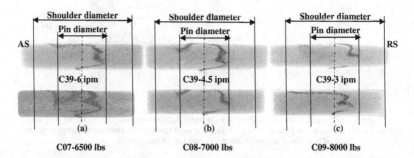

Figure 6. Minimal variation observed in an inverted x-ray of post weld copper tracer due to variations in travel (a) and plunge force (b). Weld tool rotation was held constant at 200 RPM. As the weld travel was varied (a), the plunge force was held constant at 7000 lbs and as the plunge load was varied (b), the travel was held constant at 4.5 ipm. The dashed line indicates the centerline of the former weld seam location.

In comparison, Figure 7 shows changes in the shape of the copper tracer as the RPM is varied. Figure 7 also includes corresponding transverse section metallographs of the friction stir weld mounted, polished, and etched to reveal the microstructure. Banding (onion rings) and other etch-sensitive structures that do not show up on the radiographs are visible in the metallograph. Deformation of the band by the flow field around the pin can also be seen. A rise in the band in the outer portion of the ring vortex component of the flow field is conspicuous on the RS of the pin.

The copper is observed as dark particles in the weld zone of the metallographs shown in Figure 7. Although the copper particles fall along the trace of the weld seam identified in the radiograph, their density is insufficient to mark this trace very clearly. Hence the complimentary radiographs of 0.25″ thick transverse slices of the weld, which average the copper density, provide a more continuous marking of the former seam trace.

The radiographs and metallographs reveal distortions due to the flow field around the pin. At the edge of the ring vortex component, the flow is axially upward toward the tool shoulder. An axial displacement of the band toward the shoulder increasing with the RPM is conspicuous on the RS (right) of the pin, caused by a through thickness component.

Using the flow paths predicted by the Nunes model, illustrated in Figure 1, the ring vortex circulation can be envisioned to flow inward under the tool shoulder, down the threads on the pin, outward toward the bottom of the pin, and upward on the outside to complete the circulation with conserved weld metal volume. Inward flow delays the exit of weld metal from the rotating plug flow and shifts the trace metal toward the AS of the tool, Figure 8. Outward flow hastens the exit of weld metal from the rotating plug and shifts the trace metal toward the RS of the tool. Hence, because of the interaction of the rotating plug and ring vortex flow components, one expects to see the trace of the weld seam mark out a line from the shoulder AS to the pin RS. Since the outflow does not extend all the way to the bottom of the weld, the displacement effect on the seam trace vanishes at the anvil, and the seam trace reverts back to its original position.

Using the kinematic model, the seam trace at 150 RPM (Figure 7a), the left extending advancing arm of the seam trace, starts to show a swirl characteristic. The "notch" in the "primary loop" at 150 RPM can be attributed to a ring vortex secondary distortion. As the RPM is increased from 150 to 200 RPM (Figure 7b), this notch becomes more pronounced as the arm develops an axial characteristic. As the temperatures within the weld zone increase, the amount of softened material that can be drawn into the ring vortex increases. This increases until the notch is no longer observed at 300 RPM (Figure 7c) as the ring vortex flow appears to dominate the microstructure. It is also at the 300 RPM condition that a defect is observed to open up as a classic 'wormhole', Figure 7c.

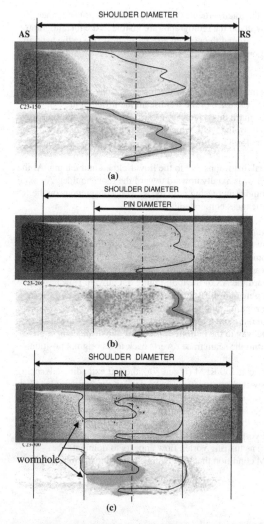

Figure 7. Transverse friction stir weld radiographs of the trace of the weld seam marked by a deposit of Cu at varying rotational speeds. The travel speed was maintained at 4.5 ipm and the plunge force at 7000 lbs with the rotation varied from: (a) 150 RPM, (b) 200 RPM, to (c) 300 RPM. All welds show the basic advancing-(shoulder-side-to-retreating-(pin)-side-and-back-to-center loop due to the interaction of the rotating plug and ring vortex flow field components. At higher RPM, greater distortions of the basic loop appear as the ring vortex flow extends out from the shear surface.

To explain the variation observed in the traced flow, Figure 8 illustrates the metal at the former seam being influenced by a ring vortex flow. This causes metal to remain in the rotating plug longer as the weld tool transverses along the initial seam. Figure 8a illustrates a mild influence of the ring vortex flow on the metal in the rotational plug, where the effect is to move the material downward, but not necessarily remaining within the rotating plug for multiple weld tool rotations. As the region of plasticized metal increases in hotter welds to where the ring vortex flow exerts a stronger influence on the rotating plug (Figure 8b), regions of the metal would stay within the rotational plug flow longer, possibly for several rotations.

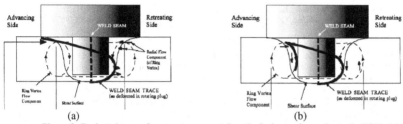

Figure 8. Projected trace of copper tracer coated on the faying surface prior to the FSW. (a) In a colder weld, the ring vortex flow offers minimal distortion to the radial deformation. (b) As the weld becomes hotter and the zone of plasticized material increases, the ring vortex would be expected to draw more material into its flow path.

Once the rotational flow of the rotating plug has moved past the trace elements embedded in the tool wake, ring vortex circulations outside the rotating plug continue to distort the seam trace. If the outer edge of a ring vortex flow passes over a metal volume, it imparts a rotary swirl to that volume in the transverse plane. If the center of the ring vortex flow passes over a metal element, it imparts an up-and-down axial motion to it in the transverse plane. There are temperature gradients in FSW [13], and plastic flow around the tool tends to be restricted to hotter regions where the weld metal flow stress is lower. Hence, in hotter (higher RPM) welds the ring vortex flow field is more broadly extended as can be observed on the RS of the metallograph in Figure 7c.

Summary

For FSW, the rotational speed affects how the shoulder interacts with the work piece. This interaction means that increasing the rotational speed results in a decreased impact on the weld metal path induced by the shoulder. As the weld heats, the shoulder and the bottom of the pin begin to have less effect on the material movement. This can also be considered to be the transition between sticking and slipping at these interfaces.

This understanding of the influence of process parameters on material flow during FSW also offers insight into defect formation. In a hotter weld, obtained by higher RPMs, the vortex dominates the metal flow causing a disruption in the continuity of flow provided by the translational component.

Acknowledgements

The assistance of Carolyn Russell, Joseph Querin, Rhonda Lash, John Ratliff and Ronnie Renfroe is gratefully acknowledged. Funding from the Mississippi Space Grant and Marshall Space Grant Consortiums contributed to this study.

References

[1]   Thomas, W.M., Nicholas, E.D., Needham, J.C., Murch, M.G., Temple-Smith, P., and Dawes, CJ. *Friction-stir butt welding*. GB Patent No. 9125978.8, International Patent No. PCT/GB92/02203 (1991).

[2]   Li, Y., Murr, L.E., McClure, J.C., "Flow visualization and residual microstructures associated with the friction-stir welding of 2024 aluminum to 6061 aluminum, *MSEA*, Vol. A271, p. 213-223, 1999.

[3]   Colligan, K., "Material flow behavior during friction stir welding of aluminum," *Welding Res. Suppl.*, p. 229s-237s, July 1999.

[4]   Seidel, T.U., Reynolds, A.P., "Visualization of the material flow in AA2195 friction-stir welds using a marker insert technique," *Met. & Mat. Trans.*, Vol. 32A, p. 2879-2884, 2001.

[5]   Guerra, M., Schmidt, C., McClure, J.C., Murr, L.E., Nunes, A.C., Jr., "Flow patterns during friction stir welding," *Mat'ls Characterization*, Vol. 49, p. 95-101, 2003.

[6]   Contreras, F., Trillo, E.A., Murr, L.E., "Friction-stir welding of a beryllium-=aluminum powder metallurgy alloy," *J. Mat. Sci.*, Vol. 37, p. 89-99, 2002.

[7]   Ouyang, J.H., Kovacevic, R., "Material flow and microstructure in the friction stir butt welds of the same and dissimilar aluminum alloys," *J. Mat. Eng. & Perf.*, Vol. 11, p. 51-63, 2002.

[8]   Schneider, J.A., Nunes, A.C., Jr., "Quantifying the material processing conditions for an optimized FSW process", *7th Int'l Conf. Trends Welding Research*, 2005.

[9]   Schneider, J.A., Nunes, Jr., A.C., "Characterization of plastic flow and resulting micro textures in a friction stir weld," *Met. Trans. B*, V. 35, p. 777-783, 2004.

[10]   Arbegast, W.J., "Modeling friction stir joining as a metal working process," *Hot Deformation of Aluminum Alloys*, ed. Z. Jin., TMS, 2003.

[11]   London, B., Mahoney, M., Bingel, W., Calabrese, M., Bossi, R.H., Waldron, D., *FSW&P* II, eds. K.V. Jata, M.W. Mahoney, R.S. Mishra, S.L. Semiatin, T. Lienert, p. 3-12, 2003.

[12]   Schneider, J.A., Beshears, R., Nunes, Jr., A.C., "Interfacial sticking and slipping in the friction stir welding process", *Mat'l Sci. & Engr. A*, Vol. 435-436, p. 297-304, 2006.

[13]   Mahoney, M.W., Rhodes, C.G., Flintoff, J.G., Spurling, R.A., Bingel, W.H., "Properties of friction-stir-welded 7075-T651 aluminum," *Metall. Mater. Trans.*, Vol. 29A, p. 1955-1964, 1998.

FRICTION STIR WELDING AND PROCESSING V

Session IV

Session Chair

Dwight A. Burford

Friction Stir Spot Welding of Advanced High Strength Steels

Yuri Hovanski[1], Michael Santella[2], Glenn Grant[1], Alan Frederick[2], Michael Dahl[1]

[1]Pacific Northwest National Laboratory;
902 Battelle Blvd.; Richland, WA 99352
[2]Oak Ridge National Laboratory; 1 Bethel Valley Road
Bldg. 4508; Oak Ridge, TN 37831-6096

Keywords: Friction Stir, Steel, Spot Welding

Abstract

Friction stir spot welding was used to join two advanced high-strength steels using polycrystalline cubic boron nitride tooling. Numerous tool designs were employed to study the influence of tool geometry on weld joints produced in both DP780 and a hot-stamp boron steel. Tool designs included conventional, concave shoulder pin tools with several pin configurations; a number of shoulderless designs; and a convex, scrolled shoulder tool. Weld quality was assessed based on lap shear strength, microstructure, and bonded area. Mechanical properties were functionally related to bonded area and joint microstructure, demonstrating the necessity to characterize processing windows based on tool geometry.

Introduction

The emergence of Friction Stir Spot Welding (FSSW) of aluminum in the automotive industry had a relatively humble beginning. To date this joining technique has managed to fill only singular niche and specialty applications for lower volume aluminum production items [1], but the appearance of FSSW in the automotive community has begun to show a more significant impact than initial implementation. This novel technology is beginning to challenge the use of the conventional standard, resistance spot welding (RSW). The attractive allure of reduced power consumption [2], zero-emissions, and greater weld performance helps to overcome traditional skepticism related to the implementation of any new technology in high production markets, and FSSW of aluminum alloys definitely delivers in these arenas.

While the argument for FSSW of steels is not as clearly defined due to the historical satisfaction of RSW performance in automotive sheet steels, ever increasing standards for safety and efficiency require implementation of new materials that challenge historical joining approaches. The introduction of advanced high strength steels (AHSS) such as dual phase, transformation induced plasticity and martensitic variations, have proven traditional RSW approaches to be ineffective [3]. While significant research in adapting existing RSW methodologies to each variation of AHSS is underway, the initial inability to join these ultra-high strength sheet steels demonstrates the need to further investigate alternative joining approaches. As FSSW had previously proven capable in joining more problematic automotive aluminum sheets, the opportunity to demonstrate the technology as an enabling joining approach for AHSS became a reality [4-8].

Project Setup and General Parameters

The U.S. DOE office of EERE Vehicle Technologies program became an ideal organization to bring together members from each of the U.S. automakers and several National Laboratories to undertake an initial effort in FSSW of AHSS. An industrial steering committee was formed with members from U.S. automakers, a friction stir tool manufacturer, and several automotive sheet steel manufacturers to guide and critique a team of researchers from the Pacific Northwest National Laboratory and Oak Ridge National Laboratory.

The project was designed to demonstrate initial concept feasibility using a series of three tool designs in both Polycrystalline Cubic Boron Nitride (PCBN) and Tungsten 25% Rhenium (W-25Re) materials. These materials were selected based on previous performance documented in linear friction stir welding of high strength materials [9]. Preliminary evaluation of as-welded material consisted of microhardness, optical metallography and lap-shear strength. The two advanced high-strength steels selected for evaluation based on availability consisted of a dual-phase steel with a 800 MPa tensile strength (DP780) and a hot-stamp boron steel (HSBS) provided by SSAB HardTech with a nominal tensile strength in excess of 1500 MPa. The chemical composition of each material is provided in Table 1.

Table 1 Chemical compositions for a Dual Phase steel and a hot-stamp boron steel in weight percentages.

Element	C	Mn	Si	Al	Cr	Ni	Ti	V	B	N	S
HSBS	0.20	1.26	0.27	0.057	0.22	0.04	0.034	0.019	0.002	0.007	0.003
DP780	0.11	1.93	0.24	0.045	0.03	0.01	0.002	0.004	<0.001	0.008	0.004

All welds were made in a lap-shear configuration using two 31 mm wide by 100 mm long sections that were fixtured in order to maintain approximately a 35 mm overlap. Friction stir spots were made at the center of the overlap with a thermocouple located at the underside of the lower sheet to provide thermal feedback during the joining process. FSSW consists of a three-step process outlined in Figure 1, designed to plunge the tool thru the sheet interface while mechanical mixing them together.

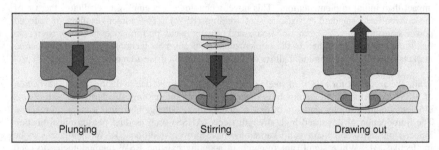

| Plunging | Stirring | Drawing out |

Figure 1 Visual Schematic of the three distinct stages of the friction stir spot welding process.

Introductory process parameters consisted of changing the plunge depth and plunge velocity while maintaining a constant 800 RPM rotational speed. Plunge depths were varied from 2.3 mm to 2.9 mm, which engaged between 50% and 90% of the lower sheet thickness. These depths were achieved via a two step plunge philosophy in which the first 90% of the total plunge depth was achieved rapidly at a rate between 0.40 mm/s and 3.0 mm/s. The final 10% of the desired plunge depth was than attained using a slower plunge rate of either 0.20 mm/s or 0.07 mm/s. Combinations of these plunge rates and depths yielded total weld times between 1.70 seconds and 9.75 seconds.

While clearly an immense discussion of the details of this mixing process is warranted, the goal of the current work is to present the unique differences between friction stir spot welds produced with a variety of tool configurations, geometries and materials, as a means of disseminating the results of the initial FSSW in AHSS.

Discussion and Analysis of Tool Performance

The division of the project scope into two phases provided an opportunity to evaluate the initial feasibility of FSSW ultra high strength steels, while avoiding a more extensive range of tool geometries, materials and process parameters. The second phase of the project was intended to provide an opportunity to reevaluate tooling based on the initial performance during the scope of the first phase.

Phase I tooling

Three initial tool designs were selected for the program including a conventional, concave-shoulder, tapered 3-flat pin design and two variations of shoulderless pin configurations. All three designs are shown in detail in Figure 2. Each of these tool designs was fabricated from PCBN, while only the two tools shown in Figure 2 (a) and (b) were fabricated from a friction stir grade W-25Re alloy.

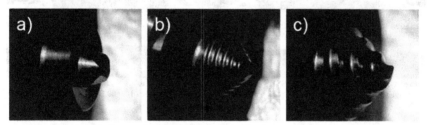

Figure 2 Phase I tool designs including (a) tapered 3-flat tool with concave shoulder designated BN77 (b) shoulderless fine thread tool designated BN86F (c) shoulderless course thread with designation BN86C.

Preliminary testing using the W-25Re tools that were precision ground to the identical shapes shown in Figure 2 (a) and (b) showed excessive wear in less than 10 plunges. While a conclusive determination about the use of tungsten based tools cannot be made based on such a limited data set, no further attempt was made to adapt the process parameters to accommodate the tungsten based tools. The PCBN tooling showed little to no wear during the initial testing, although changes in the surface were characterized by visual distortions of the tool color especially at the leading edge of sharp features and the outer portion of the tool shoulder.

An extensive range of parameters was initially examined using the conventional tapered 3-flat design designated as BN77 in Figure 2 (a). Approximately 50 different conditions changing plunge speeds and depths were characterized in the DP780 with this tool. As increased bonded area and lap shear strength clearly increased with a greater plunge depth, only the 2.9 mm plunge depth was evaluated in all additional phase I tool geometries and material combinations. After the first 13 weld conditions the PCBN tool pin showed visible indications of chipping from the surface, but no indications of wear could be measured associated with the overall length of the pin. The tool continued to make a total of 75 weld conditions in DP780 and HSBS without showing any wear to the overall length of the pin. The tool was discarded at this point however, due to the loss of approximately 10% of the pin volume on one side in the form of chipping or fracture. A second BN77 tool was then used to make more than 150 additional welds without any visible sign of fracture or wear.

All welds made using the BN77 tool in DP780 and HSBS experienced a common mode of failure consisting of a material fracture thru the hardened weld nugget between the upturned or "hooked" residual surface and the plunge cavity. Variation in the fracture distance was parameter specific as shown by comparison in Figure 3 of four joints made at a 2.9 mm plunge depth in DP780. Furthermore, in all cases using the BN77 tool lap shear tests showed a common failure method regardless of the variation in lap shear strength associated with different rotational velocities and plunge schemes.

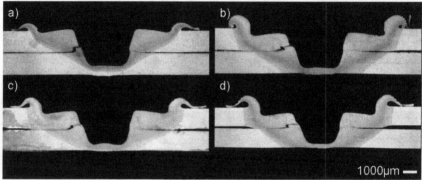

Figure 3 Optical micrographs of FSSW pulled to failure in lap shear tensile tests (a) 800 RPM - single step plunge (b) 1600 RPM - single step plunge (c) 800 RPM - two step plunge (d) 1600 RPM - two step plunge.

Shoulderless tool designs including the BN86C and BN86F shown in Figure 2 (b) and (c) were tested at a plunge depth of 2.9 mm. Plunge parameters were identical to those used for the conventional shoulder tool, including the two step plunge scheme previously described. Maximum forge loads for both the coarse and fine designs were very similar at around 33 kN in the fastest plunge condition. Lap shear strengths for the BN86C design were found to be nearly double that of the fine threaded BN86F design. Ultimately each tool was discarded due to the inability of the shoulderless design to provide sufficient down force on the upper sheet to keep both sheets in contact during the FSSW process. While bond strength for the course, shoulderless tool was similar to that of the conventional shouldered tool, the gap between sheets of greater than the sheet thickness was undesirable.

Having characterized the common modes of failure and the initial problems related to the phase I tool designs, several new tools were selected to modify the performance and failure conditions of the original designs. A shorter version of the tapered 3-flat pin was designed with the intent of further thinning the upper sheet and disrupting the "hook" effect common to joints made with the original BN77 design. Both shoulderless tools were redesigned with a shoulder that would supply the necessary down force on the upper sheet of the lap shear stack. Additionally, the fine threaded scrolled pin was narrowed to a geometry similar to that of the original tapered 3-flat design. Finally, a contemporary convex scrolled shoulder tool was introduced to the program as a means of comparison. All four phase II tools were fabricated with PCBN, and are displayed in Figure 4 showing the distinct differences between each design.

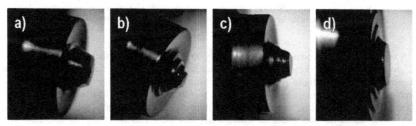

Figure 4 Phase II tool designs including (a) tapered 3-flat tool with concave shoulder designated BN97 (b) course thread tool with convcave shoulder designated BN98 (c) fine thread tool with concave shoulder designated BN99 and (d) convex scrolled shoulder tool designated BN46.

The BN97 design, a modified version of the conventional tapered 3-flat pin tool, was modified to allow the shoulder to plunge further into the upper sheet without the bottom of the pin exceeding the overall thickness of the two sheet stack. The pin length was reduced from 2.33 mm to 1.96 mm, while the pin base was increased to approximately 4.9 mm to accommodate a similar taper to the original design. The effects derived from this change significantly impacted the overall performance of the 3-flat pin design. In order to prevent over thinning the upper sheet the overall plunge depth was reduced from 2.9 mm to 2.7 mm. An immediate difference in the failure mode of friction stir spot welds made using the phase II 3-flat pin design was apparent, as the entire nugget was pulled from one of the two sheets during lap shear tests, see Figure 5, similar to a more familiar nugget pullout in RSW. No further characterization of failure between the "hooked" surface residual and pin cavity was witnessed. Other comparative differences included an increase in lap shear strength and variations in weld temperatures and forge loads, and detailed information on the performance of both 3-flat pin designs in DP780 is displayed in Table 2.

Each of the three concave shoulder tools introduced in phase II were greatly influenced by the overall surface conditions of the weld material. Each material, DP780 and HSBS, were joined in their as-received "bare" condition, as well as in several variations of surface preparation including surface ground and machined. In all cases friction stir spot welds made using the BN97, BN98 and BN99 tools performed better in lap shear when the sheet surfaces were ground or machined in lieu of the as-received condition. It should be noted that the affect of surface preparation was more significant in the HSBS than in the DP780 as previously detailed [7]. In contrast, the convex scrolled shoulder tool, BN46, showed much less impact resulting from the surface condition in either material with several conditions yielding identical mechanical performance regardless of surface preparation.

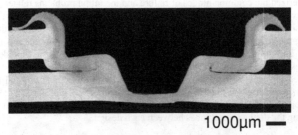

1000µm —

Figure 5 Optical micrograph of a FSSW in DP780 made with the BN97 tool. This joint was pulled to failure in lap shear and shows the crack propagation from the sheet interface to the top of the weld.

Both threaded pin tools demonstrated the ability to effectively join either AHSS. The BN98 design demonstrated a larger range of parameters that produced acceptable mechanical properties, but both of these phase II designs were hindered by similar circumstances. The combined influence of inserting a shoulder and rotating the tool such that the thread direction added an additional driving force resulted in extremely high forge loads. Table 2 shows tool loads using the 3-flat pin could be consistently maintained at or below 30 kN; however, maximum tool loading often exceeded 50 kN when using the threaded pin tools in AHSS. Some attempt was made to determine an alternative operating window specific to this type of design, yet traditional approaches only seemed to exacerbate the overall difficulties.

Table 2 Comparitive performance data for friction stir spot welds made with two PCBN tools.

Tool Designation	RPM	plunge steps (#)	Max Tool Load (kN)	Max Temp. (°C)	Avg. LSS (kN)	Failure Mode
BN77	800	1	32.5	410	4.4	Hook to Pin
		2	33.0	537	8.0	Hook to Pin
	1600	1	23.7	510	9.6	Hook to Pin
		2	31.3	641	11.2	Hook to Pin
BN97	800	1	27.6	453	12.7	Nugget pullout
		2	34.6	650	15.4	Nugget pullout
	1600	1	18.9	574	14.7	Nugget pullout
		2	21.7	767	15.2	Nugget pullout

The convex scrolled shoulder tool was fundamentally different from the other designs, and as such necessitated a different set of operating parameters. Plunge depths ranging from 1.4 mm to 1.6 mm produced welds with lap shear strengths between 5.3 kN and 25.3 kN. Tensile strength was found to be dependent on parameters such as rotational velocity of the tool, plunge scheme, weld duration and to a less extent surface preparation. The BN46 tool design demonstrated the greatest flexibility in producing joints without regard to surface preparation in both materials used in this study; however, the influence of surface conditions was slightly more prominent at shorter weld times. An example of a friction stir spot weld in HSBS using the convex scrolled shoulder tool is shown below in Figure 6.

166

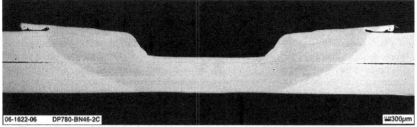

Figure 6 Optical micrograph of a friction stir spot weld made using a convex scrolled shoulder tool.

Forge loads for the BN46 tool greatly varied with operating parameters, but were most affected by the rotational tool velocity and the plunge speed. The highest loads, 47 kN in HSBS and 52 kN in DP780, were generated using an 800 RPM two-step plunge, in which the initial 90% of the desired depth was rapidly achieved in the first one second. Numerous operating conditions were found that maintained maximum forge loads at or below 30 kN. All joints produced at 1600 RPM were able to limit the maximum loading below a 30 kN threshold, although temperatures at the bottom of the weld approached 700°C. At 800 RPM, welds made with the BN46 tool produced temperatures between 400°C and 600°C.

Conclusions

Data accumulated during a multiyear program evaluating FSSW as an alternative joining technology for AHSS has demonstrated the significance of tool specific operating parameters. Hundreds of joints were successfully welded using numerous tool designs that were fabricated in PCBN. Each specific tool design evaluated in this study demonstrated the capacity to effectively join the two AHSS used herein; however, the effective operating parameters varied from tool to tool. Analysis of the weld performance and process feedback data yielded the following conclusions:

1. Each tool design has a unique operating window that is both specific to the AHSS being joined and unique to the tool geometry. This operating window is a function of the weld time, plunge depth, plunge rate(s), tool shape, tool material, tool rotational velocity, and the material being joined.
2. The highest lap shear strengths for each tool were achieved based on tool specific parameters, with no general characterization of proper operating conditions showing complete applicability to all designs. Thus, no generalized statement about proper operating parameters should be made without specific reference to the tool geometry, tool material and weld material.
3. Seemingly small modifications in tool geometry such as the change in pin length from the phase I to phase II design led to marked changes in applicable operating parameters for a tapered 3-flat pin tool with concave shoulder. In the case highlighted herein, a reduction in pin length led to variation in achievable lap shear strength, complete alteration of the generalized mode of failure, and a reduction in the applicable plunge depth for any range of operating parameters.
4. Changes in operating parameters at a fixed plunge depth seemed to have less affect on varying the failure mode prevalent to any one tool geometry than minor changes to the tool design. The use of a tapered 3-flat tool design demonstrated that changes in operating

167

parameters could affect the applicable bonded region, but that the ultimate mode of failure was unchanged by modification of rotational velocity, plunge rate, or hold duration.

5. The influence of surface preparation had a stronger impact on traditional concave shoulder pin tools than on the convex scrolled design regardless of weld material being joined.

6. Forge loads required to plunge FSSW tools to the desired depth were greatly affected by both operating parameters and tool geometry. Threaded pins showed higher tool loads than tapered pins of the same cross-sectional area, with loads increasing for more coarse threads.

Acknowledgement

The authors graciously acknowledge the assistance of members of their industrial steering committee including Susan Ward of Ford Motor Company, Bill Marttila of Chrysler Corp., Jim Chen and Jim Quinn of General Motors Corp., and Scott Packer of Advanced Metal Products for their technical input throughout the project. This work was sponsored by Dr. Joseph Carpenter in association with the U.S. Department of Energy Office of Vehicle Technologies as part of the Automotive Lightweighting Materials Program.

References

1. "Car Bodies First to be Friction Stir Welded," Machine Design, September 18, 2003.

2. R. Hancock, "Friction Welding of Aluminum Cuts Energy Cost by 99%," *Welding Journal*, 83 (2) (2004), 40.

3. K. Yamazaki, K. Sato, Y. Tokunaga, "Static and Fatigue Strength of Spot Welded Joint in Ultra-High-Strength, Cold-Rolled Steel Sheets," *Weld. Int.*, 14 (7) (2000), 533-541.

4. Z. Feng, M.L. Santella, S.A. David, R.J. Steel, S.M. Packer, T. Pan, M. Kuo, and R.S. Bhatnagar, "Friction Stir Spot Welding of Advanced-High-Strength Steels – A Feasibility Study," *SAE Technical Paper*, 2005, No. 2005-01-1248.

5. W.J. Kyffin, P.L. Threadgill, H. Lalvani, and B.P. Wynne, "Progress in FSSW of DP800 High Strength Automotive Steel," (paper presented at the 6th International Symposium on Friction Stir Welding, October 10, 2006).

6. Y. Hovanski, M.L. Santella, and G.J. Grant, "Friction Stir Spot Welding of Hot-Stamped Boron Steel," *Scripta Materialia*, 57 (2007), 873-876.

7. M.P. Miles, J. Sederstrom, K. Kohkonen, R. Steel, S. Packer, T. Pan, W.J. Schwartz, and C. Jiang, "Spot Friction Welding of High Strength Automotive Steel Sheets," (paper presented at MS&T 2006).

8. S. Packer, T. Nelson, C. Sorensen, R. Steel and M. Matsunaga, "Tool and Equipment Requirements for Friction Stir Welding Ferrous and Other High Melting Temperature Alloys," (paper presented at the 4th Int. FSW Symp., Park City, Utah, 2003).

Friction Stir Welding and Processing V
Edited by: Rajiv S. Mishra, Murray W. Mahoney, and Thomas J. Lienert
TMS (The Minerals, Metals & Materials Society), 2009

MECHANICAL EVALUATION OF FRICTION STIR SPOT WELDED ADVANCED HIGH STRENGTH STEELS

Jeff Rodelas, Rajiv S. Mishra, Greg Hilmas, and Wei Yuan

Center for Friction Stir Processing and Department of Materials Science and Engineering
Missouri University of Science and Technology
McNutt Hall, 1870 Miner Circle, Rolla, MO 65409

Keywords: friction stir spot welding, advanced high strength steel, cermet tools, sinterbonding

Abstract

Friction stir spot welding (FSSW) was used to join dual phase (uncoated DP 590, hot-dipped galvannealed (HDGA) DP 600 and DP 780) and martensitic (uncoated M190) advanced high strength steels (AHSS). The process featured a hybrid WC-Co/W-Ni-Fe bi-material FSSW tool capable of withstanding process temperatures in excess of 1000°C. Some geometric changes in the WC-Co tool face were observed due to plastic deformation. For welds produced with this tool, the lowest strength grade AHSS (DP 590) possessed the highest average lap shear strength (LSS) of 11.4 kN. Higher strength AHSS grades (DP 780 and M190) possessed lower average LSS, likely due to higher material flow stresses at elevated temperatures. Although HDGA DP 600 possesses similar mechanical properties to DP 590, its LSS was 55% lower than that of DP 590 due to the formation of a Zn-rich faying surface layer. Some welds produced in this study exceeded the maximum LSS presented in other published work for similar materials. In addition, some welds exceeded the minimum design shear strength mandated by the American Welding Society and Military Handbook for resistive spot welded steel by as much as 150% and 200%, respectively.

Introduction

In the automotive industry, increasingly strict crash safety standards and more stringent C.A.F.E. regulations mandate that vehicles retain sufficiently high crash integrity without significantly increasing curb weight. Incorporating AHSS, such as dual phase (DP) steel, in vehicle structures is one way of achieving this goal [1]. However, care must be exercised when joining these materials to ensure properties do not degrade as a result of the joining process.

When AHSSs are joined using the conventional resistance spot welding (RSW) method, solidification problems such as cracks and defects degrade weld strength and performance, especially for AHSS with relatively high carbon content [2]. A small number of studies evaluating the mechanical properties of AHSS, such as dual phase (DP) steel [1,3-6] and martensitic steels [1-2,7], have been reported. These studies, however, offer no analysis relating the resulting weld mechanical properties to base steel strength and sheet thickness. The relationship between RSW lap shear strength and base metal strength for steels is established in existing design guidelines issued by the American Welding Society [8] and MIL Handbook [9]. However, the relationships between base metals strengths for steels joined by FSSW are not currently well understood.

The objective of this work is to evaluate and compare the lap shear strength (LSS) performance of various AHSS grades joined by FSSW welds using a hybrid bi-material tool. Tensile lap shear testing of FSSW welds was performed to characterize weld mechanical performance in relation to existing published data and RSW design guidelines. Additionally, an analysis was performed of processing, wear, and deformation behavior of a hybrid WC-Co/W-Ni-Fe bi-material FSSW tool used to produce the welds performed in this work.

Experimental Procedure

Development of Hybrid Bi-material Tools for Steel FSSW

Material selection for aluminum FSSW tools is straightforward. Tools made from age-hardened tool steels, nickel alloys, tungsten alloys, and some ceramics have been used with success [10-12]. Widespread application of FSSW to steel, however, is limited in part by the lack of widely available economical tool materials that are mechanically and chemically robust to withstand the plunge forces when exposed to process temperatures in excess of 1000°C [13]. The application of FSSW to steel, therefore, demands the development of tool materials and designs that can withstand these temperatures to produce steel joints with satisfactory properties.

Welds in this work were produced using a hybrid bi-material tool with a Co-cemented WC face and a W-Ni-Fe heavy alloy (WHA) shank. This approach provided a combination of the advantageous properties of both materials to produce a more effective tool overall. The high elevated temperature strength/hardness of the WC-Co portion of the tool is restricted to the tool face/pin features, allowing the complex geometry necessary for interfacing with the FSSW machine from high-toughness W-Ni-Fe alloy rather than from WC-Co.

Fabrication of the hybrid tool is a unique process in that the WC-Co tool face is produced by sinterbonding WC-Co powder to a dense WHA shank under uniaxial load. Tools produced in this manner possess a consolidated interface while retaining the machinability of the WHA shank, a property that would otherwise be lost by sinterbonding WC-Co powder to WHA powder [14]. Simple tool features were used in this work. The tool had a right-angle cylindrical pin 4 mm in diameter, 1.2 mm in length, and 10 mm in shoulder diameter. Figure 1 shows a photograph of the as-fabricated tool used in this study.

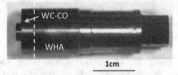

Figure 1. Photograph of hybrid bi-material tool.

FSSW Experimental Procedure

Both dual phase and martensitic grades of AHSS were selected for this study. Table I lists the types and grades of AHSS welded. All welds were produced from 1.0 mm thick sheet. Sheet metal coupons used for weld strength characterization were 127 mm long and 38 mm wide. Coupons were welded in a lap configuration with a 38 mm square overlap. Prior to welding, the coupons were cleaned ultrasonically in acetone to remove any surface contamination. The LSS was determined by loading the spot weld coupon in tension at a constant extension rate of 0.02 mm/s and recording the peak load at failure.

Table I. Material properties of AHSS joined by FSSW in this study

Material Designation	Condition	YS (MPa)	UTS (MPa)	Elongation (%)
DP 590	uncoated	370	602	27
DP 600	HDGA	375	640	26
DP 780	HDGA	460	838	17
M190	uncoated	1188	1392	6

Friction stir spot welds were produced with a RoboStir™ spot welder (Friction Stir Link Inc., Waukusha, WI). This particular machine uses a linear actuator capable of applying axial loads up to 22.2 kN with a maximum spindle rotation speed of 3000 RPM. For this study, weld parameters were held constant for all AHSS types tested. Welds were performed using a constant tool rotation speed of 1500 RPM, a tool plunge rate of 0.1 mm/s, and a dwell time of 490 ms. For all welds, a constant tool plunge depth was used such that the measured distance from the pin tip to the underside surface of the bottom sheet was 0.4 mm

Results and Discussion

Mechanical Evaluation of AHSS Spot Welds

The LSS for four AHSS grades was evaluated. Figure 2 shows the measured LSS. DP 590 exhibited the highest LSS of 11.4 kN. The average LSS for M190 nearly equaled that of DP590, despite an ultimate tensile strength (UTS) more than double that of DP 590. For the welding conditions explored in this work, both HDGA DP 600 and DP 780 exhibited nearly equal average LSS, however, both average LSS values are significantly lower than those of uncoated DP 590 and M190 steel. Scanning electron micrographs (Figure 3) revealed a Zn/ZnO layer forming at the faying surface that these authors believe reduces the bonded area and therefore the LSS of the FSSW joint. Further work is necessary to determine the exact role played by this Zn-rich layer in decreasing the LSS of HDGA DP steels.

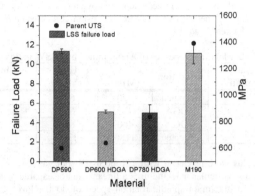

Figure 2. FSSW LSS for materials evaluated in this study.

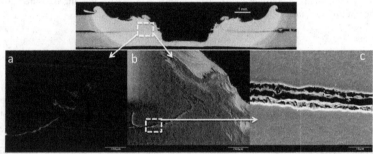

Figure 3. SEM micrographs of FSSW HDGA DP 600 a.) Energy dispersive spectroscopy compositional map showing presence of Zn in weld nugget. b.) Bonding ligament approximately 0.5 mm. c.) ZnO forms at faying surface interface inhibiting bonding.

Similar LSS values for DP 590 and M190 did not to follow trends generally expected for RSW welds. Figure 4 shows the minimum design shear strengths for steel joined by RSW as a function of base metal UTS for various thicknesses. As the base metal UTS and workpiece thickness increased, the expected LSS strength increased as well. In FSSW, the size and quality of the bonded region directly influences joint strength [15]. The results shown in Figure 2 suggest that the base metal UTS had some effect on the size/quality of the bonded region. The observation that both DP 590 and M190 have similar LSS but drastic UTS differences may be related to the disparity in high-temperature flow stresses for the various AHSS grades. For constant FSSW parameters, the higher UTS of M190 lead to a higher high-temperature flow stress [16]. The higher high-temperature flow stress reduced the extent of material flow under the tool and appeared to reduce the size/quality of the bonded region.

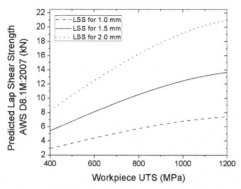

Figure 4. Minimum RSW LSS guidelines for steel [7]

Because FSSW is considered an alternative to RSW for joining of AHSS in the automotive industry, it is useful to compare the strength of welds produced in this study to existing FSSW data and RSW design guidelines. Commonly used design guidelines for RSW-joined steel are given by the American Welding Society (AWS) [7] and Military Handbook (MIL-HDBK-5H) [8]. Published data for similar materials with the same minimum design specifications may also be usefully compared.

Six published papers have reported the strength of AHSS joined by FSSW (DP 590/600 [1,3,5], DP 780/800 [2,4], and M190 [1,2,6]). Because sheet thicknesses and material strengths of AHSS grades addressed in these papers differ from those used in the current work, a comparison can be made by normalizing the maximum reported LSS to the ultimate tensile strength (UTS) of the sheet material. All the welds in the aforementioned studies were performed with different tool materials including W-25Re, Si_3N_4, and PCBN. The tool shoulder diameters used in these studies varied between 9.8mm and 10.2 mm.

Figure 6 is a plot of UTS-normalized LSS (LSS_{UTS}) versus material thickness values found in both the present work and work reported by others. Normalizing the maximum LSS for welds performed in this study to UTS reveals no observable relationship between obtained LSS and material thickness. These results contrast with the relationships established by the AWS and MIL-HNDBK specifications for RSW. Additionally, relatively higher strength materials (e.g., DP780, DP800, and M190) have low LSS_{UTS} values. Again, this behavior is likely due to the lower extent of material flow in high UTS materials due in part to an increased high-temperature flow stress. Lastly, DP 590 welds on a 1.0 mm sheet produced by the hybrid bi-material tool yielded the higher LSS_{UTS} than to all values reported in the technical literature.

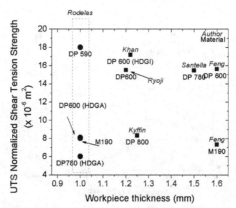

Figure 5. UTS-normalized LSS for welds produced in this study compared to results reported in technical literature

To further evaluate the efficacy of FSSW, the LSS values obtained in this work, and reported in the technical literature, were plotted along with existing design guidelines. Figure 7 compares the maximum LSS obtained by FSSW to minimum design guidelines for RSW. For welds performed in this study, DP 590, M190, and HDGA DP 600 exceeded AWS/MIL-HDBK requirements by 150/200%, 45/80%, and 10/35%, respectively. HDGA DP 780 failed to meet the minimum LSS guidelines for AWS; however, it exceeded the MIL-HNDBK minimum value. Reported LSS values in other works show that the lower strength materials (e.g., DP 590/600) overall met or exceeded design minimum guidelines, whereas stronger materials such as DP 780/800 and M190 failed to meet one or both guidelines. Additionally, LSS_{UTS} does not appear to increase overall with increasing sheet thickness. This trend, again, is contrary to RSW LSS expectations. The FSSW process for steels is not as refined as RSW; however, further refinement of tool geometries and FSSW process parameters that increase bonded region size will likely show further improvement in LSS values with respect to minimum design guidelines.

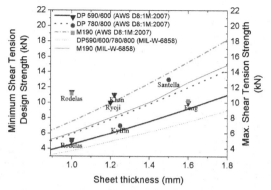

Figure 6. Maximum FSSW LSS and minimum design guidelines for RSW welded steels as per AWS [7] and MIL-HDBK [8]

Hybrid Bi-materials Tool: Wear and Deformation

Figure 7 demonstrates the tool geometry changes from the initial tool geometry shown in Figure 1. Plastic deformation of the WC-Co tool face material occurring at the process temperature is the primary mechanism for geometric changes. Within the first 10 plunges, pin height decreased and pin diameter increased as the pin deformed during the FSSW process. The once cylindrical pin (Figure 1) chipped/wore and deformed to a tapered cone. A slight increase in shoulder diameter accompanied the pin dimensional changes. After approximately 20 plunges, material from the face of the tool around the pin was pushed towards the end of the pin during welding. The movement of material created a tapered 45 degree conical pin with accompanying shoulder concavity. After approximately 20-25 plunges of the tool, no further significant tool dimension deformation had occurred. Welds produced with this tool, and used for LSS characterization, were performed after major tool deformation had occurred to ensure unchanging tool dimensions. Finally, there was no observable degradation of the WC-Co/WHA bimaterial interface after performing over 70 welds with this tool with average tool normal forces and spindle torque exceeding 15 kN and 16 N-m, respectively.

Figure 7. Progression of hybrid bi-material tool wear/deformation

Although deformation of the WC-Co face ceased after approximately 20 plunges, deformation of future hybrid WC-CO/WHA tools can perhaps be mitigated by refinement of the WC-Co properties. Deformation at the FSSW process temperature could be minimized in the future by altering the WC grain size to prevent deformation by grain boundary sliding. Additionally, further research is required to determine suitable metallic binder compositions that will optimize the high-temperature stability of the WC based tool.

Conclusions

This work successfully joined two different types of AHSS of varying grades using a WC-Co/WHA hybrid bimaterial tool. Although the tool experienced chipping/wear and plastic deformation during the weld process, the deformation ceased after approximately 20 plunges. The lap shear strength for the strongest grade AHSS tested, M190, exhibited LSS similar to that obtained for DP 590 despite the 230% higher UTS of M190. For the welding conditions explored in this work, the HDGA sheet exhibited drastically lower LSS than the uncoated varieties. The presence of a Zn layer at the faying surface appeared to decrease the extent of the bonded region significantly.

Based on the LSS results obtained in this study, and the results currently published in technical literature, there appears to be no direct relationship between the workpiece UTS and the FSSW LSS. These results suggest that higher strength materials under constant welding conditions have correspondingly higher flow stresses than lower strength materials, which affects the size of the bonded region. Improvements to the FSSW process, including changes in tool design and geometry, may produce greater weld strength.

Acknowledgements

The authors would like to acknowledge Sylvia Loo (Arcelor Mittal), Roger Kekeis (Arcelor Mittal), Ming Shi (US Steel), and the Intelligent Systems Center (Missouri University of Science and Technology) for generously providing materials and financial support. Additionally, the help of Jeremy Watts, Steve Webb, and Jacob Johnson at the Missouri University of Science and Technology is greatly appreciated.

References

1.   S. Subramaniam et al., "Considerations for Spot Welding of Advanced High Strength Steels," *Society of Automotive Engineers Technical Papers*, 2006-01-0089, 2006.

2.   Z. Feng et al., "Friction Stir Spot Welding of Advanced High Strength Steels: a Feasibility Study," *Society of Automotive Engineers Technical Papers*, 2005-01-1248, 2005.

3.   M.L. Santella et al., "Friction-Stir Spot Welding of Advanced High-Strength Steel: FY 2006 Progress Report," Automotive Lightweighting Materials. Oak Ridge National Laboratories. 2006, 208-214.

4.   M.I. Khan et al., "Resistance and friction stir spot welding of DP600: A Comparative Study," *Science and Technology of Welding and Joining*, 12 (2007), 175-182.

5.   W.J. Kyffin et al., "Progress in FSSW of DP800 High Strength Automotive Steel," *Proceedings of the 6th International Symposium on Friction Stir Welding*. St. Sauveur, Canada. October 10-13th, 2006.

6.   O. Ryoji et al., "Study on Friction Stir Spot Joining of High Strength Steel Sheet," *Preprints of the National Meeting of the Japan Welding Society*. Spring 2007, 4.

7.   Y. Hovanski, M.L. Santella, G.J. Grant, "Friction Sitr Spot Welding of Hot-Stamped Boron Steel," *Scripta Materialia*, 57 (2007), 873-876.

8.   American Welding Society. AWS D8.1M:2007: Specification for Automotive Weld Quality: Resistance Spot Welding of Steel, 8.

9.   Military Handbook, "Metallic Materials and Elements for Aerospace Vehicle Structures" MIL-HDBK-5H-1998, section 8.2.2.3.1

10.   C.B. Fuller. "Friction Stir Tooling: Tool Materials and Design," *Friction Stir Welding and Processing*, ed. R.S. Mishra and M. Mahoney (Materials Park, OH, ASM International, 2007), 9.

11.   O.T. Midling, G. Rorvik, "Effect of Tool Shoulder Material on Heat Input During Friction Stir Welding," *Proceedings of the 1st International Symposium on Friction Stir Welding*, Thousand Oaks, CA, USA, 14-16 June, 1999.

12.   M. Santella, G. Grant, W. Arbegast, "Plunge Testing to Evaluate Tool materials for Friction Stir Welding of 6061+20wt%Al$_2$O$_3$ Composite," *Proceedings of the 4th International Symposium on Friction Stir Welding*, Park City, UT, 14-16 May, 2003.

13.   C.B. Fuller, "Friction Stir Tooling: Tool Materials and Designs," *Friction Stir Welding and Processing*, ed. R.S. Mishra and M. Mahoney (Materials Park, OH, ASM International, 2007), 11.

14.   J. Rodelas, G. Hilmas, R.S. Mishra, "Sinterbonding of Cobalt-cemented Tungsten Carbide to Tungsten Heavy Alloys," *Int. J. of Refractory Metals and Hard Materials*, (in review).

15.   T. Freeney, S. Sharma, R.S. Mishra, "Effect of Welding Parameters on Properties of 5052 Al Friction Stir Spot Welds," *Society of Automotive Engineers Technical Papers*, 2006-01-0969, 2006.

16.   J.N. Harris. *Mechanical Working of Metals: Theory and Practice* (New York, NY, Pergammon Press, 1983), 54.

Friction Stir Welding and Processing V
Edited by: Rajiv S. Mishra, Murray W. Mahoney, and Thomas J. Lienert
TMS (The Minerals, Metals & Materials Society), 2009

Low Z-force Friction Stir Spot Welds – Conventional Tool & Process Development Approach

Tze Jian Lam[1], Christian A. Widener[1], Jeremy M. Brown[1],
Bryan M. Tweedy[2], Dwight A. Burford[1]

[1] Wichita State University, 1845 Fairmount, Wichita, Kansas, 67260, USA
[2] H.F. Webster Engineering and Professional Services, Rapid City, SD 57702 USA

Keywords: Friction Stir Spot Welding, Low Z-force, Process Development, Pin Tool
Geometries, Aluminum Alloy 2024-T3

Abstract

An investigation was conducted to develop low normal load (Z-Force) Friction Stir Spot Welds (FSSW) using conventional tooling and process development approaches. Low Z-forces can be achieved by studying the relationship between pin tool features, geometries, processing parameters, and the resultant strengths of coupons produced by FSSW. Effects of geometrical changes of pin tool design, including shoulder diameter and probe features, on joint properties in 1mm (0.040") thick 2024-T3 Al alloy were evaluated. Weld tools included the Psi™, Counterflow™ and Trivex™ tools. A design of experiments approach was used to investigate the effects of process parameters, which included spindle speed, plunge depth, plunge rate, plunge load, travel rate and tilt angle. The goal of the program to maintain the ultimate shear strength tested in tension of unguided lap shear coupons was achieved while reducing the normal force required for producing a sound joint. In addition to single spot unguided lap shear tests, the performance of low Z-force FSSW joints were evaluated by optical metallographic cross section analyses, which were correlated with process parameters, process forces, UTS and pin tool designs.

Introduction

Friction stir spot welding (FSSW) development work has commonly used gantry-type systems because of the wide range of Z-forces (aka "forging force" or "normal force") required to produce desirable FSSW. However, articulated robots, which are limited to lower Z-forces, are preferred for implementation in manufacturing plants because of their potential to produce three dimensional structures with more flexibility and lower costs than a conventional gantry system. Thus, an investigation into low Z-force FSSW using conventional tool and process development is crucial for the development of this technology for robotic applications. Lower Z-forces can be achieved by studying the relationship between pin tool features, geometries, process parameters, UTS and optical metallographic cross sections. This research helps to indentify the portability issues associated with moving this technology from gantries to robots and provides a path for implementation of FSSW utilizing articulated robots in the automotive and aerospace industries.

Existing fastening methods such as rivets and resistance spot welds have been widely applied in the automotive and aerospace industries for decades. FSSW has been introduced recently as an alternative fastening method for thin gauge materials. The simplest type of FSSW, referred to as a plunge FSSW or poke FSSW, is an attractive replacement for existing discrete fastening methods because it can be produced rapidly and with a simple motion. Plunge FSSW has shown a lot of benefits and has already been implemented in the automotive industry [1]. Besides plunge spots, refill FSSW can fill the exit hole and leave a nearly flush surface with an opposing pin and shoulder [2, 3]. Swept FSSW, such as the Squircle™ disclosed by TWI and developed at Wichita State University (WSU) as an Octaspot™, has been shown to be up to 250% stronger than rivets and resistant spot welds in single spot lap shear [4]. Plunge and refill FSSW differ from swept FSSW. Swept FSSW has an additional closed loop translation movement (Figure 1). This additional closed loop translation increases the joint shear area and has been demonstrated by WSU to have better mechanical properties compared to plunge or refill FSSW [5,6].

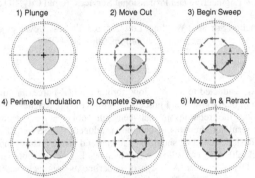

Figure 1. Octaspot™ Travel Path

Plunge FSSW cross sections tend to exhibit an upward flow of material from the bottom sheet causing an uplift of the faying surface, called hooking. The hooking caused by the vertical translation creates a thinning of the top sheet effective thickness. Whereas, swept FSSW consumes the hook by sweeping around the perimeter, giving it better control of the faying surface geometry and increasing the effective shear area of nugget (Figure 2).

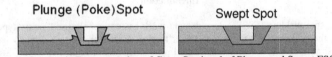

Figure 2. Schematic Representation of Cross Sectional of Plunge and Swept FSSW

Procedure

Five pin tool designs were initially considered for this project. Three pin tools were extensively investigated for linear lap FSW [4], plunge and swept FSSW. Two preferred pin tools for swept FSSW are the Counterflow™ and Psi™ tool designs developed at WSU [5], whereas a modified Trivex™ tool has been shown to be successful for plunge FSSW, (Figure 3). In addition, two more new pin tool designs will be included in a future investigation. Two pin tool shoulder diameters of 10.2 mm (0.400") and 7.6 mm (0.300") were included in the study. The pin tool probes had base diameters of 3.45 mm (0.136") and a 7 degree taper angle. All the pin tools included in this project had a 5 degree concave shoulder.

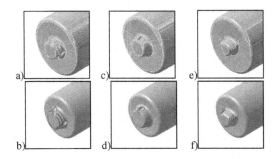

Figure 3. Counterflow™ Tool with concave shoulder diameter a) 10.2 mm (0.40") and b) 7.6 mm (0.30") Psi™ Tool with concave shoulder diameter c) 10.2 mm and d) 7.6 mm Modified Trivex™ Tool with concave shoulder diameter e) 10.2 mm and f) 7.6 mm

The weld coupon used in this study was a lap joint configuration with 25.4 mm (1.0") overlap in 1 mm (0.040") thick, bare 2024-T3 material. The specimen coupon configuration is shown in Figure 4, with W = 25.4 mm (1.0") and L = 101.6 mm (4.0") with 2024-T3 for both the top and bottom sheets.

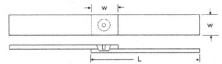

Figure 4. Single Spot Unguided Lap Shear Specimen

All FSSWs were made with a 5-axis ISTIR™ Process Development System (PDS) friction stir welding machine from MTS Systems Corporation. The welding was supported with a 12.5mm (0.50") thick steel backing plate with a 1.0 mm (0.040") machined step for lap welds (Figure 5). Steel bars were spaced 19 mm (0.75") apart clamped with finger clamps spaced 152.4 mm (6.0") apart, and were tightened with a torque wrench to 54.2 N-m (40 ft-lbs), providing approximately 4 kN (900 lbf) down force. The process parameters evaluated in this study included spindle speed (rotational speed), travel speed, plunge rate, plunge force (forge load) and dwell time. The test matrix used a Box Behnken design of experiments (DOE) approach to determine the process windows and significance of each process parameter.

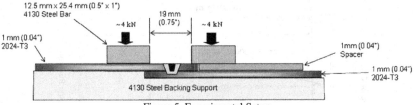

Figure 5. Experimental Setup

Weld programs used on the MTS FSW machine were written using a hybrid weld program that combined load control with position control. This capability of the MTS software gives an advantage to researchers to further investigate FSSW with low Z-force with innovative weld schedules tested in this work. The first weld program written was using position control, with specific plunge depths determined during previous work. The second program utilized a hybrid weld program with a partial initial plunge using position control and then switched to load control for the remainder of weld. In addition to controlling maximum weld forces, load control FSSW has been shown to have more consistent ultimate tensile strength results with lower standard deviations [5].

Mechanical tension testing of single spot unguided lap shear coupons was used to evaluate the mechanical properties of low Z-force FSSW. Besides the tensile strength test, optical metallographic analyses of FSSW cross-sections were used to qualitatively evaluate the welds.

Results and Discussion

Previous research had been performed using a 10.2 mm (0.4") shoulder. This data was beneficial in taking steps towards effective low Z-force FSSW. Swept FSSW using the hybrid weld program consisted of position control in the plunge process and switched to load control in the swept stage. The feedback from the forge force (also known as the normal force or Z-force) of the position control welds had two distinctive force spikes, the probe spike and the shoulder spike as the material was in contact with the pin tool during the initial plunge, which reached up to 13.34 kN (3000 lbf) (Figure 6). The high Z-force spike created by the pin tool shoulder was undesirable for this low Z-force study because it is beyond the force capability of most robotic arms.

Therefore, a few possible solutions were considered, such as pre-drilling before FSSW, pre-heating before FSSW, modifying initial plunge depth, increasing spindle speed, increasing cycle time, decreasing rate of plunge. Most of these possible solutions were tested using the existing hybrid weld program and compared directly with existing FSSW data. Pre-drilling and pre-heating before FSSW were not investigated. The remaining solutions were unsuccessful when implemented with the existing hybrid weld program. The last option was to modify the hybrid weld program to be fully under load control. Existing data from welds using a 10.2 mm (0.4") shoulder were used as a benchmark for Z-force comparison; however, existing data did not utilize the same fully load control program. The fully load control weld program successfully produced FSSW with low Z-force. The Z-force was lowered to 4.44 kN (1000 lbs). This is a commanded force of 4.00 kN (900 lbs) with additional shoulder spike of 0.44 kN (100 lbs) (Figure 7).

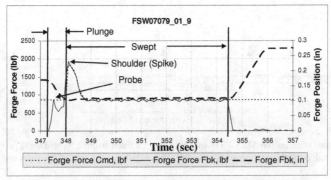

Figure 6. Command and Feedback Plot for Typical Swept Spot

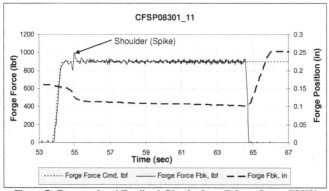

Figure 7. Command and Feedback Plot for Low Z-force Swept FSSW

The weld program that was fully load control, or low Z-force weld program, successfully created a swept FSSW with low normal load 4.00 kN (900 lbs), desirable joint interface, and fully consolidated weld nugget. The weld joint interface of low Z-force swept FSSW is shown in Figure 8, using a Psi™ tool with a 7.6 mm (0.30") diameter shoulder, and corresponds to the Z-force feedback shown in Figure 7. The weld also exhibited a desirable joint interface or faying surface interface with minimal or no hooking, shown in Figure 9.

A weld program that was fully position control was used to determine an appropriate Z-force for a corresponding hybrid program. The position control program was also used to estimate the required Z-force to maintain the tool depth while in the swept stage of the weld. The 10.2 mm (0.40") shoulder created a spike of 13.34 kN (3000 lbs) which decreased to an average of 7.56 kN (1700 lbs) during the swept stage of the FSSW. Whereas, the 7.6 mm (0.30") shoulder spiked up to 15.57 kN (3500 lbs) and continuously dropped to an average of 3.56 kN (800 lbs) by the end of swept stage. This reduction of normal load for the 7.6 mm (0.30") shoulder showed that lower Z-force could be achieved simply by reducing the shoulder diameter. The data also suggests that all position control aspects of the weld can be eliminated and performed under load control in order to achieve the lowest Z-force spot welds.

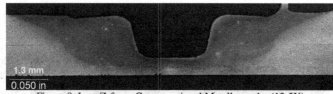

Figure 8. Low Z-force Cross-sectional Metallography (12.5X)

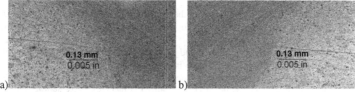

Figure 9. Joint Interface of Figure 8 (100X) a) Left Side b) Right Side

185

To further investigate strategies for reducing Z-force, the three essential process parameters, i.e. Z-force, spindle speed and travel speed, were studied in further detail while all other parameters selected were held constant. Process parameters were investigated using Box Behnken design of experiments to show correlations between these three process parameters. The first DOE had a process parameter range of 1.7 – 3.4 mm/s (4-8 IPM) for travel speed, 3.1 – 4.9 kN (700-1100 lbf) for Z-force, and 800-1200 RPM for spindle speed. The ultimate tensile strength (UTS) of unguided single spot lap shear was used to correlate the process parameters.

Low Z-force specimens were welded with two pin tool diameters, 7.6 mm (0.30") and 10.2 mm (0.40"), using a concave shoulder with a Psi™ tool probe. UTS increased for the 7.6 mm (0.30") shoulder specimens as the forge force load decreased. On the other hand, as the UTS decreased the forge load decreased for 10.2 mm (0.40") shoulder diameter. The Counterflow™ pin tool produced similar trends with the same process parameters. Cross-sectional metallographic analysis using optical microscopy provided more evidence to support this trend. The 7.6 mm (0.30") concave shoulder with the Psi™ pin welded at 4.89 kN (1100 lbs) over-plunged which created sheet thinning in top sheet and exit hole (Figure 10). The 10.2 mm (0.40") concave shoulder of Psi™ Tool welded at low Z-force 3.11 kN (700 lbs) created an unconsolidated weld nugget which showed insufficient Z-force (Figure 11). Additionally, the nugget is smaller where the shoulder had a lack of contact and created a U-shaped nugget instead of a V-shaped nugget. The 10.2 mm (0.40") Counterflow™ tool was unable to plunge at 3.11 kN (700 lbs) Z-force because the tip of probe has a larger surface area as compared to the Psi™ tool. The 7.6 mm (0.30") shoulder pin tool showed a better performance and achieved comparable UTS to the 10.2 mm (0.40") shoulder pin tool with low Z-force. Since the shoulder diameter investigation confirmed that the smaller shoulder can achieve lower Z-force, a study of shoulder features is planned for future work. The second phase of this project was an investigation into different probe designs affecting the Z-force and mechanical properties of swept FSSW.

Figure 10. Low Z-force Swept FSSW with 7.6 mm (0.30") Psi™ Tool at 4.9 kN (1100 lbs)

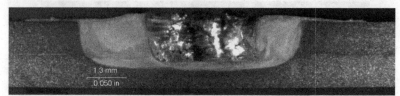

Figure 11. Low Z-force Swept FSSW with 10.2 mm (0.40") Psi™ Tool at 3.1 kN (700 lbs)

In the second phase probe design study, another 7.6 mm (0.30") concave shoulder with the modified Trivex™ pin tool was included in the study and welded with the low Z-force weld program with similar weld parameters as Psi™ and Counterflow™ tools. The different probe designs have differences in nugget joint area and joint interface morphology. The Counterflow™ probe has aggressive threads and flutes, promoting more flow of material (Figure 12) compared with the Psi™ tool (Figure 13). Similarly, the modified Trivex™ tool weld nugget (Figure 14) is slightly wider than the Psi™ tool weld nugget (Figure 13) because of its aggressive triangular

186

probe design. The effects of probe design were more obvious in plunge FSSW, where the probe features determine the nugget joint area.

Figure 12. Low Z-force Swept FSSW with 7.6 mm (0.30") Counterflow™ Tool at 3.1 kN

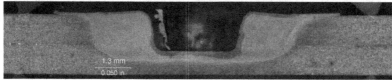

Figure 13. Low Z-force Swept FSSW with 7.6 mm (0.30") Psi™ Tool at 3.1 kN

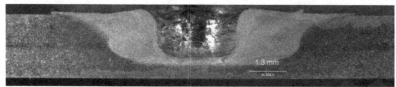

Figure 14. Low Z-force Swept FSSW with 7.6 mm (0.30") Modified Trivex™ Tool at 3.1 kN

In addition, at high Z-force, 4.9 kN (1100 lbs), the Counterflow™ and the Modified Trivex™ tools over-plunged completely (Figure 15 and 16) whereas the Psi™ tool still had material at bottom of exit hole (Figure 10). The Modified Trivex™ tool has a smaller probe tip area and combination of aggressive movement of material around the probe. Whereas, the Counterflow™ tool draws the material up through the flutes and pushes it back down with the threads. This creates a symmetrical and wider nugget joint area. Another benefit of the load control weld program with low Z-force FSSW is that the probes seem to experience even less wear, presumably due to the lower forces during the plunge stage. The little probe wear that was observed was near the tip and was associated with the higher Z-force 4.9 kN (1100 lbs) and hotter welding parameters. Welding with the tool at an angle off of the normal to the plane of the weld surface, called a tilt angle, can also lead to wear of the probe tip. Zero degree tilt angle tools such as Wiper™ shoulder [7] might be able to avoid direct loading to the edge of probe tip and therefore reduce wear. Another shoulder feature convex shoulder will also be beneficial for low Z-force FSSW will be included in further study.

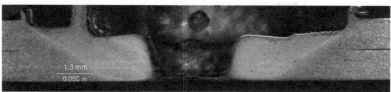

Figure 15. Low Z-force Swept FSSW with 7.6 mm (0.30") Counterflow™ Tool at 4.9 kN

187

Figure 16. Low Z-force Swept FSSW with 7.6 mm (0.30") Modified Trivex™ Tool at 4.9 kN

Again, using a Box Behnken design, a first DOE was created to find the process window and a second DOE was created to achieve maximum ultimate tensile strength (UTS) for each tool. Each DOE consisted of 15 coupons. The coupons were naturally aged for a minimum of 4 days before tensile testing. The Psi™ tool was able to match the average UTS of Counterflow™ tool on second DOE. Whereas, the average UTS of modified Trivex™ was significantly lower compared to Psi™ tool and Counterflow™ tool on both DOEs (Figure 17).

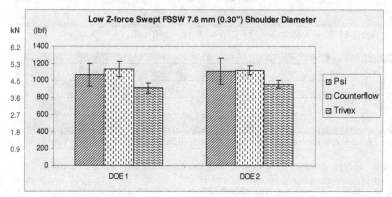

Figure 17. Chart Low Z-force 7.6 mm (0.30") Concave Shoulder Swept FSSW DOE

Further investigation of the modified Trivex™ tensile coupons and cross-sectional metallographic inspection revealed the primary cause of the low UTS. The failure mode of both DOEs was entirely due to nugget pullout (Figure 18). Additionally, metallography of the weld cross-section revealed that the faying surface interface had an upturn on each side (hooking) (Figure 19) due to the aggressive triangular probe. In the swept stage, the hooking created from plunge was not consumed but instead was remade, this time by the passing of the probe. The low UTS correlates well with the nugget pullout failure mode and the cross-sectional metallographic investigation of the faying surface interface. Clearly, the probe design can have a significant effect on the mechanical properties and faying surface interface.

Figure 18. Nugget Pullout Failure Mode

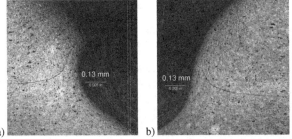

a)   b)

Figure 19. Joint Interface of Figure 14 (100X) a) Left Side b) Right Side

Low Z-force FSSW was able to maintain good mechanical properties using the Psi™ and the Counterflow™ tools using a low Z-force weld program. The modified Trivex™ tool with an extreme probe design requires further study to achieve good mechanical properties for low Z-force FSSW. A variety of pin tool designs have been proven to created sound welds in linear FSW using different process parameters and process windows [8]. Additionally, a modified Trivex™ tool may still have the potential to achieve good mechanical properties if an appropriate process window and process parameters could be found. The low Z-force FSSW data obtained from this research takes a step closer to transferring FSSW from gantry systems to articulated robots.

Conclusions

The results of this study shows that programmable load control enables low Z-force FSSW. Conversely, a position control weld program cannot be controlled to repeatably produce low Z-force FSSW. This is due to the sudden increase in normal force when the tool shoulder comes into contact with the material. Different possible solutions were tried individually with a hybrid weld program without success. However, the solution using a fully load control weld program provided significant reduction in Z-force and maintained FSSW quality. The reduction of shoulder diameter significantly reduces the required Z-force while simultaneously maintaining good mechanical properties. The combination of pin tool design coupled with a low Z-force weld program and appropriate process parameters created a sound FSSW. Mechanical properties of low Z-force FSSW were investigated using unguided single spot lap shear. Statistical analysis software (Statgraphics®) was used to correlate the UTS of the lap shear coupons with process parameters. The faying surface interface and the weld nugget quality were also investigated using cross-sectional metallography to correlate process parameters and pin tool designs. These low Z-force FSSW results indicate that is possible to produce sound FSSW joints within the force capability range of a typical articulated robot. Since the automotive and aerospace industries are moving toward automation to improve production and quality control simultaneously, the investigation of low Z-force FSSW will accelerate and bridge the implementation of FSSW for articulated robots in those industries.

Acknowledgements

This work was performed as part of the National Science Foundation's Center for Friction Stir Processing which is part of their Industry/University Cooperative Research Program. The authors would like to recognize the hard work of students in the Advanced Joining and Processing Lab in the National Institute for Aviation Research at Wichita State University, especially James Gross who performed much of the welding.

189

References

1.   Kallee, S.W. Mistry, A., "Friction Stir Welding in the Automotive Body in White Production," Proceedings of the 1st International Conference on Friction Stir Welding, California, USA 1999.
2.   Allen, C.D., Arbegast, W.J. "Evaluation of Friction Spot Welds in Aluminum Alloys," Proceedings of the Spring 2005 SAE World Congress, Detroit, MI, 2005.
3.   C. Shilling and J. dos Santos, "Method and Device for Joining at Least Two Adjoining Work Pieces by Friction Welding," US Patent App. 2002/0179 682.
4.   Merry, J., Tweedy, B., Widener, C., Burford, D. "Static Strength Comparison of Discontinuous Friction Stir Welded Stiffened Panels," 7th AIAA Aviation Technology, Integration and Operations Conference (ATIO), Belfast, Ireland 2007.
5.   Tweedy, B., Widener, C., Merry, J., Brown, J., Burford, D. "Factors Affecting the Properties of Swept Friction Stir Spot Welds," SAE International, SAE 2008 World Congress, Detroit, MI, 2008.
6.   Burford, D., Tweedy, B., Widener, C., "Fatigue Crack Growth in Integrally Stiffened Panels Joined Using Friction Stir Welding and Swept Friction Stir Spot Welding," Journal of ASTM International, Vol. 5, No. 4, Paper ID JAI101568
7.   Burford, D., Tweedy, B., Widener, C., "Influence of Shoulder Configuration and Geometric Features on FSW Track Properties," 6th International Symposium on Friction Stir Welding, Saint-Sauveur, Nr Montréal, Canada, October 10-13, 2006
8.   Widener, C., Dwight, B., "FSW Path Independent Study- Evaluation of the Sensitivity of FSW Butt Joints to Weld Tool Design," 12th MMPDS Coordination Meeting, PIMWG FSW Meeting, Orlando, FL, 2007.

Friction Stir Welding and Processing V
Edited by: Rajiv S. Mishra, Murray W. Mahoney, and Thomas J. Lienert
TMS (The Minerals, Metals & Materials Society), 2009

FRICTION STIR FORM WELDING OF ALUMINUM TUBES

K. Gupta [1], R.S. Mishra[1], Y. L. Chen[2], B. Carlson[2] and X. Q. Gayden[2]

[1]Center for Friction Stir Processing, Materials Science and Engineering, Missouri University of Science and Technology, Rolla, MO 65409, USA
[2]GM Research and Development Center, Warren, MI 48090, USA

ABSTRACT

The objective of the present study was to establish the feasibility of friction stir welding (FSW) rectangular aluminum tubes. Partial penetration welds were made for 6063 Al tubes having rectangular cross sections. FSW runs were made on a conventional machine as well as on a robotic machine. A comparative study of tool plunge force variation was made between unsupported and plug-supported welds. Effect of penetration depth (PD) was also examined for both weld categories. Multiple weld cross sections were taken across the weld length to correlate the exerted plunge force with the resulting tube deformation. Also, the effect of paint-bake cycle was studied on the load bearing capability of the welds. Unsupported welds exhibited higher load bearing capability than plug-supported welds. Optimum heat input required for the weld was quantified by developing a process map. This was done for unsupported welds on a robotic welding machine. The unsupported weld was used as it was better than plug-supported weld. Runs made on the robotic machine were more uniform with respect to their load bearing capabilities. A discussion on the failure mode for the two different weld types is included. Overall joint efficiency was higher for the paint bake condition. A penetration depth of 56% gave higher failure loads as compared to specimens made with 67% penetration depth.

Keywords: Unsupported weld, Plug supported, Process map

1. INTRODUCTION

Friction stir welding is a solid state joining process that involves plunging and traversing a rotating tool along the weld line. Improved weld quality, low power requirements, no filler material, no shielding gas, and environmental friendliness are some of the significant advantages that FSW has to offer over conventional welding technology [1,2]. Industrial implementation of FSW technology is hindered because of a lack of industry standards, and design guidelines [1,3]. Design for manufacturing (DFM) issues related to the fabrication of built-up structures by FSW method are addressed to some extent in the development of an aluminum rail car hopper [1,4]. Fixturing issues, process parameter optimization and design modification for FSW are considered thoroughly for the fabrication of built-up beam assemblies [5].

Aerospace and automotive industries have focused on the development of various aspects of FSW [6]. Automotive application of this technology depends primarily on the types of joints possible. The variety of joint designs possible by FSW are full penetration and partial penetration butt joints, lap joints [7], T-joints, edge joints etc. and are indicated in Fig. 1 [1,2]. These joint designs are explored for solid components where the work-piece is supported by a backing plate during the weld.

There is not much literature on the FSW of hollow structures and as such this work focused on exploring the possibility of joining hollow structures using the FSW technique. The possibility of making partial penetration butt welds with hollow tubes was explored. Such a joint is observed mostly in vehicle frames. The forces experienced by the work-piece during welding should be kept as low as possible to minimize tube distortion. Plunge force in the downward direction was studied as this governs the deformation of the hollow structure. A method to control the tube distortion was explored in which a plug was used as a backing piece. Performance of the welds is analyzed in the as-welded (AW) and paint bake (PB) conditions in order to understand the effect upon FSW properties, if any, that the typical vehicle frame thermal processing in a paint shop has. FSW runs were successfully made on conventional as well as on a robotic machine thereby exploring the feasibility of industrial implementation.

The effect of process parameters on the weld quality was determined for process parameter optimization. The temperature attained by the work-piece during FSW affects the quality of the weld. The amount of heat input into the work-piece during FSW depends on the process parameters and could be measured by the pseudo heat index (PHI) defined as below [8]:

$$PHI = \frac{rpm^2}{ipm \times 10000} \tag{1}$$

where rpm is the tool rotation speed in revolutions per minute, and ipm is the tool traverse rate in inches per minute.

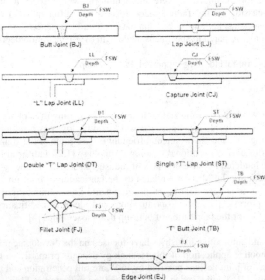

Fig. 1. Joint designs possible with FSW [1,2].

2. EXPERIMENTAL PROCEDURE

Rectangular tubes of 6063-T56 Al were welded in two configurations indicated in Fig. 2. Tube dimensions were 76.2mm X 25.4mm in cross section with a wall thickness of 3.175 mm. Supporting plugs of 6061-T6 Al alloy with an outer dimension of 76.2mm X 25.4mm and 3.175 mm thickness were machined to form a 1.0 mm thick flange leaving an area of 69.85 X 19.05mm of original thickness in the center which was used to press fit into the hollow of the tube to be joined for support, refer to Fig. 2b. Single pass welds of 64.8mm were made along the horizontal surface of each joint. The weld is shorter than the tube width in order to contain the initial plunge and pullout within the tube material. Conventional machine Z-forces experienced by the tool were recorded and analyzed for both weld types and correlated with the macrostructural variations along the weld length.

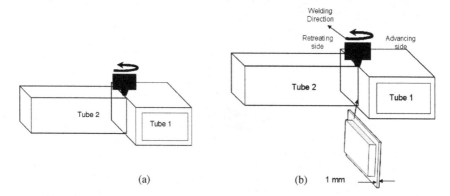

(a)   (b)   1 mm

Fig. 2. Tube to tube weld configurations, a) unsupported (TT), and b) plug supported (TTP).

The first objective was to do a preliminary work aimed at determining what one might do to optimize the penetration depth (PD) for unsupported and plug supported welds. Welds were made on a conventional FSW machine keeping all other process variables constant and run in position control. These runs were made with a tool rotation speed of 750 rpm, tool traverse rate of 1.058 mm/sec and tool tilt angle of 2.5°. Two tools were used with differing pin heights targeted for 56% and 67% penetration of the total thickness for both unsupported and plug-supported tubes. The tools had a conical stepped spiral pin and concave shoulder with specific geometries summarized in Table 1.

Table 1. Tool geometries used in this study.

Penetration Depth, %	Pin Height mm	Pin Diameter mm	Shoulder Diameter mm
67	2.1	4	12
56	1.5	4	12

For both the penetration depths used, effect of paint bake cycle on failure load was determined by pulling tensile samples having undergone thermal cycling. The thermal cycling consisted of being placed in a table-top oven at 175 °C for 30 min and air cooled which is representative of a commercial paint-bake cycle. Fig. 3 presents the tensile test coupon geometry. An MTS frame with 99 kN load capacity was used for tensile testing. All coupons were subjected to pulling in tensile mode at the rate of 0.02 mm/sec up to yielding and subsequently the cross-head speed was increased to the rate of 1 mm/sec beyond yielding.

A process map was developed for the unsupported weld type using an ABB robotic machine. For these runs a constant penetration depth of 1.4 mm was used with tool tilt angle of 3° and work angle of 2°. The process parameter matrix used in making the runs is presented in Table 2. Transverse weld sections were taken at different distances starting from the weld beginning in order to study the effect of process parameters on failure load. Keller's reagent was used as the etching reagent.

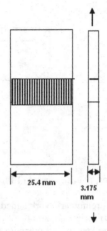

Fig. 3. Coupon used for tensile testing.

Table 2. Unsupported weld process matrix.

		Tool Rotation Speed (rpm)		
		750	900	1000
Tool Traverse	0.42	Y	Y	Y
Speed (mm/s)	1.06	Y	Y	Y
	2.12	Y	Y	Y

3. RESULTS AND DISCUSSION

3.1 Plunge force analysis

The welds for this study were made using position control of equipment and as such the plunge forces varied during both the initial plunging of the tool as well as traversing the tool along the weld. Variations in the Z-force during the tool plunge and tool traverse stages are presented in Figs. 4a and 4b, respectively. For this analysis unsupported and plug-supported welds were made using 56% PD. During the tool plunge stage the Z-forces increase with the vertical progression of the tool into the workpiece in order to overcome the greater material resistance needed to continue its rotation. The Z-force is maximum when the tool shoulder comes in contact with the material surface and was 4.72 kN for unsupported and 6.45 kN for plug-supported welds.

Temperature increases are observed in the work-piece as a result of the heat generated by the extensive plastic deformation of the material and frictional heat between the tool and work-piece during welding. This leads to a drop in the Z-force during traverse. Z-forces during traverse drop from 5.34 kN to 3.11 kN (averaging 3.93 kN) for plug-supported weld and from 4.45 kN to 2.67 kN (averaging 2.95 kN) for unsupported welds as the tool moves along the weld line. For plug-supported welds, Z-force decreases continuously traversing along the length of the weld whereas for unsupported welds, the Z-force first decreases from the weld start (4.45 kN) to the middle of the weld length (2.67 kN) beyond which a marginal increase in the Z-force is observed towards the end (2.9 kN). This difference can be explained by the fact that welds made in the middle of the unsupported tubes correspond to maximum deflection of the tube wall whereas the deflection is minimal at the ends. The shape of the Z- force curve follows a similar trend. However, forces at the end of the weld do not attain the same level as the beginning Z-force which is attributed to the high temperature attained by the work-piece during the run causing softening of the material and a lower required Z-force.

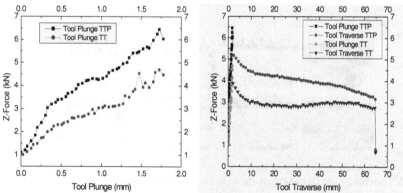

a) Z-force variation during tool plunge   b) Z-force variation during tool traverse
Fig. 4. Plunge force analysis during tool plunge and traverse.

Transverse weld sections were taken at different distances from the weld start along the weld line as indicated in Fig 5a. Figures 5b and 5c represent transverse weld sections for unsupported and plug-supported welds respectively. Several observations can be made from these figures. First for the unsupported welds, it can be seen in Fig. 5b that there was greater bending of the retreating side tubes at the center of the tubes as compared to the ends. This becomes apparent by comparing the relative position of the original top surface in the macrographs labeled 1-4 of Fig. 5b. The middle two locations (#2&3) have extra material deposited on top of the retreating side surface. The extra material is built up there because there is maximum deflection of the retreating side tube in this area because of a lack of supporting structure underneath. Fig. 5b macrographs #1 & 4 for the unsupported weld show relatively small build up of material on the retreating side, i.e., little deflection of the tube because of the supporting tube wall underneath the weld. Subsequently, the transverse weld sections for the plug-supported welds which are supported along the entire length of the weld do not show much bending of the supported tube because of the support provided by the plug as evidenced by macrographs #1–4 of Fig. 5c. However, relatively large local deformation of the other tube under the weld is observed (Fig. 5c).

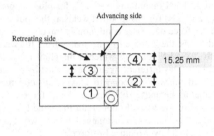

a) Location of samples

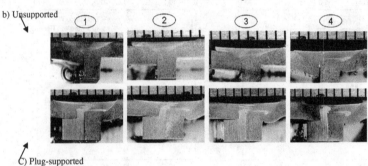

Fig. 5. Transverse weld sections along the weld length with the advancing side on the right hand side of each macrograph.

3.2 Effect of paint bake (PB) and penetration depth (PD)

Except penetration depth, all other process variables like tool rotation speed, tool traverse rate, and tool tilt angle were kept constant. The nomenclature used for different samples is indicated in Fig. 6. Welds were made on both the sides of the tubes. Side welded first is indicated

196

as side 1 and side 2 was welded last. The time after welding and before heat treatment was not recorded. However, after making the welds, tubes were kept in the freezer until the paint bake cycle for the PB samples. As welded (AW) tubes were also retained in the freezer until the tensile test was done. Note that Sample I was positioned closer to the weld stop position and Sample II positioned closer to the weld start position. Each coupon was one inch wide and the load bearing cross sectional area was based upon the penetration depth used to make the weld. Figs. 7(a-d) show the effects of paint bake cycle on the load bearing capabilities of the welds for the two different penetrations of 56% and 67% of the total thickness under tensile loading condition. These tests were done with two different expectations. First, the paint bake heat treatment is expected to increase the strength of 6XXX alloys because of ageing. Second, it was assumed that higher penetration depth may lead to larger weld nugget and therefore higher joint strength. The analysis of the results suggests that:

1. Effect of PB is inconclusive especially given only 1 sample for each location & condition. Lack of a trend suggests that the joint strength does not just depend on the material strength, but also on other geometrical factors not considered in these figures.
2. There is a general trend for the plug-supported samples to have a higher failure load at higher % penetrations. This is consistent with the initial expectation that a larger stir zone would provide better properties.
3. The unsupported samples exhibited noticeable bending after welding at higher % penetration. These runs were made in position control. So, this suggests that there might be a critical penetration depth up to which the resultant process loads are adequately supported. Further work is needed to define this.

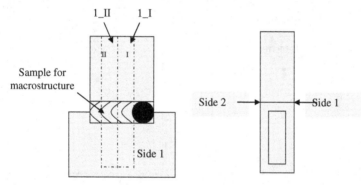

Fig. 6. Specimen details for weld categories.

Maximum load bearing capabilities for all welds made during this study are summarized in Table 3. Plug-supported welds show lower load bearing capability than unsupported welds both in AW and in PB conditions. This is likely to be a result of the geometrical nature of the remnant joint line leading to a reduced cross-sectional thickness for the plug-supported welds. Another possibility is that the plug material was a different 6XXX alloy than the tube material. The differences in alloy chemistry and material strength could lead to difference in material flow during FSW and subsequent strength. One idea for future work would be to use a plug of the same alloy as the tube. The 6XXX alloys are heat treatable and different studies done to analyze the effect of paint bake on the properties of FSW 6XXX series Al alloys indicate an improvement in weld strength after paint bake. The current results obtained do not agree with the results of the earlier studies [1]. However, in our study we are comparing weld strengths of two different samples. If we could analyze the same coupon in AW condition and in PB conditions, different results might be expected.

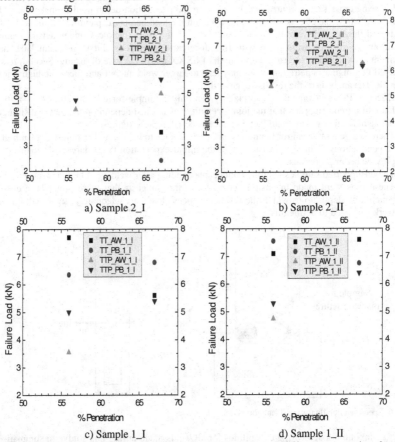

a) Sample 2_I b) Sample 2_II

c) Sample 1_I d) Sample 1_II

Fig. 7. Effect of paint bake on failure loads for welds with two different penetration depths, AW= as-welded, PB = with paint bake.

Analysis of failure load against % penetration curves indicates that there is no regularity in the data for different samples. There might be some internal variations in the weld quality along the weld length which could be responsible for the randomness in the results of the tensile testing. Overall weld efficiency (weld strength/parent material strength for same cross section) was determined for different weld types as indicated in Table 3. Minimum weld efficiency of 66% was recorded for an optimum penetration depth of 56% for unsupported welds. The unsupported weld exhibited higher weld efficiency than plug-supported weld. Welds subjected to PB treatment showed higher overall weld efficiency than AW condition.

Table 3. Summary of results of tensile testing.

Weld Type	% Penetration	Condition	Failure Load (kN)					% Weld Efficiency
			1_I	1_II	2_I	2_II	Minimum	
TT	56	As Welded (AW)	7.7	7.1	6.05	5.95	5.95	65.88
		Paint Bake (PB)	6.35	7.55	7.9	7.6	6.35	70.30
	67	As Welded (AW)	5.6	7.6	3.5 (Unbonded Area + SLOF)	6.25	5.6	52.29
		Paint Bake (PB)	6.8	6.75	2.4 (SLOF)	2.65 (SLOF)	6.75	63.03
TTP	56	As Welded (AW)	3.55	4.75	4.4	5.4	3.55	39.30
		Paint Bake (PB)	5	5.3	4.75	5.6	4.75	52.59
	67	As Welded (AW)	Traverse TD X	Traverse TD X	5	6.3	5	46.13
		Paint Bake (PB)	5.4	6.4	5.55	6.2	5.4	49.82

3.3 Process map development

Visual inspection of the weld length was done for each tool rotation and tool traverse combination. These results are indicated in Figs. 8(a-j). Scratch marks observed on some welds are a result of the filing of flashes to facilitate the welding on the second side.

All welds are free from wormholes. A significant digging of the material was observed on the advancing side. Material on the retreating side was bent during welding due to lack of support. The advancing side was supported by the tube wall. The difference in the support on advancing and retreating sides was responsible for the material dig on advancing side. This extra material moved with the tool and was deposited on the retreating side. Material deposition was higher for higher heat indices as can be seen in Figs. 8a and 8d. Full cup was observed at all the tool rotation rate – tool traverse rate combinations. Shoulder-workpiece contact on the retreating side was not observed in the middle of the weld length for heat index values less than 40. For almost similar heat index values of 32.4 and 36 the appearance of the weld was different as observed in Figs. 8e and 8j, respectively. The PHI term defined in eqn. (1) does not encompass all the process variables that control heat input in the weld. Some other parameters could also be responsible for this variation, like tool geometry, clamping, temperature of the backing piece during the run, etc.

Fig. 9 presents the transverse weld sections for the welds shown in Fig. 8. The tool rotation rate-traverse rate combination of 1000:0.423 represents the hottest weld and 750:2.12 represents the coldest weld. The weld section for 1000:0.423 tool rotation tool traverse combination was bent during cutting. The first parameter for comparison is the bending of the tube on the retreating side. Bending of retreating side tube was higher for hotter welds (1000:0.423, 1000:1.058 and 1000:2.12). Within the nugget region, an arm indicating microstructural variation is observed on the advancing side for some heat indices. PHI of 100, 56.25, and 40 showed presence of this microstructurally distinct region in the nugget. Relatively colder welds for PHI of 32.4, 22.5, 16.2 and 11.25 showed a nugget which was macroscopically uniform.

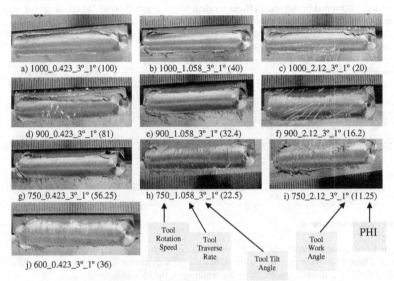

a) 1000_0.423_3°_1° (100)   b) 1000_1.058_3°_1° (40)   c) 1000_2.12_3°_1° (20)

d) 900_0.423_3°_1° (81)   e) 900_1.058_3°_1° (32.4)   f) 900_2.12_3°_1° (16.2)

g) 750_0.423_3°_1° (56.25)   h) 750_1.058_3°_1° (22.5)   i) 750_2.12_3°_1° (11.25)

j) 600_0.423_3°_1° (36)

Tool Rotation Speed   Tool Traverse Rate   Tool Tilt Angle   Tool Work Angle   PHI

Fig. 8. Comparison of weld beads for different process parameters.

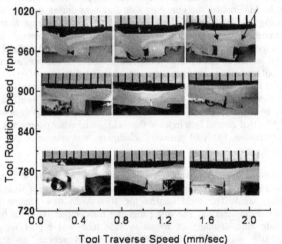

Retreating side   Advancing side

Fig. 9. Process map for unsupported tube to tube welds.

200

Two samples per Fig. 3 were subjected to tensile testing. Sample No. I was taken from the end of the weld length and sample No. II was from the middle of the weld length. Failure load for these coupons was plotted against the corresponding PHI. All other parameters were kept constant. Results of tensile testing are indicated in Fig. 10a for sample I and in Fig. 10b for sample II, respectively. PHI varied in the range of 11-100. For sample I, failure load varied from 5.3 kN to 7.7 kN for various heat indices. Minimum failure load of 5.3 kN was obtained for sample I for a colder weld corresponding to 11.25 PHI. For same PHI, sample II failed at a load of 6.4 kN.

Welds made on the robotic machine exhibited much less variation in load bearing capability than the welds made on the conventional machine. Heat index values in the range of 15-40 showed higher failure loads for both samples I and II respectively and exhibit an optimized process parameter window for these welds. PHI of 100 also gave failure loads of the same order as heat indices in the range of 15-40.

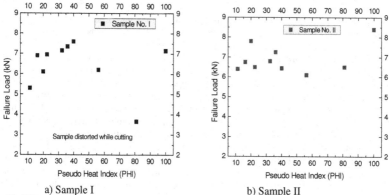

a) Sample I   b) Sample II

Fig. 10. Effect of PHI on tensile properties of unsupported tube to tube welds.

3.4 Failure path

The observed failure paths for unsupported and plug-supported weld coupons when subjected to tensile loading are indicated in Fig. 11. In both cases, failure occurred by the propagation of cracks on the retreating side. Partial penetration butt welds were made as the hollow structures were joined. In the joint, partially welded thickness acts as the weakest load bearing cross section. When load is applied, the gap between two tubes continued to widen with consequent thinning of the welded area along the indicated line. This crack propagated from bottom to the top surface. For the plug-supported weld, a crack initiated on the retreating side at the interface between the plug and retreating side tube. The sample was able to withstand the applied load till the critical thickness was reached, beyond which the sample failed.

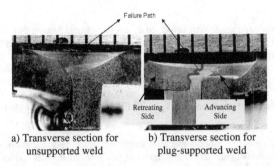

a) Transverse section for   b) Transverse section for
unsupported weld   plug-supported weld

Fig. 11. Failure path for a) Unsupported and b) Plug supported tube to tube welds.

4. CONCLUDING REMARKS

The FSW method could be successfully used for making partial penetration butt welds for rectangular 6063 Al tubes. However, unsupported welds exhibited better performance than plug-supported welds. Differences in the materials between the plug and the tube could be responsible for poor performance of plug-supported welds in comparison to unsupported welds. A FSW joint was possible for both unsupported and plug-supported tubes. The plunge force pattern was different for unsupported and plug- supported welds corresponding to the nature of the backing plate. A paint bake cycle improved overall performance of the weld. A stray randomness was observed in tensile properties of samples taken along the same weld length with same process parameters. Welds made for 56 % target depth exhibited better properties for both the weld types. Maximum weld efficiency of around 70% was obtained for unsupported welds. Macroscopically defect free welds were possible on both a conventional and a robotic welding machines, thereby demonstrating feasibility for industrial implementation. Process parameters corresponding to the PHI values in the range of 15-40 gave optimized weld performance. Samples failed on the retreating side of the weld by gradual decrease in the thickness of the region of partial penetration weld.

5. REFERENCES
1. R. S. Mishra, M. W. Mahoney, *Friction Stir Welding and Processing, ASM International, 2007, Chapters 1, 2, 13, and 15.*
2. C. B. Smith, J. F. Hinrichs and P. C. Ruehl, " Friction Stir and Friction Stir Spot Welding-Lean, Mean and Green", *Friction Stir Link, Inc. W227 N546 Westmound Dr., Waukesha, WI 53186*, pp.1-8.
3. W. J. Arbegast, "Friction Stir Welding: After a Decade of Development", *Friction Stir Welding and Processing IV, Edited by Mishra, Mahoney, Lienert & Jata TMS* (2007), pp.3-17.
4. C. Allen, C. Oberembt, W. J. Arbegast, D. Medlin and H. Mercer, "Friction Stir Welding of an Aluminum Coal Hopper Railcar" , *Friction Stir Welding and Processing IV, Edited by Mishra, Mahoney, Lienert & Jata TMS* (2007), pp.19-28.

5. W. J. Arbegast and A. K. Patnaik, "Process Parameter Development and Fixturing Issues for Friction Stir Welding of Aluminum Beam Assemblies" , *SAE International* (2005)
6. A. Scafe, A. Joaquin, "Friction Stir Welding of Extruded Aluminum for Automotive Applications" , *SAE International* (2004)
7. L. Cederqvist and A. P. Reynolds, " Factors Affecting the Properties of Friction Stir Welded Aluminum Lap Joints", *Proceedings of the second international conference on Friction Stir Welding, The Welding Institute (TWI), Gothenburg, Sweden, June 2000.*
8. S. Kandukuri, W. J. Arbegast, A. K. Patnaik, and C. D. Allen, "Development of Design Curves for Tensile Strength and Fatigue Characteristics of 7075 T73 Aluminum FSW Butt Joints" , *Friction Stir Welding and Processing IV, Edited by Mishra, Mahoney, Lienert & Jata TMS* (2007), pp.29-38.

203

Friction Stir Welding and Processing V
Edited by: Rajiv S. Mishra, Murray W. Mahoney, and Thomas J. Lienert
TMS (The Minerals, Metals & Materials Society), 2009

FRICTION STIR SPOT WELDING OF AA6016 ALUMINUM ALLOY

W. Yuan[1], R.S. Mishra[1], S.S. Webb[1], Y. L. Chen[2], X. Q. Gayden[2], B. Carlson[2], and G. J. Grant[3]

[1]Center for Friction Stir Processing, Materials Science and Engineering, Missouri University of Science and Technology, Rolla, MO 65409, USA
[2]General Motors Research and Development Center, Warren, MI 48090, USA
[3]Pacific Northwest National Laboratory, Richland, WA 99356, USA

Keywords: Friction stir spot welding, AA6016, cross-tension tests, separation modes

Abstract

Friction stir spot welds of 1 mm thick 6016-T4 aluminum alloy sheets were made using conventional pin (CP) tool and off-center feature (OC) tool. Different tool rotation speeds were employed for maximizing the bonded region: 1500 rpm for CP tool and 2500 rpm for OC tool. Effects of penetration depth and plunge speed on cross-tension separation load were investigated and a maximum separation load of about 1.8 kN was obtained. Results indicated that the cross-tension separation load did not change after thermal processing i.e. paint-bake cycle for both tools. It appears that overageing in the nugget region after welding leads to the lack of thermal treatment response. Two separation modes were observed: shear separation and pull-out under cross-tension loading condition. Based on the experimental observation of separation path, load-displacement curve, and microhardness profile, the effect of paint-bake cycle on cross-tension strength and the separation mechanisms are discussed.

Introduction

Mass reduction in order to improve fuel economy and reduce emissions without affecting the safety performance is a great challenge in the automotive industry. Lightweight materials like aluminum alloys are promising candidates for replacing equivalent steel assemblies and its use in the automotive industry has increased in recent years.

For replacing steel with aluminum in the body of automobiles, it is necessary to explore joining methods that can be used efficiently. Friction stir welding (FSW) is a solid state welding technique developed as a novel method for joining aluminum alloys. Friction stir spot welding (FSSW) is a derivative of this process and incorporates a tool consisting of a shoulder with a protruding pin which is rotated and plunged into the work piece to a predetermined depth. After the weld is made, it is retracted and a key hole is left. The frictional heat generated at the tool-workpiece interface softens the surrounding material, and the rotating and moving pin causes material flow around it. The forging pressure applied by the tool shoulder and the mixing of the plasticized material result in the formation of a solid bond region [1,2].

The property of spot welds is typically evaluated by the joint strength or the maximum load the weld bears prior to separation. Previous work has shown that a large bonded region with sufficient thickness of top sheet under shoulder indentation and dispersed oxide layer (hooking defect) produces high shear strength spot welds [3] Tool features and process parameters determine the size of bonded area, the remaining thickness of top sheet under shoulder and the morphology of the hooking defect. Most of the current literature focuses on the lap shear strength of spot welds and how the process parameters influence it [1-6].

The present work expands upon the process – microstructure literature by investigating the cross-tension properties of AA6016 aluminum alloy spot welds as a function of two different tool designs. AA 6016 is typically used for automobile body outer panels due to its combined properties of good dent resistance, relatively high formability, medium strength and superior surface finish [7]. The aim of this work was to study the effects of tool features, plunge depth and speed on the cross-tension property as well as the heat treatment response of welds.

Experimental Procedure

AA6016 sheets of 1 mm thickness in T4 temper were used in this study. Aluminum sheets were sheared to a dimension of 6 x 1.5 inch. Figure 1 presents a schematic of the cross-tension specimen and testing fixture used to evaluate the weld strength. Friction stir spot welds were produced using a RoboStir™ spot welder (Friction Stir Link Inc., Waukusha, WI) with axial load capacity of 22.2 kN and spindle rotation speeds up to 3000 rpm. In this study, the spot welder was operated under position control mode and the tool rotation speed was varied from 1000 rpm to 2500 rpm in steps of 500 rpm. Plunge speeds of 0.5 mm/s and 2.5 mm/s were applied with a constant dwell time of 490 ms. Because of the stiffness of the motor, actual plunge depth was always less than the predetermined plunge depth. Actual plunge depth was measured from the weld cross-section.

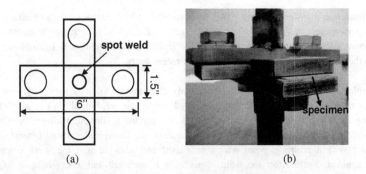

(a) (b)

Figure 1. Schematic of (a) cross-tension specimen, and (b) cross-tension fixture.

Two tools were used in this work; a conventional pin, CP, and an off-center pin feature, OC, and are presented in Figure 2. The reasons for designing this off-center pin feature were elucidated in earlier work [5]. Both tools were machined from Densimet tungsten alloy and have the same 10 mm diameter concave shoulder. The CP tool had a 1.51 mm long step spiral pin with a root diameter of 4.5 mm and tip diameter of 3 mm. The OC tool had three off-center 0.8 mm long and 2 mm diameter hemispherical pin features.

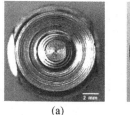 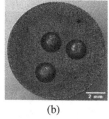

(a)   (b)

Figure 2. Marco images of (a) conventional pin, CP, tool, and (b) off-center pin feature tool, OC.

Cross-welding and cross-tension tests were performed after selecting the specific tool rotation speed for two tools based on its effect on size of bonded region. Cross-tension specimens were first made using a plunge speed of 0.5 mm/s and varying penetration depths to determine which penetration depth for each tool produced the maximum cross-tension separation load. Three cross-tension specimens were made for each penetration depth and the samples were tested to obtain the 'optimum' or maximum average cross-tension separation load by using an MTS testing machine. Specimens were pulled at a rate of 0.02 mm/s. Three additional samples were made at these optimum parameters and were subsequently thermally processed (170 °C for 20 minutes and air cooled) in order to evaluate any effect a thermal paint-bake cycle, PB, may have upon the cross-tension separation load. Furthermore, three additional runs were completed using the 'optimum' penetration depth for each tool although using an elevated plunge speed of 2.5 mm/s in order to investigate the effect of plunge speed on cross-tension properties. Microhardness tests were performed on the cross-section of welds. Vickers microhardness measurements were made at 0.4 mm below the upper sheet surface with 0.5 mm indent spacing, a 0.5 kgf load and 10 s dwell time. The samples were stored in a freezer after FSSW for days, and then allowed to equilibrate to room temperature and then put directly into the furnace. In addition to mechanical tests, welds were cross-sectioned, mounted and etched using a 5% HF reagent for metallographic studies.

Results and Discussion

Figures 3a & b present the cross-section and microstructure of a weld made using the CP tool at 1500 rpm tool rotation speed. A large stir zone is notable. A typical hooking defect, which curved up into the stir zone and turns toward the pin, is clear in Figure 3b. It has been reported that the curved up hooking defect is deleterious to the weld strength [8]. Figure 3c shows the cross-section of a weld made with a large bonded region using the OC tool at a 2500 rpm tool rotation speed similar to what was produced with the CP tool although a larger exit hole was produced. Though a stop-then-retraction mode was used for the OC tool, the motor inertia caused softer material around the pin to be pulled out with the rotating tool.

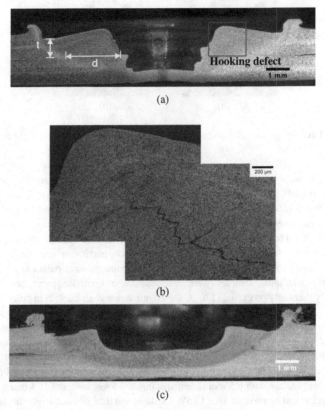

(a)

(b)

(c)

Figure 3. Cross-section and microstructure of spot welds made using: (a) CP tool with a close-up of a hooking defect with secondary cracking towards the upper sheet surface (b), and (c) OC tool.

The strength of FSSWs depends on the microstructure and size of the stir zone (schematically indicated as "d" in Figure 3a) as well as the thickness of the top sheet (schematically indicated as "t" in Figure 3a) in the area under the shoulder indentation, and the hooking effect. Strong welds are achieved if the thickness of the top sheet is not much reduced due to the hooking defect as well as if the hooking defect is relatively flat, dispersed or discontinued.

Figure 4 shows the size of bonded region as a function of tool rotation speed for both tools at 0.5 mm/s plunge speed, 490 ms dwell time, and a specific penetration depth. For the CP tool, the bonded region size increases with tool rotation speed from 1000 up to a maximum at 1500 rpm after which it decreases with increasing tool rotation speed up to 2500 rpm. For the OC tool, the size of bonded region increases monotonically with the tool rotation speed in the range used. This observation matched pervious shear-tensile results and the reasons for these drastically different behaviors were also detailed [3]. The tool rotation speeds achieving the maximum bonded region size for each tool were adopted for characterizing cross-tension properties. It should be noted that it is not reliable to compare strength of welds made using

208

different tools by comparing the sizes of bonded region. For different tools, same size of bond region does not guaranty the same weld strength, since thickness of top sheet under shoulder indentation may vary with tool designs.

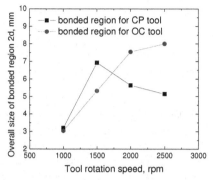

Figure 4. Size of bonded region as a function of tool rotation speed.

Figure 5 shows the average cross-tension separation load and standard deviation for specimens produced at different process parameters including thermal processing i.e. a simulated paint-bake cycle. Both tools show a parabolic increase and subsequent decrease of separation load as a function of penetration depth in the range explored. The maximum cross-tension separation load is 1.84 kN for CP tool and 1.25 kN for OC tool at 0.5 mm/s plunge rate. The plunge rate exhibits opposite effects on cross-tension strength for the CP and OC tools. The cross-tension separation load for the CP tool reduces to a value of about 1.37 kN with increase in the plunge rate from 0.5 mm/s to 2.5 mm/s whereas, the cross-tension separation load for the OC tool increases to about 1.62 kN with increase in the plunge rate. Results for both tools indicate that cross-tension separation load is unchanged after the paint-bake cycle. This observation is quite different from the response of lap-shear strength, which increases after a paint-bake cycle [3]

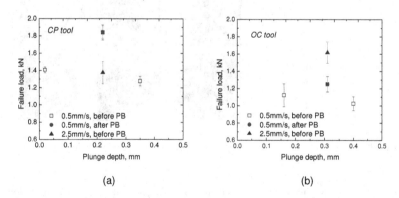

Figure 5. Effects of shoulder penetration depths, plunge speeds and paint-bake on cross-tension load of welds made using, (a) CP tool @ 1500 rpm, and (b) OC tool @ 2500 rpm. Note that in both figures 0.5 mm/s data before and after PB are superimposed at the intermediate plunge depth.

The trend of separation load curves with penetration depth can be explained by the variations in thickness of top sheet under shoulder indentation, "t", and the size of bonded region of welds, "d". The "d" becomes larger with increasing plunge depth because of the increase in the frictional heat between the tool and workpiece. So the bond strength between top and bottom sheets increases. However, as the tool plunge depth increases, the "t" reduces, hence the outer circumference of the shoulder indentation becomes weakest region and that is where separation occurs. It is of interest to note here that Tozaki et al. [9] showed a similar observation although, induced by increasing tool holding time rather than plunge depth. So "d" determines the bond strength and "t" influences the strength required to break through the top sheet.

Two different separation modes were observed for both tools under cross-tension loading: shear separation and pull-out separation. The spot welds produced with large "t" value (shallow plunge depth) separated in a shear mode and generated a relatively low separation load of 1.41 kN. As the plunge depth increased, "t" decreased and "d" increased, separation occurred with a pull-out separation mode, which gave a relatively higher separation load. However, with further decrease in "t", the separation load decreased with pull-out separation mode. The impression from tool shoulder may impose a stress concentration which facilitates earlier crack initiation. Figure 6 shows the photographs of the fractured samples made using CP tool under cross-tension loading.

t, mm	Separation load, kN	Mode	Top sheet top view	Top sheet bottom view	Bottom sheet top view
0.98	1.41	Shear separation	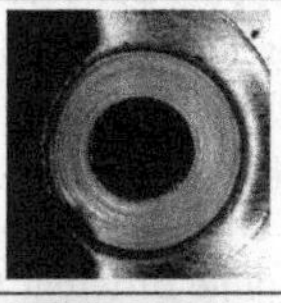	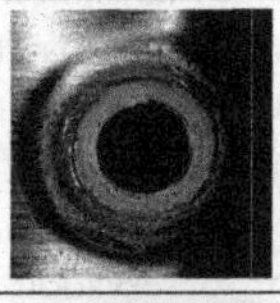	
0.78	1.84	Pull-out			
0.65	1.28	Pull-out	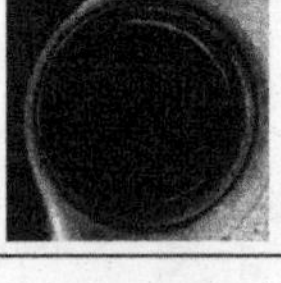		

Figure 6. Photographs of the CP tool specimen separation surfaces at various penetration depths.

Figures 7a and 7b present the cross-sections of separated samples under both shear separation and pull-out separation modes. For the shear separation mode, the crack initiates near the tip of the unwelded part between two sheets and propagates along the faying surface into the nugget until separation occurs with limited bending of upper and bottom sheets. For the pull-out separation mode, the crack passes through the thermomechanical affected zone

210

(TMAZ) and propagates through the top sheet with a fracture surface parallel to the loading direction. At the same time, another crack propagates into the bottom sheet up to certain depth. In this case, much more bending of both sheets is observed. Figure 8 shows dimpled fracture of welds made at 0.5mm/s & 2.5mm/s plunge rates.

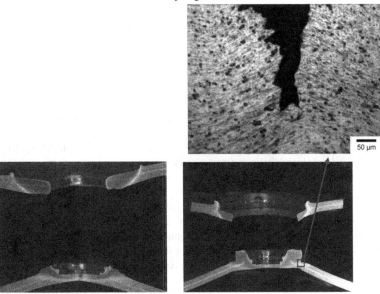

Figure 7. Cross-sections of welds separated at (a) shear mode and (b) pull-out mode.

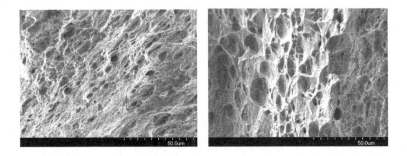

Figure 8. SEM images of fracture surface.

Figures 9a & 9b present the load-displacement curves for welds separated by shear and pull-out modes, respectively. Two different regions are observed: (1) crack growth through the bonded area, and (2) crack shear through nugget (shear mode) or propagation through the top sheet (pull-out mode). It was observed that there was not much difference in the maximum load in region (1), 1.10 kN for shear separation and 1.25 kN for pull-out mode. However, in region (2), the increment of load for pull-out condition is two times that for shear separation condition. A continuous hook defect in the nugget induces the shear separation and lowers separation load.

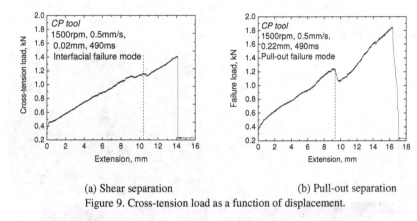

(a) Shear separation   (b) Pull-out separation

Figure 9. Cross-tension load as a function of displacement.

Figure 10 shows the microhardness of the weld cross-section before and after the paint-bake cycle. It can be seen that the microhardness increases greatly in the parent material region, however, the variation in the welded region is not obvious, especially for the region (between ±2mm and ±5mm) where separation occurs. Since the hardness is almost unchanged in this region (±2mm to ±5mm), it can be inferred that the local property (like, yield strength) did not change. This may be the reason for achieving same cross-tension separation load before and after paint-bake cycle.

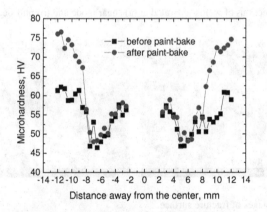

Figure 10. Effect of paint-bake cycle on welds microhardness.

Based on the cross-tension separation results and associated separation paths observed in this study, there appears to be a relationship between the material property, material thickness and maximum cross-tension separation load when a specific tool is used. It is useful to compare current results with other published data. Table I shows a general comparison of present results with previous literature results. It seems that the possible maximum cross-tension load is proportional to the material thickness and material property, when suitable tool is used.

However, this is a general comparison, and further research is required to verify this relationship.

Table I. Comparison of current results with literature results on cross-tension properties.

	Current result	Literature result [10]
Thickness (T)	1mm 6016-T4	2 mm 6061-T4
Tool (pin, shoulder)	1.5mm, 10mm	3.1mm, 10mm
Process parameter	1500rpm, 0.2mm shoulder Pd	2000rpm, 0.2mm shoulder Pd
Fc,max (max separation load)	1.8 kN	3.8 kN
Material yield strength (δy)	130 MPa	145 MPa
Fc,max/(T* δy)	*0.0138*	*0.0131*

Conclusions

Friction stir spot welds of 1 mm 6016-T4 aluminum alloy were made using conventional pin tool and off-center feature tool. Maximum bonded region was achieved at 1500 rpm for CP tool and 2500 rpm for OC tool. Maximum separation load of about 1.8 kN was obtained at 0.22 mm shoulder penetration depth for CP tool. Cross-tension separation load did not change after paint-bake cycle for both tools. Two different separation modes were observed: shear separation and pull-out under cross-tension loading condition. Based on the experimental observation of separation path, load-displacement curve, a possible relationship between cross-tension separation load, material thickness and yield strength is proposed.

Acknowledgments

This work was performed under the NSF-IUCRC for Friction Stir Processing and the additional support of NSF, GM, PNNL and Friction Stir Link for the Missouri University of Science and Technology site is acknowledged.

References

[1] S. Lathabai et al., "Friction spot joining of an extruded Al-Mg-Si alloy," *Scripta Materialia*, 55 (2006), 899-902.

[2] T. Freeney, S. R. Sharma, R. S. Mishra, "Effect of Welding Parameters on Properties of 5052 Al Friction Stir Spot Welds," Society of Automotive Engineers Technical Papers. 2006-01-0969; 2006.

[3] W. Yuan, "Friction stir spot welding of aluminum alloys" (M.S. thesis, Missouri University of Science and Technology, 2008), 20-42.

[4] H.J. Liu et al., "Tensile properties and fracture locations of friction-stir-welded joints of 2017-T351 aluminum alloy," *Journal of Materials Processing Technology*, 142 (2003), 692-696.

[5] R.S. Mishra et al., "Friction stir spot welding of 6016 aluminum alloy," TMS, (2007).

[6] H. Badarinarayan et al, Friction Stir Spot Welding. In: Mishra RS, Mahoney M, editors. Friction Stir Welding and Processing, Materials Park: ASM International; 2007, 235-272.

[7] S.M. Hirth et al., "Effects of Si on the aging behavior and Formability of Aluminum Alloys based on AA6016," *Materials Science and Engineering. A*, 319-321 (2001), 452-456.

[8] P. Su, A. Gerlich and T.H. North, "Friction Stir Spot Welding of Aluminum and Magnesium Alloy Sheets," Society of Automotive Engineers Technical Papers. 2005-01-1255; 2005.

[9] Y. Tozaki, Y. Uematsu and K. Tokaji, "Effect of process parameters on static strength of dissimilar friction stir spot welds between different aluminum alloys," *Fatigue Fract Engng Mater Struct*, 30 (2006), 143-148.

[10] Y. Tozaki, Y. Uematsu and K. Tokaji, "Effect of tool geometry on microstructure and static strength in friction stir spot welded aluminum alloy," International Journal of Machine Tools & Manufacture, 47 (15) (2007), 2230-2236.

Friction Stir Welding and Processing V
Edited by: Rajiv S. Mishra, Murray W. Mahoney, and Thomas J. Lienert
TMS (The Minerals, Metals & Materials Society), 2009

Evaluation of Swept Friction Stir Spot Welding in Al 2219-T6

Jeremy Brown[1], Christian Widener[1],
Gary Moore[2], Ken Poston[2], Dwight Burford[1]

[1]National Institute for Aviation Research Advanced Joining and Processing Laboratory, Wichita
State University, Wichita, KS 67260, USA
[2]Short Brothers PLC/ Bombardier Aerospace, Belfast, Northern Ireland, United Kingdom

Keywords: friction stir, FSSW, anodized, 2219, sealant, gasket

Abstract

The purpose of this investigation was to evaluate the effects of swept Friction Stir Spot Welding
(FSSW) on tensile strength and fatigue life in 2219-T62 material with a faying surface gasket
compound. The sheets were 0.100-in. (2.5 mm) thick. The top sheet was chromic acid anodized
while the bottom sheet was sulfuric acid anodized. A polyurethane based non-setting and non-
hardening gasket compound was placed at the faying surface. The first round of testing involved
exploratory bounding of the process windows for three tools. The bounding spots were
evaluated through macroscopic inspection of spot cross sections. One tool was eliminated in the
first round; however, the remaining two tools were evaluated for coupon tensile strength. The
coupons were pulled to failure in a single spot unguided lap shear configuration. Weld
parameters for each tool were varied in a Box-Behnken design of experiment (DOE). Coupons
were also produced for limited fatigue testing from the three best welding parameters for each
tool. The coupons were made in the 100% load transfer configuration per the NASM 1312-21
specification. A single tool was then chosen based on the previous tensile and fatigue results.
Another set of DOEs were performed to evaluated tensile strength and fatigue life. These DOEs
again used the NASM 1312-21 100% load transfer coupons. Select FSSW coupons were then
compared to riveted coupons at equal fatigue load levels. The rivets used in this experiment
were MS20426E5-7 flush countersink. The FSSW coupons were able to outperform the riveted
coupons in regards to tensile strength and for fatigue life at high load levels. At lower load
levels FSSW coupon results were comparable to riveted coupons.

Introduction

Several types of Friction Stir Spot Welding (FSSW) are available. These include the simple
plunge type spot developed by Mazda [1] to the more complex refill type developed by GKSS
[2]. Another type of FSSW, the swept spot, was initially developed by TWI in their Squircle™
design [3].

Wichita State University (WSU) has done extensive development of the TWI swept FSSW
concept with their own version called an Octaspot™ pattern, shown in Figure 1. The potential of
swept FSSW was first demonstrated successfully with thin gauge 1 mm (0.040-in.) 2024-T3
sheet [4]. The pattern has also been used to weld through surface treatments [5, 6] as well as a
couple of industrial faying surface sealants [6]. While authors like Li et al. have shown that
continuous FSW lap welds can be made successfully through sealants [7], WSU has
demonstrated that it is just as viable for swept FSSW as well.

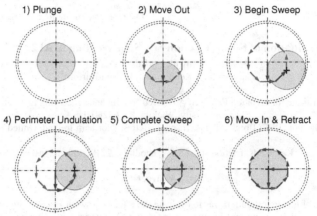

1) Plunge   2) Move Out   3) Begin Sweep

4) Perimeter Undulation 5) Complete Sweep 6) Move In & Retract

Figure 1: Typical Octaspot Pattern

Approach

<u>Goal</u>

Until now, the Octaspot™ swept spot has only been used in thin gauge material. Experimentation was needed to determine if it was applicable though thicker material. The goal of this experiment was to determine the viability of swept FSSW through sealants and surface treatments in medium gauge sheet.

<u>Material</u>

The material chosen for this experiment was 0.100 inch (2.5 mm) thick, 2219-T62 sheet. The sheets were surface treated with anodized coatings. In all cases tested in this study the top sheet was chromic acid anodized (CAA) and the bottom sheet was sulfuric acid anodized (SAA). A gasket compound was also placed at the faying surface. Unlike a previous experiment [6], this sealant does not cure. The gasket compound was a polyurethane based non-setting and non-hardening gasket compound. A side view schematic of the layup is show in Figure 2.

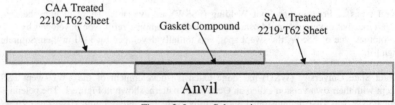

Figure 2: Layup Schematic

<u>Tool Bounding</u>

Initially, three tools were considered to investigate swept spots through the chosen material. The three tools are called the Psi™, the Counterflow™ with threads, and the Counterflow™ with tapered flats. The tools are shown, respectively, in Figures 3a, 3b, and 3c. The three tools have common pin and shoulder geometries, which are:

- ¬ 0.400 inch diameter shoulder
- ¬ 7 degree shoulder concavity
- ¬ 0.137 inch pin length
- ¬ 0.158 inch pin base diameter
- ¬ 7 degree pin taper

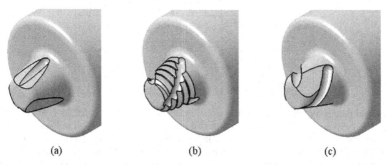

(a)   (b)   (c)

Figure 3: (a) Psi™ tool (b) Counterflow™ tool with threads
(c) Counterflow™ with tapered flats

The tools underwent weld parameter bounding trials. The swept spots had a path diameter of 0.160 inch (4.0 mm) in this and subsequent sections. Other weld parameters were also held constant in this experiment. Plunge rate was 7 IPM (178 mm/min). Initial plunge depth was 0.005 inch (0.13 mm). A tilt angle of 0.25 degrees was also used for most of this experiment.

Metallographic cross-sections were made from some of the spot welds. After examining the cross-sections, tensiles were created and pulled to failure. The coupons were single spot unguided lap shear coupons. A schematic of this coupon is shown in Figure 4. Any tools which failed to produce an acceptable level of results were eliminated from the program after this point.

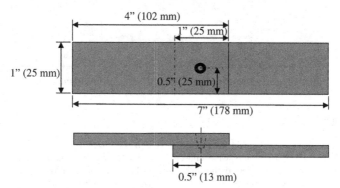

Figure 4: Single Spot Unguided Lap Shear coupon configuration

First Design of Experiment (DOE)

The remaining tools were tested in a DOE. The DOE was a Box-Behnken design. Table 1 shows the variables tested and their ranges. The response studied in this DOE was the ultimate tensile strength of the coupon. The coupons used were of the single spot unguided lap shear configuration.

Table 1: First DOE variable ranges

	Low	Medium	High
Rotation Speed (RPM)	1,400	1,700	2,000
Travel Speed (IPM) [mm/min]	5.0 [127]	8.0 [203]	11.0 [279]
Forge Load (lbs) [kN]	1,150 [5.12]	1,300 [5.78]	1,450 [6.45]

Some weld parameters that resulted in good strengths were chosen to perform a limited fatigue study. The coupons tested were four spot guided lap shear coupons, shown in Figure 5, and are based on the NASM 1312-21 coupon. The distance, D, is 1.2 inches (30.5 mm) for this experiment. The coupons were fatigue tested at a load ratio of 0.1 and a frequency of 10 Hz. Three coupons were made from each selected weld parameter. One was pulled for ultimate strength. Two were fatigued at 50% average strength of the selected weld parameters. A single tool was selected to be used in the next section of this experiment.

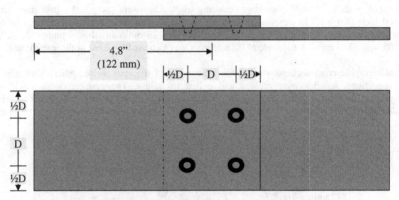

Figure 5: Four Spot Guided Lap Shear coupon configuration

Second DOE

A Box-Behnken DOE was again used in the next step. The four spot guided lap shear coupon was used. The weld parameters were chosen based on the results from the previous section. Table 2 shows the variables tested and their ranges. The DOE was performed 7 times; once for ultimate tensile strength, 3 times for 66.7% load fatigue life, and 3 times for 50.0% load fatigue life. The fatigue load levels are based on results from the tensile section of this DOE. The fatigue load ratio was 0.1. The frequency was increased to 40 Hz to reduce test time. The tilt angle was changed in this section to 0.50 degrees to reduce weld flash.

218

Table 2: Second DOE variable ranges

	Low	Medium	High
Rotation Speed (RPM)	1,700	1,850	2,000
Travel Speed (IPM) [mm/min]	8.0 [203]	9.5 [241]	11.0 [279]
Forge Load (lbs) [kN]	1,200 [5.34]	1,300 [5.78]	1,400 [6.23]

Rivet Comparison

The final section of this experiment involved a direct comparison between flush countersunk riveted coupons and FSSW coupons. The rivets used were MS20426E5-7, with the four spot guided lap shear coupon. The FSSW coupons were fatigue tested at the same load levels as the riveted ones. The load levels were 66.7% and 50.0% of the rivet coupon ultimate strength. The welded parameters were selected from the previous DOE. Three of each tensile and fatigue coupon were tested. The fatigue ratio was 0.1. The frequency was 40 Hz. Three riveted single spot unguided lap shear coupons were also tested. These results can be directly compared to the single spot FSSW coupons tested in the first DOE.

Results

Tool Bounding

Only two of the three tools produced fully consolidated welds, the Psi™ and the Counterflow™ with threads. The Counterflow™ with flats consistently left voids at the base of the nugget, over the tested parameter ranges. An example of this is shown in Figure 6.

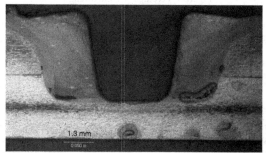

1.3 mm

Figure 6: Typical unconsolidated nugget from the Counterflow™ with flats

Table 3 and Table 4 show the results of the tensile coupons from this section. As both the Psi™ and Counterflow™ with threads were able to make welds free of voids, both tools were used in the next section.

Table 3: Initial Psi™ tensile results

Rotation Speed (RPM)	Travel Speed (IPM) [mm/min]	Forge Load (lbs) [kN]	Tensile Strength (lbs) [kN]
1,800	5 [127]	1,450 [6.50]	1,561 [6.94]
1,800	5 [127]	1,300 [5.78]	1,522 [6.77]

219

1,800	5 [127]	1,150 [5.12]	1,580 [7.03]

Table 4: Initial threaded Counterflow™ tensile results

Rotation Speed (RPM)	Travel Speed (IPM) [mm/min]	Forge Load (lbs) [kN]	Tensile Strength (lbs) [kN]
600	5 [127]	1,150 [5.12]	965 [4.29]
600	5 [127]	1,000 [4.45]	1,043 [4.64]
1,800	5 [127]	1,300 [5.78]	1,413 [6.29]
1,800	5 [127]	1,150 [5.12]	1,499 [6.67]

First DOE

Both of the down-selected tools performed well. The Psi™ tool had an average ultimate strength of 1,489 lbs (6.62 kN) with a standard deviation of 116 lbs (0.52 kN). The Counterflow™ with threads had an average strength of 1,374 lbs (6.11 kN) with a standard deviation of 125 lbs (0.56 kN).

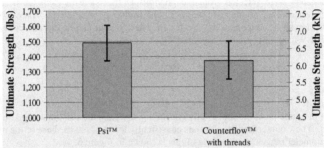

Figure 7: Average Tensile Strengths

Three weld parameters were selected for each tool and welded in the four spot configuration for ultimate strength and fatigue. The 3,073 lbs fatigue load, shown in Tables 5 and 6, was 50% of the average of the 4 spot coupon's strength in guided lap shear. The limited fatigue data did not reveal any conclusive differences between the two tools; however, since the Psi™ had higher strengths and lower standard deviation. The Psi™ was selected to be used in the next sections of this experiment.

Table 5: Psi™ first DOE 4 spot coupon results

Rotation Speed (RPM)	Travel Speed (IPM) [mm/min]	Forge Load (lbs) [kN]	Tensile Strength (lbs)	Fatigue Load (lbs) [kN]	Fatigue Percent	Fatigue Life (cycles)
1,700	8.0 [203]	1,300 [5.78]		3,073 [13.67]	50.9%	33,400
1,700	8.0 [203]	1,300 [5.78]	6,038			
1,700	8.0 [203]	1,300 [5.78]		3,073 [13.67]	50.9%	41,400
2,000	11.0 [279]	1,300 [5.78]		3,073 [13.67]	49.2%	38,491
2,000	11.0 [279]	1,300 [5.78]	6,245			
2,000	11.0 [279]	1,300 [5.78]		3,073 [13.67]	49.2%	35,841
2,000	8.0 [203]	1,150 [5.12]		3,073 [13.67]	48.9%	35,678
2,000	8.0 [203]	1,150 [5.12]	6,288			
2,000	8.0 [203]	1,150 [5.12]		3,073 [13.67]	48.9%	39,042

Table 6: Threaded Counterflow™ first DOE 4 spot coupon results

Rotation Speed (RPM)	Travel Speed (IPM) [mm/min]	Forge Load (lbs) [kN]	Tensile Strength (lbs)	Fatigue Load (lbs) [kN]	Fatigue Percent	Fatigue Life (cycles)
1,400	8.0 [203]	1,450 [6.50]		3,073 [13.67]	54.6%	51,286
1,400	8.0 [203]	1,450 [6.50]		3,073 [13.67]	54.6%	32,190
1,400	8.0 [203]	1,450 [6.50]	5,631 [25.05]			
2,000	5.0 [127]	1,300 [5.78]		3,073 [13.67]	48.1%	24,323
2,000	5.0 [127]	1,300 [5.78]		3,073 [13.67]	48.1%	33,278
2,000	5.0 [127]	1,300 [5.78]	5,394 [23.99]			
2,000	8.0 [203]	1,450 [6.50]	6,275 [27.91]			
2,000	8.0 [203]	1,450 [6.50]		3,073 [13.67]	49.0%	21,142
2,000	8.0 [203]	1,450 [6.50]		3,073 [13.67]	49.0%	28,140

Second DOE

The second DOE tensile results, shown in Table 7, had an average ultimate strength of 5,754 lbs (25.59 kN) with a standard deviation of 270 lbs (1.20 kN). The fatigue testing results of this second DOE are shown in Figure 8. Based on the tensile and fatigue data, three weld parameters were chosen for comparison to rivets. They were I, L, and G. These are the three strongest in tension and range from medium to low fatigue life for the set.

Table 7: Second DOE Tensile Results

Weld Parameter	Rotation Speed (RPM)	Travel Speed (IPM) [mm/min]		Forge Load (lbs) [kN]		Ultimate Strength (lbs) [kN]	
A	1,700	8.0	[203]	1,300	[5.78]	5,676	[25.25]
B	1,700	9.5	[241]	1,200	[5.34]	5,612	[24.96]
C	1,700	9.5	[241]	1,400	[6.23]	5,634	[25.06]
D	1,700	11.0	[279]	1,300	[5.78]	5,791	[25.76]
E	1,850	8.0	[203]	1,200	[5.34]	5,692	[25.32]
F	1,850	8.0	[203]	1,400	[6.23]	5,483	[24.39]
G	1,850	9.5	[241]	1,300	[5.78]	5,812	[25.85]
G	1,850	9.5	[241]	1,300	[5.78]	5,737	[25.52]
G	1,850	9.5	[241]	1,300	[5.78]	5,722	[25.45]
H	1,850	11.0	[279]	1,200	[5.34]	5,528	[24.59]
I	1,850	11.0	[279]	1,400	[6.23]	6,617	[29.43]
J	2,000	8.0	[203]	1,300	[5.78]	5,862	[26.08]
K	2,000	9.5	[241]	1,200	[5.34]	5,580	[24.82]
L	2,000	9.5	[241]	1,400	[6.23]	5,954	[26.48]
M	2,000	11.0	[279]	1,300	[5.78]	5,604	[24.93]

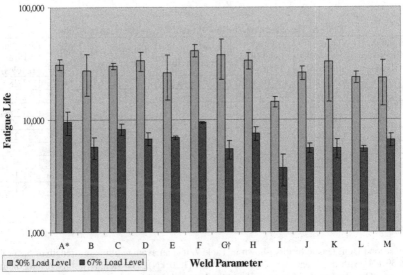

Figure 8: Fatigue Results
*2 data points for 66.7% load level
†9 data points per load level

Rivet Comparison

The 4 spot riveted coupons carried an average load of 3,719 lbs (16.5 kN) with a standard deviation of 177 lbs (0.79 kN). The single spot riveted coupons carried an average load of 959 lbs (4.3 kN) with a standard deviation of 12 lbs (0.05 kN). These strengths are much less than the FSSW are capable. Figure 9 compares the single spot riveted coupons to the first DOE Psi™ results. It also compares the four spot riveted coupons to the second DOE tensile results. The fatigue results at both high and medium load ranges are shown in Figure 10. Weld parameters L and G give comparable results to the riveted coupons. L and G were medium life fatigue coupons from the second DOE. The low life fatigue coupons from weld parameter I were unable to match the fatigue life of the riveted.

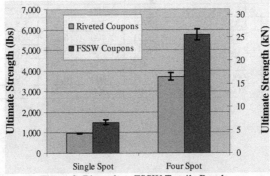

Figure 9: Riveted vs. FSSW Tensile Results

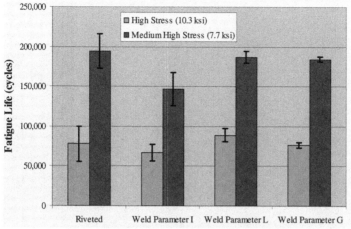
Figure 10: Riveted vs. FSSW Fatigue Comparison

Conclusions

Based on the results of this study, swept FSSW is capable of producing sound joints through 0.100 inch 2219-T62 with a faying surface gasket compound and surface treatments. It was also shown that strength alone is not a reliable predictor of fatigue performance. When compared to riveting, swept FSSW has the potential to be much stronger while maintaining comparable fatigue properties. Further work is planned to confirm the findings of this study.

References

1. T. Iwashita, U.S. Patent 6,601,751
2. C. Schilling and J. dos Santos, U.S. Patent 6,722,556 B2
3. A.C. Addison and A.J. Robelou, "Friction Stir Spot Welding: Principle Parameters and Their Effects," Proceedings of the 5th International Symposium on Friction Stir Welding (Metz, France), TWI, Sept 14-16, 2004
4. B. Tweedy et al, "Factors Affecting the Properties of Swept Friction Stir Spot Welds," *Proceedings of the 2007 SAE World Congress*, (Detroit, MI, USA), April 14-17, 2008
5. B. Tweedy, C. Widener, and D. Burford, "The Effect of Surface Treatments on the Faying Surface of Friction Stir Spot Welds," *Friction Stir Welding and Processing. TMS 2007 Annual Meeting*, (Orlando, FL, USA), Feb 25 – March 1, 2007.
6. J. Brown et al, "Evaluation of Swept Friction Stir Spot Welding through Sealants and Surface Treatments," 8th *International Conference on Trends in Welding Research*, (Pine Mountain, GA, USA), ASM International, June 1-6, 2008.
7. T. Li et al. "Friction Stir Lap Joining of Al 7075 with Sealant," *Friction Stir Welding and Processing. TMS 2007 Annual Meeting*, (Orlando, FL, USA), Feb 25 – March 1, 2007.

Friction Stir Welding and Processing V
Edited by: Rajiv S. Mishra, Murray W. Mahoney, and Thomas J. Lienert
TMS (The Minerals, Metals & Materials Society), 2009

Energy Generation during Friction Stir Spot Welding (FSSW) of Al 6061-T6 Plates

Mokhtar Awang[1] and Victor H. Mucino[2]

[1] Senior lecturer, Mechanical Engineering Department, Universiti Teknologi Petronas, Malaysia
[2] Professor, Mechanical and Aerospace Engineering department, West Virginia University, USA

Keywords: friction stir spot welding, finite element model, thermo-mechanical, explicit.

Abstract

Effective and reliable computational models would greatly enhance the study of energy dissipation during the friction stir spot welding (FSSW) process. Approaches for the computational modeling of the FSSW process, however, are still under development and much work is still needed, particularly the application of explicit finite element codes for a verifiable simulation. The objectives of this work are to develop a finite element modeling of FSSW of 6061-T6 aluminum alloy and analyze energy generation during the welding process. In this work, a three dimensional (3-D) finite element (FE) coupled thermal-stress model of FSSW process has been developed in Abaqus/Explicit code. The rate dependent Johnson-Cook material model is used for elastic plastic work deformations. Temperature profile and energy dissipation history of the FE model have been analyzed. The peak temperature at the tip of the pin and frictional dissipation energy are in close agreement with the experimental work by Gerlich et al. [1], with a difference of only about 5.1%.

Introduction

Friction stir spot welding is a solid state joining process that uses a non-consumable tool to generate frictional heating and produce a plasticized region at the bonding interface as a result of a strong compressive forging pressure. The process consists of three phases, i.e, plunging, stirring and retraction. The process starts with spinning the tool at a high rotational velocity and plunging it into a weld spot until the shoulder contacts the top surface of workpiece. Then, the stirring phase enables the materials of two work pieces to mix. Lastly, once a predetermined penetration is reached, the process stops and the tool retracts from the workpiece. The welding operation is completed remarkably quickly since cycle times are within a few seconds.

As a relatively new manufacturing process, there is very limited published research on the FSSW process. Lin et al. [2] investigated microstructures and failure mechanisms of FSSW in aluminum 6111-T4 based on experimental observations. Feng et al. [3] reported about their feasibility study of FSSW in advanced high-strength steels such as AHSS. Mitlin et al. [4] conducted experiments to investigate structure-properties relations in spot friction welded 6111 T4 Aluminum. Other works on FSSW were presented by Pan et al. [5] on sheet aluminum joining and Sakano et al. [6] on development of FSSW robot system for the automobile industry.

Recently, some researchers also studied the thermal and heat generation aspects of FSSW. Gerlich et al. [1] measured peak temperature in friction stir spot welds of aluminum and magnesium alloys. Two thermocouples were embedded in a welding tool assembly in the experiment. They found that the peak temperatures during FSSW of Al 5754 and Mg alloy AZ91D were 565°C and 462°C, respectively. Energy utilization during FSSW process has been studied by Su et al. [7]. The group reported that less than 4.03 % of the energy generated during the FSSW was required for stir zone formation in Al 6111 welds.

Thermo-mechanical Simulation

In this work, two thin aluminum alloy 6061-T6 plates are modeled as two separate workpieces. The bonding of the workpieces and retracting phase, however, have not been modeled. The FE analysis has been conducted by prescribing plunge rate and angular velocity of the pin tool, and by imposing appropriate boundary conditions.

The coupled thermomechanical analysis performed in Abaqus/Explicit uses 376.6 MB of memory for solving the 517902 unknowns. The CPU time is 14 days and 12 hours on a 3.60 GHz Intel Pentium 4 processor for the simulation time of 1.505 seconds.

Mesh and Geometry

The FE model is comprised of three main parts, i.e., workpieces, tool and backing anvil as depicted in Fig. 1. It only includes a limited part of the workpieces to optimize the resolution close to the tool and minimize the computational expenses.

In the FE model, two workpieces are stacked on top of each other with a dimension of 25 mm by 25 mm by 1 mm. The mesh is dense at the center of the workpiece to reduce the hourglassing[3] effect as depicted in Fig. 2. The element size is approximately 0.1 mm in the region surrounding the tool. A total of 132162 elements and 142816 nodes have been generated in the model. The workpieces have been modeled using thermal coupled element C3D8RT. This element type has 8-node tri-linear displacement and temperature degrees of freedom and reduced integration with hourglass control.

Since this work is not intended to study the mechanical responses of the tool and backing anvil, they are modeled as analytical rigid surfaces with prescribed motion at a reference node. The rigid body surfaces carry thermal response, and are assumed to be isothermal. The tool has a shoulder and an unthreaded pin as shown in Fig. 3. The diameters of the shoulder and the pin are 10 mm and 3 mm, respectively. The backing anvil has a diameter of 15 mm. The reference node is defined for each rigid body that has translation, rotation, and thermal degree of freedoms.

---

[3] Hourglassing is an element deformation that does not cause any strains at integration points.

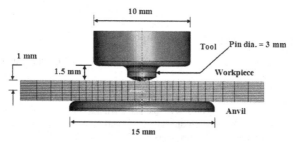

Fig. 1: Mesh representation of two layers of work piece with a pin and an anvil.

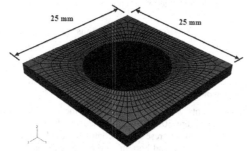

Fig. 2: Mesh scheme for workpieces.

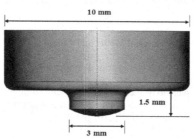

Fig. 3: Geometry of the welding tool.

Assumptions

To model the actual physics phenomena of the FSSW process is complicated. Therefore, several simplifying assumptions have been made to the FE model.

a) Workpieces are assumed to be made of a deformable material (Al 6061-T6) having displacement and temperature degree of freedoms.

b) The workpieces are assumed to behave as an elastic-plastic Johnson-Cook material model, both temperature, and strain rate dependent.
c) The workpieces and pin are assumed to experience frictional sliding contact described by Coulomb's friction law.
d) The tool and the backing anvil are considered rigid bodies having translation, rotation and thermal degree of freedoms.
e) The tool and the backing anvil are assumed isothermal.
f) The friction coefficient, μ is assumed to be a function of surface temperature.
g) Initial temperatures for both workpieces, pin and anvil are assumed to be 22^0C.
h) It is assumed that 100% of the dissipated energy caused by friction is transformed into heat and 90 % of the heat goes to the workpiece.
i) The heat is distributed evenly at the top and bottom workpieces interface.
j) 90 % of the energy due to plastic deformation is converted into heat.

Boundary Conditions

Boundary conditions imposed on the current FE model are described in the following paragraphs. The edges of the top and the bottom workpieces are restrained in the X direction. The tool can have only translation and rotation in the Y-direction. The backing anvil is fixed in all degrees of freedom to avoid rigid body motion.

Heat convection coefficients, h on the top surface of the upper workpiece and the bottom surface of the lower workpiece are 30 W/m^2-^{0}C as used in Chao and Qi's paper [8], with an ambient temperature of 22^0C. In this research, a contact thermal gap conductance of 100,000 W/m^2-^{0}C [9] is introduced at the top and bottom workpieces interface to simulate the heat loss through conduction between the two workpieces. There are no data or theory to predict precisely the heat loss through the bottom surface of the upper workpiece and the top surface of the bottom workpiece.

Material Model and Properties

Aluminum alloy 6061-T6 was chosen as a material for workpieces. The temperature dependent material properties for Al 6061-T6 are taken from Chao and Qi [8]. However, the thermal conductivity was found to be incorrect in the paper. To simulate the workpieces material behavior in the analysis, a temperature and strain rate dependent material law was used using the elastic-plastic Johnson-Cook material model [10] given by

$$\sigma_y = [A + B(\varepsilon^{pl})^n]\left[1 + C\ln\frac{\dot{\varepsilon}^{pl}}{\dot{\varepsilon}_0}\right]\left[1 - \left(\frac{T - T_{ref}}{T_{melt} - T_{ref}}\right)^m\right]$$

where σ_y is the yield stress, ε^{pl} is the effective plastic strain, $\dot{\varepsilon}^{pl}$ is the effective plastic strain rate, $\dot{\varepsilon}_0$ is the normalizing strain rate (typically, 1.0 /s). A, B, C, n, T_{melt}, and m are material constants. T_{ref} is the ambient temperature, which is 22 ^{0}C in this case. This material model is a type of Mises plasticity model with analytical forms of the hardening law and rate dependence. Moreover, it is suitable for high-strain-rate deformation of many materials.

Discussion of the Results

Thermal Verification

Fig. 4 shows the comparison of temperature history between the FE simulation results and the experimental results reported by Su et al. [7]. Both analyses were performed on Al 6061-T6 material using 3000 rpm rotational speed and 2.5 mm/s plunge rate. The tool geometry and the thickness of the workpiece, however, were different. Based on the results, the FE simulation shows the maximum tool temperature is 514.3^{0}C and the experiment shows a peak temperature at the tip of the tool of about 542^{0}C. From this, it can be concluded that both results are in good agreement with a difference of about 5.1 %.

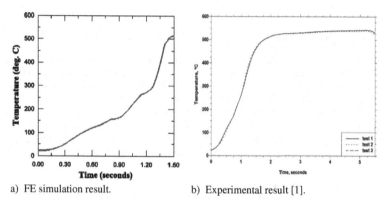

a) FE simulation result.   b) Experimental result [1].

Fig. 4:   FE simulation and experimental results of temperature profile.

Energy Dissipation during the FSSW Process

FSSW relies on the heat generated during the process to join the workpieces together. There are three possible heat sources during FSSW, i.e., friction work at the tool and top workpiece interface, friction work at the interface of top and bottom workpiece, and plastic deformation of the workpiece material. In this work, the friction and plastic work have been investigated using frictional energy dissipation and plastic energy dissipation histories, respectively. The angular velocity and plunge rate were 3000 rpm and 1 mm/s, respectively. Table 1 shows the summary of energy dissipation during FSSW.

Su et al. [7] have conducted an experiment on energy generation of 6.3 mm Al 6061-T6. In the experiment, rotational speed and plunge rate were set at 3000 rpm and 2.5 mm/s, respectively. The average energy generated by a 10 mm diameter tool with shoulder at five different plunge depths was 1.964 kJ. Based on the FE results shown in Table 1, the total energy dissipation at a plunge depth of 1.505 mm is 1.519 kJ. The discrepancy between the two results is due to the different thickness of the workpiece and different plunge rate.

Table 1: Summary of energy dissipation during FSSW process.

	Energy Dissipation (kJ)	Percentage (%)
Friction at tool and workpiece interface	1.471	96.84 %
Friction at top and bottom workpieces interface	0.000375	0.02 %
Plastic deformation	0.0478	3.14 %
Total Energy Dissipation	1.519	100 %

Heat Generation due to Friction Work at the Tool and Workpiece Interface

The simulation results show that friction work at the tool and top workpiece interface contributes the most heat to the welding process. The frictional energy contributes 96.84 % of the total energy as shown in Fig. 5. This high percentage of energy generation is expected due to the presence of high differential velocities (slip rate) on the workpiece surface caused by the high rotational speed and pressure from the tool have created.

Fig. 5 also shows that the frictional dissipation energy increases drastically after the time reaches 0.9450 seconds. The increase of frictional energy is due to the additional friction from the interface of the shoulder of the tool and the surface of the top workpiece.

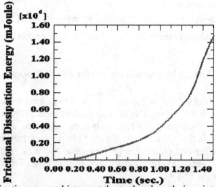

Fig. 5: Frictional dissipation energy history at the tool and workpiece interface.

Heat Generation due to Friction Work at the Top and Bottom Workpieces Interface

Based on the results in Table 1, about 0.02% of the total energy is due to frictional force at the interface between the top and bottom workpieces. Even though the amount of energy generated at this interface is almost negligible, the interface is important in FSSW because the actual welding occurs here. Consequently, it is relevant to trace how much energy is used at the interface and how this is affected by the process parameters.

Fig. 6 shows the frictional dissipation energy history at the top and bottom workpieces interface is about 0.000375 kJ (375 mJ) at 1.505 seconds. The friction work at the interface between the top and bottom workpieces gives the least contribution to the total heat generation because the interface has a very small relative motion due to friction.

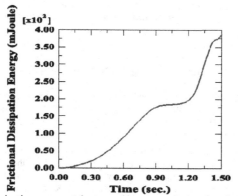

Fig. 6: Frictional dissipation energy at the top and bottom workpiece interface.

Heat Generation due to Internal Friction/Plastic Work

As shown in Fig. 7, plastic dissipation energy in the material is 47.8 kJ, which is about 3.14% of the total energy dissipated. This energy is due to the presence of internal frictional forces that tend to resist the motion of the material. As can be seen from the figure, the plastic dissipation energy increases drastically after 1.3 seconds. This is because, in addition to the plastic straining process due to the pin penetration, a large plastic deformation occurs when the shoulder of the tool touches the surface of the workpiece at the end of the cycle time.

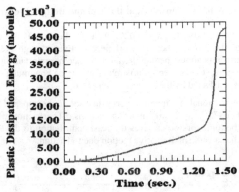

Fig. 7: Plastic dissipation energy history.

Effect of Welding Tool's Rotational Speed

A parametric study has also been conducted for various rotational speeds. Three different welding tool rotational speeds, 3000 rpm, 2500 rpm, and 2000 rpm, have been run on the FE models in order to study the effect of welding tool's rotational speed on heat generation. Plunge rate was set to be 1 mm/sec.

Fig. A-2(a-c) shows frictional dissipation energy histories for 3000 rpm, 2500 rpm, and 2000 rpm welding tool rotational speeds, respectively. Based on the results, the higher the rotational speed, the higher the dissipation energy. The frictional dissipation energy is reduced by about 8.4% when the rotational speed is reduced from 3000 rpm to 2500 rpm and reduced by about 19.2% when it is reduced from 3000 rpm to 2000 rpm. This is because higher rotational speed will result in higher relative velocity of the material, consequently higher energy will be produced. Table 2 summarizes frictional dissipation energy for different tool rotational speeds.

Table 2: Summary of frictional dissipation energy for different tool's rotational speeds.

Tool's Rotational Speed (rpm)	Frictional Dissipation Energy (kJ)	Percentage Reduction (%)
3000	1.471	100
2500	1.347	91.6
2000	1.188	80.8

232

Effect of Welding Tool Plunge Rate

Parametric studies have been conducted to determine the effect of tool penetrating speed on frictional and plastic dissipation energies. Three different tool penetration speeds of 1 mm/s, 5 mm/s, and 10 mm/s, are modeled with contact pressure friction coefficient dependent. The welding tool rotational speed was set at 3000 rpm. The targeted welding tool displacement was 1.505 mm. Fig. A-3 shows the frictional and plastic dissipation energy history for three different welding speeds. Based on the curves, frictional and plastic dissipation energy increases as the tool velocity decreases. The frictional dissipation energy of 5 mm/s plunge rate reduces to about 29% of the 1 mm/second frictional dissipation energy. The same decreasing trend is observed with the 10 mm/s plunge rate (14% decrement). This is because the slower the tool velocity, the more time it spends to spin on the workpiece thus more energy is produced.

The plastic dissipation energy also follows the same trend of frictional dissipation energy curves. In short, the slower the penetration speed, the higher the plastic dissipation energy. Table 3 summarizes the frictional and plastic dissipation energy for different tool plunge rates.

Table 3: Summary of friction and plastic dissipation energies for different tool's plunge rate.

Plunge Rate (mm/s)	Frictional Dissipation Energy (kJ)	%	Plastic Dissipation Energy (kJ)	%
1.0	1.471	100	0.0478	100
5.0	0.425	28.9	0.0237	49.6
10.0	0.207	14.1	0.0193	40.4

Conclusions

A fully coupled thermomechanical 3-D FE model of FSSW has been developed using an Abaqus/Explicit code. The following conclusions can be drawn from this work:

a) The peak temperature obtained from the simulation, equivalent to $0.95T_s$ (T_s is solidous temperature of Al 6061-T6), is consistent with the theory reported by Su, et al. [1].

b) Friction work at the interface of the tool and the workpiece generates the most energy, which is about 96.84 %, for FSSW. The rest of the energy comes from the friction work between the plates (0.02%) and the plastic deformation (3.14%)

c) The simulation results also show a significant affect of welding tool's rotational speed on frictional dissipation energy. In general, lower rotational speed yields lower frictional dissipation energy. The frictional dissipation energy is reduced by about 8.4% when the rotational speed is reduced from 3000 rpm to 2500 rpm and reduced by about 19.2% when it is reduced from 3000 rpm to 2000 rpm.

d) The affect of plunge rate on frictional dissipation energy is noted. The results show that the lower the plunge rate, the higher the energy. The frictional dissipation energy for a plunge rate of 5 mm/s reduces to about 29% of the 1 mm/s frictional dissipation energy. The same decreasing trend is observed with the 10 mm/s plunge rate (14% decrement).

REFERENCES

[1]   Gerlich, A., Su, P., and North, T., 2005,   "Peak Temperatures and Microstructures in Aluminum and Magnesium Alloy Friction Stir Spot Welds", Science and Technology of Welding and Joining, vol. 10, p.647-652.

[2]   Lin, P., Lin, S., Pan J., Pan, T., Nicholson, L., and Garman, M., 2004, "Microstructures and Failure Mechanisms of Spot Friction Welds in Lap-Shear Specimens of Aluminum 6111-T4 Sheets", Proceeding of the 2004 SAE World Congress, Michigan.

[3]   Feng, Z., Santella, M., David, S., Steel, R., Packer, S., Pan, T., Kuo, M., and Bhatnagar, R., 2005, "Friction Stir Spot Welding of Advanced High-Strength Steels-A Feasibility Study", Proceeding of the 2005 SAE Congress, Detroit, MI.

[4]   Mitlin, D., Radmilovic, V., Pan, T., Feng, Z., Santella, M., 2005, "Structure-Properties Relations in Spot Friction Welded 6111 T4 Aluminum", TMS 2005 Annual Meeting, San Francisco.

[5]   Pan, T., Joaquin, A., Wilkosz, E., Reatherford, L., Nicholson, J., Feng, Z., and Santella, M., 2004, "Spot Friction Welding for Sheet Aluminum Joining", 5th International Symposium on Friction Stir Welding, P. Threadgill, ed., The Welding Institute, Metz, France, Paper No. 11A-1.

[6]   Sakano, R., Murakami, K., Yamashita, K., Hyoe, T., Fujimoto, M., Inuzuka, M., Nagao, Y., and Kashiki, H., 2001, "Development of Spot FSW Robot System for Automobile Body Members", Proceedings of the 3rd. International Symposium of Friction Stir Welding, Kobe, Japan.

[7]   Su, P., Gerlich, A., North, T., and Bendzsak, G., 2006, "Energy Utilization and Generation during Friction Stir Spot Welding", Science and Technology of Welding and Joining, p. 163-169.

[8]   Chao, Y., and Qi, X., 1998, "Thermal and Thermo-Mechanical Modeling of Friction Stir Welding of Aluminum Alloy 6061-T6", Journal of Materials Processing & Manufacturing Science, vol. 7, p. 215-233.

[9]   Khandkar, M., Khan, J., and Reynolds, A., 2002, "Input Torque Based Thermal Model of Friction Stir Welding of Al-6061", Proceeding of the 6th International Trends in Welding Research Conference Proceeding, Pine Mountain, GA.

[10] Johnson, G., and Cook, W., 1983, "A Constitutive Model and Data for Metals Subjected to Large Strains, High Strain Rates and High Temperatures", Proceeding of the 7th Int. Symp. On Ballistics, The Hague, the Netherlands, p. 1-7.

FRICTION STIR WELDING AND PROCESSING V

Session Session V

Session Chair

William J. Arbegast

Friction Stir Welding and Processing V
Edited by: Rajiv S. Mishra, Murray W. Mahoney, and Thomas J. Lienert
TMS (The Minerals, Metals & Materials Society), 2009

Aging Weapons Systems Repair using Friction Stir Welding

Bryan M. Tweedy[1], William A. Arbegast[2], Robert H. Hrabe[1] and James M. Hutto[1]

[1] H.F. Webster Engineering and Professional Services, Rapid City, SD 57702 USA
[2] Advanced Materials Processing Center, South Dakota School of Mines and Technology, Rapid City, SD 57701, USA

Keywords: Friction stir, repair, aging aircraft, FSW

Abstract

Friction stir welding and processing (FSW&P) was identified in the FY07 Aging Aircraft Study conducted by the South Dakota School of Mines and Technology as a technology that is ready to enter into a qualification process for use as standard repair technique on aging weapons systems. FSW has been widely investigated as a manufacturing process with successes reported in the commercial and government sectors; however, little is reported in the literature on the qualification of FSW&P for repair applications. Preliminary analysis in this study utilizing FSP for repair of several components showed technical feasibility. In addition, the demonstration of FSP to refurbish an actual part was successful. Radiographic inspection showed that the volumetric defects and fatigue cracking were processed from the candidate component. The study produced a cost benefits analysis which estimated $9.4M annual savings to the USAF alone for components repairable with FSW&P.

Introduction

The purpose of this FY07 Aging Aircraft Study [1] was to investigate the use of advanced processing technologies such as friction stir welding and processing, laser deposition and cold spray in repair/remanufacturing applications of aging weapons systems. In particular, friction stir welding and processing presented some unique opportunities to repair worn or cracked components. One of the chosen parts for this remanufacturing study is the ruddervator fitting for the KC-135 re-fueling boom (Figure 1). This structural component is responsible for connecting the re-fueling boom to the rudder shaft. The component has two ears, each with a bearing race. When the bearings are installed, it is common to use a punch and hammer to deform the part and stake the bearing in place. Unfortunately, these stake marks are stress concentration points and form fatigue cracking when placed into service. A picture of the typical stake marks in the ruddervator fitting is shown in Figure 2.

 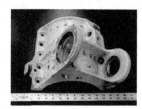

Figure 1. KC-135 Re-Fueling Boom (left) and Ruddervator Fitting (right)

Figure 2: Typical Stake Damage to a Ruddervator Ear

The ruddervator project was completed in three phases. Phase I, a defect processing study, focused on testing the ability of friction stir processing (FSP) to repair volumetric defects (holes) from the aluminum test part. Phase II, welding of surrogate test parts, proved friction stir welding (FSW) can be used to install a plug into a bore to eliminate defects and add material, then be re-machined to part tolerances. Phase III was the application of knowledge learned from Phases I and II in the demonstration of FSP to refurbish the ruddervator fitting.

Experimental Procedure

Phase I: Defect Processing Study

The defect processing study "stake-marks" were simulated with holes drilled into a piece of 0.125" 7075-T6 sheet stock. These 0.065" diameter holes are similar in diameter to the defects found on the defective parts and were made to be 50%, 100%, and 150% of the observed defect depth (0.030", 0.060", and 0.090" respectively). Hole spacing and depth were controlled using a vertical mill with digital readout.

All defect locations remained constant in the coupon specimens. Two evenly spaced defects were made in a 1.0" lap-shear specimen. Defect size for lap-shear specimen locations was ordered using design of experiments (DOE) to identify any trend in specimen location. A metallographic specimen was taken for all processing parameters with the 0.030", 0.060", and 0.090" defects 0.5", 1.25" and 2.25" into the specimen, respectively (Figure 3).

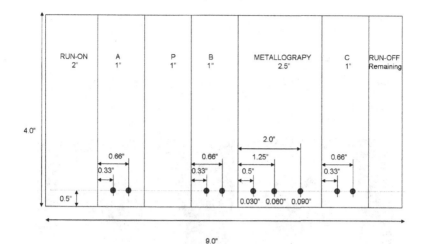

Figure 3. Defect Location for Process Development

The base sheet was aligned in a jig and clamped in place with a 0.040" 7075-T6 sheet on top in a lap-weld configuration. The fixture allowed all parts to be welded with the same weld program and set-up. Two sets of welding parameters were down selected from previous weld schedule development: 600 RPM and 11 IPM yielding a Pseudo Heat Index (PHI) of 2.27, Parts 4, 5, and 6 used 800 RPM and 9 IPM yeilding a PHI of 7.57. Weld runs were randomized and both sets of parameters used a 2° lead angle and heel plunge of -0.005". All welds were made with adjustable pin tool with scrolled shoulder and tapered, tri-flute, threaded pin profile as shown in Figure 4. Welds were centered over the defects 0.5" from the edge of the part and were 8.2 inches in length.

Figure 4. Pin Tool and Shoulder

Welds were allowed to naturally age for a minimum of 24 hours, then all lap-shear specimens were cut to approximately 1" wide, randomly loaded into the MTS 858 mini tensile machine with shims and pulled at 0.060" per minute. This load rate allowed for all specimens to be broken in approximately one minute. All specimens broke at the sheet-thinning defect in the lap weld.

Metallographic specimens for parts 1 and 4 were sectioned at the defect location and parts 2 and 5 were sectioned 0.125" up-weld of the defect locations which is approximately equivalent to 1 pin diameter. Metallography was conducted to inspect for remaining defects. Selected

239

metallographic specimens were radiographically inspected at Ellsworth Air Force Base (AFB) in Rapid City, SD. Defects were found to be fully processed from the specimen coupons.

<u>Phase II: Welding of Surrogate Parts</u>

The pathfinder experiment utilized two surrogate parts, similar in geometry to the Ruddervator fittings, and two plugs machined from 7075-T753 barstock (Figure 5). Four defects, 0.065" in diameter and 0.090" deep were installed in the four quadrants of each bore. Defects were installed approximately 0.060" on center, radially from the edge of the bore.

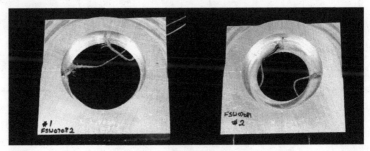

Figure 5. Surrogate Plugs

Weld path plans were developed in MasterCAM and modified with South Dakota School of Mines and Technology (SDSM&T) Advanced Materials Processing Center's (AMP) 3D path planning software. FSW tooling and parameters were identical phase I development. Pin penetration into bore surrogate was approximately 0.120". Thermocouples were mounted in the surrogate parts at three locations to capture comparative information between a 2 and 4 pass weld and to compare the thermal conditions between the surrogate part and the prototype (Figure 6). Thermocouples were placed in the quadrants of the surrogate parts where the bore meets the bearing stop. Plugs were pressed into the bores using a manual hydraulic press (Figure 7).

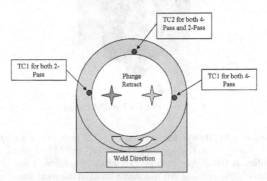

Figure 6. Thermocouple Placement Schematic

Figure 7. Installation of Plugs and Thermocouples

Parts were fixtured on the FSW table and were elevated on 1.5" tubing from a tooling system from Marino Gauge. The only clamping holding the parts in position was a single 5/8" bolt through the center of the plug (Figure 7). To test the weld program, a 4-pass test-weld was made on a piece of 0.25", 6061-T6 aluminum fixtured in a similar manner to the surrogate parts. A two-pass weld (Figure 8) was completed on the first surrogate part with success. On the exit radius however, a lack of consolidation (LOC) defect formed through the radius (Figure 9). A four-pass weld (Figure 10) was completed on the second surrogate part with success. The travel speed was reduced through the exit radius, which reduced the LOC defect, but did not totally eliminate it.

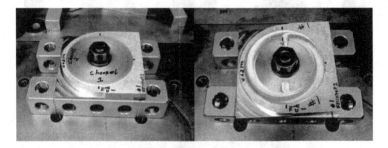

Figure 8: Surrogate 1 (2-Pass Weld)

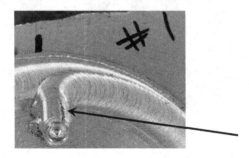

Figure 9: Lack of Fill on Exit Radius

Figure 10: Surrogate 2 (4-Pass Weld)

Parts were re-machined to remove the plugs from the bearing bores and radiographically inspected at Ellsworth AFB, Rapid City, SD (Figures 11-12).

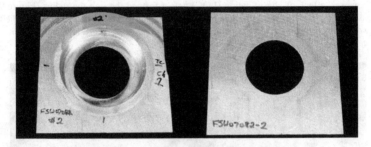

Figure 11: Surrogate 2 Post-machining

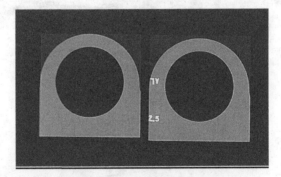

Figure 12: X-ray Inspection of Surrogate Parts

Phase III: Prototype Production Study

Paint was removed from the working surfaces of the ruddervator fittings using a pneumatic abrasive wheel to expose the bare aluminum. The bearing race was then submerged in a NaOH solution (10% by mass) prior to conducting a dye-penetrant inspection. The ruddervator fittings were measured with the Faro Coordinate Measuring Machine to determine the parallelism of the two part faces to be welded. Measurements indicated that Parts 1 and 4 were parallel to within 0.005" and 0.0025" respectively. It was determined that these features were adequate to use as datums.

Four plugs, two large and two small, were machined from 7075-T753 bar stock. Thermocouples were mounted to the part in quadrants (Figure 13) at the base of the bearing stop and were placed with high temperature thermocouple cement.

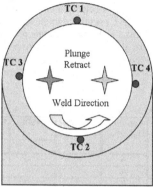

Figure 13: Thermocouple Placement Schematic for Ruddervator Prototype Fitting

The prototype ruddervator fitting was cleaned with a solvent wipe and the machined plugs were pressed into place (Figure 14). Standard tooling was used and was similar to the setup used in welding the surrogate parts in the previous section. This was important to avoid differences in the thermal and mechanical conditions used in the development of the repair process to aide portability between surrogate and prototype production. Welds were made using the parameters developed in the previous section (800 rpm, 8 ipm) while thermal data was being collected via a data acquisition system. Welds were made counter-clockwise about the center of the plug, with two inner passes followed by two outer passes. Following welding, the resultant plugs were machined from the ruddervator fittings and the repaired part was radiographically inspected which revealed that no volumetric defects remained in the part.

Figure 14: Dry-fit of Repair Plug (left) and Post Weld Repair (right)

Results and Discussion

During the defect processing study, the traverse (x) forces showed a force spike whenever the pin tool encountered a defect. The largest force spike corresponded to the defect of the largest size (deepest hole) although the difference was relatively small at 100-200 lbf. Metallographic inspection was unable to find remaining volumetric defects, further suggesting the defect was repaired. Radiographic inspection supports the conclusion that defects were processed from the parts. Lap-shear testing results suggest that weld strength varies slightly with specimen location and "hotter" or higher PHI welds may do a better job of processing larger defects (Table 1). The specimen location identification (A, B, C) is related to the location on the welded plate as shown in the previous section. Defect size and position were placed randomly using a design of experiments approach. It appears, though the evidence is not conclusive, welds later in the coupon are usually stronger than those welds earlier in the coupon. This may argue for a longer run-on than currently is being used. In any case, position A seems to yield the lowest strength across defect size.

Table 1: Unguided Lap Shear Test Results

Part Number	Specimen Location	Specimen #	Width	Peak Load
			in	lbf
1	A	20	1.032	1568.0
2	A	9	0.945	1443.3
3	A	7	0.922	1493.5
4	A	24	1.018	1349.3
5	A	14	0.954	1218.4
6	A	18	0.921	1058.9
1	B	6	0.983	1374.3
2	B	17	0.955	1549.7
3	B	2	0.942	1561.9
4	B	11	1.003	1349.0
5	B	13	0.935	1412.2
6	B	23	0.984	1505.5
1	C	16	0.920	1424.9
2	C	21	0.958	1540.2
3	C	4	0.968	1616.8
4	C	5	0.963	1130.6
5	C	3	0.973	1422.2
6	C	10	0.977	1222.9
1	P	12	0.940	1246.3
2	P	22	0.967	1433.6
3	P	19	0.957	1619.7
4	P	8	0.944	1396.5
5	P	1	0.963	1446.5
6	P	15	1.010	1587.3
		Mean	0.96392	1415.47345
		Std. Dev.	0.02999	151.82884

In the welding of the surrogate parts, the test weld used for path planning in the 6061 aluminum alloy was marginally thin to support the process loads induced by the FSP. Because of the thickness of this test piece and the inconsistent backing support, there was some material "blistering" on the back side. The actual surrogate parts additional thickness was adequate and this indication was not present. The exit radius LOC seen in the first surrogate part was lessened in the second article by utilizing a reduction in the programmed travel speed in the transition zone to the termination of the weld. The article with four passes was only slightly hotter than the article with two passes. It is possible that steady state temperature was reached and that peak temperatures will not exceed 600°F with more than four passes.

Utilizing the information and procedures developed in the defect processing study and the surrogate parts welding, the prototype production parts were processed with some confidence in the quality of the repair. A LOC indication was still present even with the increased radii employed for the exit procedure. The stake defects were observed to be successfully processed from the part once the plug was machined from the bore (Figure 15). The plug machined out of the ruddervator part, however, slight concentricity issues exist between the re-machined bore and the pre-welded bore. The radiographic inspection of the part showed no remaining volumetric defects. The maximum temperature for both the surrogate and prototype part as measured by the thermocouples was approximately 550°F.

Figure 15: Pre and Post Repair of Ruddervator staking damage

For the FSW/P repairs investigated in this study a cost benefits analysis estimated a $9.4M annual savings to the Air Force (AF) alone accounting for approximately 25% of the AF inventory. The cost estimates were projected by identifying likely candidates for repair from items that are currently non-repairable. Potential repair applications were analyzed and in some cases, demonstrated. The cost of purchasing new items verses repairing items was then analyzed to determine the projected cost savings on an annual basis for that component. The estimated savings was then extrapolated to other airframes with similar components. For example, the B-1, F-16 and F-15 have similar engines so a component that can be repaired on a B-1 engine was extrapolated to account for the additional number of F-16 an F-15 engines. The combined analysis accounted for the Air Force inventory and did not account for Navy, Marine or Army hardware.

Conclusions

Overall, the repair demonstration on the ruddervator fittings was a success. Radiographic inspection showed that the defects were fully processed from the parts. The parameter development welds showed a defect size of 150% could be processed with a single pass. While the defects have been removed, the heat input to the part alters the material temper of the aluminum part (W condition in the weld nugget and overage in the HAZ); this could be a problem since the ruddervator fitting is a structural component and may require a harder/stronger material to hold the bearing in place. However, the fatigue defects causing the scrapping of the parts may not be as critical in the heat-affected-zone of the FSW for the general reason that softer materials are less susceptible to fatigue failure. More development work is warranted to resolve this issue. Volumetric defects and fatigue cracking have been successfully processed out of the ruddervator fittings with little difficulty using friction stir technology. This demonstration suggests the repair of structural aircraft components is feasible via friction stir processing.

Investigation of forge (z-force) controlled welding is recommended; it may be beneficial to use forge force control over the position control method currently being used. Forge force control would reduce the force spikes occurring at the defects and could potentially increase the defect size capable of being processed.

References

1. Arbegast, W., Pillay, G., "Aging Aircraft Repair Feasibility Study," Alion Science and Technology Contract number #1919KR through the United States Air Force, 2007.

Acknowledgements

This study was supported by the Air Force and accomplished under a contract through Alion Science and Technology contract number FY2007 - #1919KR. Thanks to the students at the Advanced Materials Processing Center at SDSM&T who participated in this study.

Friction Stir Welding and Processing V
Edited by: Rajiv S. Mishra, Murray W. Mahoney, and Thomas J. Lienert
TMS (The Minerals, Metals & Materials Society), 2009

THE ROLE OF FRICTION STIR WELDING IN NUCLEAR FUEL PLATE FABRICATION

Douglas Burkes[1], Pavel Medvedev[1], Michael Chapple[1], Amit Amritkar[2], Peter Wells[3], Indrajit Charit[3]

[1]Idaho National Laboratory
P.O. Box 1625, Idaho Falls, Idaho 83415-6188, U.S.A.
[2]University of Utah, 50 S. Central Campus Dr. Room 2110, Salt Lake City, Utah 84112, U.S.A.
[3]University of Idaho, P. O. Box 440902, Moscow, Idaho 83844-0902, U. S. A.

Keywords: Friction Stir Welding, Aluminum Alloys, Nuclear Fuel, Fabrication

Abstract

The friction bonding process combines desirable attributes of both friction stir welding and friction stir processing. The development of the process is spurred on by the need to fabricate thin, high density, reduced enrichment fuel plates for nuclear research reactors. The work seeks to convert research and test reactors currently operating on highly enriched uranium to operate on low enriched uranium without significant loss in reactor performance, safety characteristics, or significant increase in cost. In doing so, the threat of global nuclear material proliferation will be reduced. Feasibility studies performed on the process show that this is a viable option for mass production of plate-type nuclear fuel. Adapting the friction stir weld process for nuclear fuel fabrication has resulted in the development of several unique ideas and observations. Preliminary results of this adaptation and discussion of process model development are discussed.

Introduction

Friction Stir Welding

Friction stir welding (FSW) has been in existence since the early 1990s and has routinely been developed for joining applications in the aerospace and automotive industries. These developments have been encouraged because FSW is a non-melting joining technology that typically produces higher strength and ductility, increased fatigue life and toughness, lower distortion, less residual stress, less sensitivity to corrosion, and essentially discontinuity-free joints when compared to more conventional arc welding techniques [1]. A sufficient review of the FSW process can be found in References. 2 and 3.

Research and Test Reactors

The friction bonding (FB) process, a modified FSW process, allows the fabrication of nuclear fuel plates for research and test reactors containing thin, monolithic fuel alloys. The FB fabrication concept is of great importance for the U.S. National Nuclear Security Administration-sponsored Global Threat Reduction Initiative. This initiative seeks to enable research and test reactors throughout the world that currently operate with highly enriched uranium (HEU, $\geq 20\%$ ^{235}U) fuel to operate with low-enriched uranium (LEU, $<20\%$ ^{235}U) fuel, which is desirable for nuclear nonproliferation reasons [4]. The new fuels must behave in a manner without significant penalties in reactor or experiment performance, economics, and safety.

HEU-fueled research and test reactors have given rise to advanced and novel fuel concepts because the reduction of enrichment in the fuel requires an increase in the overall uranium density. This has currently been accomplished through use of a uranium-molybdenum alloy in a monolithic form [5]. Monolithic fuel contains a single foil to replace multiple fuel particles comprising a dispersion fuel plate, provides the highest possible uranium loading, and provides a smaller contact surface area with the aluminum cladding to minimize reaction. Ultrasonic test scans of a typical dispersion fuel and a typical monolithic fuel are provided in Figure 1.

Figure 1. Ultrasonic test scans of a typical dispersion fuel plate and a new monolithic fuel plate proposed for core conversion of high power research reactors.

Current Process Design

The current application involves sandwiching a monolithic uranium-molybdenum fuel foil between two thin pieces of 6061-T6 aluminum alloy used as the cladding to fabricate fuel plate assemblies. The cladding is subjected to the friction bonding process, over its entire surface, both top and bottom. The process bonds the top and bottom pieces of thin material to one another without the movement of material across the joint interface, resulting in sufficient bonding between both the aluminum cladding and monolithic fuel alloy.

The tool pin is designed so that penetration across the joint interface is prohibited, minimizing any potential interference with the nuclear fuel foils. Sufficient bonding between the AA6061-T6 cladding plates and between the uranium-molybdenum monolith and AA6061-T6 cladding is required, accomplished primarily by mechanical bonding at the interface that is driven mainly by process load and temperature. There is minimal material movement between the pin face and the foil surface, meaning that the stir action used in the conventional sense is not present or desired. Figure 2 provides an example schematic of the weld tool design being employed. Some of the main features of the tool are the short pin that extends down from the shoulder, a recessed region between the shoulder and pin that increases the stability of the process, beveled edges along the outer radius of the shoulder to aide in warp reduction, and an annular plenum that allows heat removal from the weld face via an appropriate coolant, e.g., ethylene glycol. The FB process produces a surface finish that closely meets most reactor specifications and, therefore, requires minimal post-processing surface treatment. Most fuel plates are rectangular and are no thicker than 1.4-mm (0.055-in). A photograph of a typical monolithic fuel plate prior to insertion into the fuel assembly is provided in Figure 3.

Friction Bond Process Modeling

Concept

A three dimensional model was created using the COMSOL software as a means of modeling the thermal aspects of the unique friction bond process developed at the Idaho National Laboratory. The FB model is designed to allow changes in the operational parameters, including weld traverse speed, tool rotation rate, and applied force. It does not model the initial tool plunge or

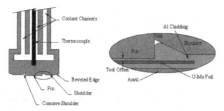

Figure 2. Schematic of a typical tool employed for the Friction Bonding process at INL.

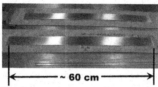

Figure 3. Photographs of a typical, finished monolithic fuel plate assembly. The reactor core will consist of tens to hundreds of such fuel plates.

the tool removal and many other real operational aspects have been simplified for ease of modeling and computation.

The model is designed as a moving continuum, in which the aluminum plate moves under the tool rather than the tool moving along the plate. This allows the plate material under the tool to be modeled as a viscous fluid that can have rotational and translational properties applied as needed. This region or zone is labeled as viscous zone and is assumed to be cylindrical right beneath the tool.

Currently, the model only applies to the first pass across the plates; however, it can be adjusted to consider additional passes. The model does not assume symmetry across the weld line because additional passes overlap with previous passes and include material with different microstructures on either side of the weld line. In addition, the differences between the advancing and retreating tool sides suggest that the heat generation is not symmetrical across the tool, even on the first pass.

Model Geometry

The model includes the following geometrical components. The tool is a composite object of two cylinders, the pin and the shoulder. The tool is not precisely the same design as the actual tool because it lacks rounded edges. The height of the shoulder is arbitrary because the shank to which the shoulder is attached is not included in the model. Instead, a heat sink equation is applied to the top of the shoulder to reflect the heat loss due to coolant running through the tool. The value of this heat loss is modeled based on the experimental results [6]. There is only one subdomain for this component. The complete plates are simulated (to the dimensions of 30 in. × 4 in. × 0.038 in.) to emulate the actual process. This plate width includes the portion of the plate under the clamps during a pass. Geometrically, the top plate is not a perfect rectangular prism, as a small volume has been replaced by the pin. The plates are further subdivided into two subdomains each, making it four subdamains total, in which the subdomain directly under the

249

tool is the viscous zone for each plate. The insert is a large piece of steel that supports the plates from beneath. In the model it primarily serves as a heat sink for the bottom of the plate. In order to improve computational efficiency, the rigid anvil is modeled with convective flux. There is only one subdomain for this part of the process.

COMSOL application modules used

The quasi-static friction bonding model makes use of four application modes, applied as needed to the subdomains. The General Heat Transfer module is applied to all subdomains. This mode allows for the heat transfer calculations which are the basis of the thermal model. The Weakly Compressible Navier-Stokes module applied to the viscous region in both the top and the bottom plate. This mode allows the plate to behave as a continuum by modeling it as a highly viscous fluid. The plate can then "flow" past the tool at a given weld-speed, achieving applied rotational velocities as it passes under the tool. The plate "flows" in the positive x-direction and rotates clockwise below the tool (as viewed from above). The Solid, Stress-Strain module is applied to all subdomains. This mode allows a force to be applied to the top of the tool and transferred down through the plate. The resulting stresses vary due to tool geometry and are used to calculate the heat distribution due to friction. The viscoplastic material model for AA6061 is applied to the simulation through this module. The Moving Mesh mode is applied to all subdomains. This mode allows efficient coupling of the Solid, Stress-Strain mode and the Weakly Compressible Navier-Stokes mode. Also, the visualization of the tool and plate movement can be obtained using transient analysis in this mode.

Process Model Output

The model can be solved as either time-dependent or steady-state. A near-steady-state condition is reached rapidly in the time-dependent model and should accurately reflect the conditions somewhere in the middle of the weld. Considering that the initial plunge area and some of the nearby weld are sheared from the plate before it is used, much of the initial non-steady-state weld is inconsequential. Thus the quasi static part of the bonding process is modeled in the simulations.

Temperature is the most significant output variable. Cross-sections through the plate are useful in visualizing the effects of the pin and shoulder (Figure 4 and Figure 5) but line-sections most closely resemble the type of data recorded experimentally. Line sections along the length of the weld simulate thermocouples attached at the joint of the assembly during an experimental run. The x-axis of the model can be equated to time (using the weld-speed) such that the temperature of material a certain distance away from the tool, say "X", correlates to the temperature of material that passed under the tool X divided by weld-speed seconds ago. This is because the modeled plate moves as a continuum with the weld-speed as its constant velocity. These line plots can be compared to experimental data recorded by thermocouples attached beneath or between the aluminum plates (Figure 6 to Figure 8). It is also possible to analyze the temperature of individual points using at any point and, in a time-dependent analysis, at any desired time.

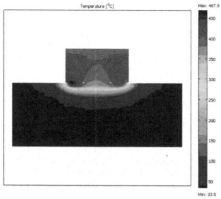

Figure 4. Results of the COMSOL calculation of the cross-sectional temperature distribution during FB process. Process parameters: feed rate 35 in/min, load 6000 lbf, tool rotation 400 rpm.

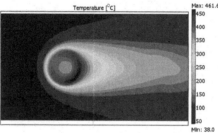

Figure 5. Results of the COMSOL calculation of the temperature distribution during FB process at the interface between the aluminum plates. Process parameters: feed rate 35 in/min, load 6000 lbf, tool rotation 400 rpm.

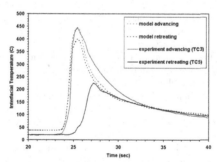

Figure 6. Comparison of the model prediction and the thermocouple reading at the tool shoulder. Process parameters: feed rate 35 in/min, load 6000 lbf, tool rotation 400 rpm.

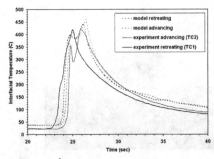

Figure 7. Comparison of the model prediction and the thermocouple reading at the tool pin edge. Process parameters: feed rate 35 in/min, load 6000 lbf, tool rotation 400 rpm.

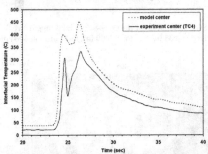

Figure 8. Comparison of the model prediction and the thermocouple reading at the center of the weld pass. Process parameters: feed rate 35 in/min, load 6000 lbf, tool rotation 400 rpm.

Future Process Modeling Efforts

As of the writing of this paper, there were still several unanswered questions concerning the model behavior. Below are several untested ideas which may improve the model fidelity:

- Improve the heat generation equation. The current friction-based equation functions well but is based on an arbitrary friction coefficient. An equation based on viscous dissipation may be more accurate.
- The change in the contact area for various conditions can be modeled based on the experimental data obtained. Similarly the viscous zone size and shape can be varied as per experimental results.
- Incorporate the friction coefficient variation with temperature based on an external study.
- The material modeling used in COMSOL needs to be appropriately done in order to model the heat transfer through the AA6061 plates correctly.

Simulate successive passes by including a microstructural adjustment in the model. Each new pass after the first overlaps a welded area so the microstructure will be asymmetrical across the weld line

Process-parameter relationships

252

Surface Finish

Surface finish of the as-fabricated fuel plates is extremely important in terms of economics, since excessive flash or ridges as a result of fabrication must be eliminated prior to insertion in the reactor. Process heat is the most important factor in controlling the surface finish of the plate. Studies were conducted to examine the effects of tool rotational rate, tool traverse rate, and tool applied load on the as-fabricated surface finish of surrogate fuel plates. Photographs demonstrating these results are provided in Figure 9. A low tool applied load results in a very rough surface finish that would be unacceptable for commercial fabrication of fuel plates. Higher applied loads (6 kips or greater) for a given tool rotational and traverse rate will alleviate this problem significantly. At applied loads of 10 kips a flash is developed that results in ridges between bond passes that must be removed by sanding. Similarly, low tool rotational rates for a given tool traverse rate and applied load will produce a relatively rough surface finish. This can be greatly alleviated by increasing the tool rotational rate, although above 400 rpm excessive flash is generated and ridges begin to develop between each bond pass. The surface finish is much less sensitive to tool traverse rate than for tool rotational rate or applied load. In each of the seven conditions investigated, the combination 35 ipm tool traverse rate, 400 rpm tool rotational rate, and 6 kips applied load yielded the best surface finish with the least amount of post-processing work required.

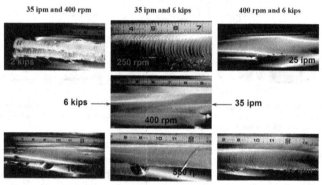

Figure 9. Photographs of as-fabricated surface finish for different tool traverse rates, tool rotational rates, and tool applied load. A combination of 35 ipm, 400 rpm, and 6 kips (center) produced the best surface finish of all conditions investigated.

Aluminum to Foil Bonding

Another important factor for plate-type nuclear fuels is bonding not only of the aluminum, but also between the two dissimilar metals. Studies were carried out based on the experimental matrix employed in Figure 9 to investigate bonding between the AA6061 cladding and a surrogate SS304 foil 0.010 in. thick. A single pass was made over the 3.25 in. long and 0.75 in. wide foil to evaluate bonding non-destructively. A single pass was made to better understand the distribution of heat across the contact surface of the tool, since fewer passes across the fuel plate to bond the Al and foil can result in increased time and cost savings and minimize the potential for error that may result in a destroyed foil. Again, process heat is the most important factor for adequate bonding between the cladding and foil. Ultrasonic scans of the foil region demonstrating these results are provided in Figure 10. Through observation of these ultrasonic scans, it becomes apparent that a low tool applied load will result in insufficient bonding, represented by the very large black void in Figure 10. Increasing the tool applied load for a given tool rotational and traverse rate will significantly improve the degree of bonding. However, increasing the tool load to 10 kips will result in too high of a process temperature, and actually minimize the amount of aluminum that can mechanically adhere to the foil surface. This effect is evidenced for the condition of 400 rpm, 35 ipm, and 10 kips where gouges in the plate surface caused by recessed region between the advancing edge and pin. Similarly, a low tool rotational rate for a given tool traverse rate and applied load will result in inadequate bonding on the retreating edge of the tool. This is alleviated by improving the process temperature by increasing the tool rotational rate. Similar to the surface roughness studies, bonding is somewhat independent of tool traverse rate, and the best combination of parameters for bonding was a tool traverse rate of 35 ipm, tool rotational rate of 400 rpm, and tool applied load of 6 kips.

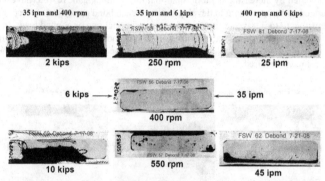

Figure 10. Ultrasonic scans of surrogate SS304 foils for different tool traverse rates, tool rotational rates, and tool applied load. A combination of 35 ipm, 400 rpm, and 6 kips (center) produced the most acceptable bonding determined by this technique of all conditions investigated.

Tool Temperature Distribution

Experiments were carried out to investigate the temperature distribution across the tool (from the advancing to the retreating edge) by placing thermocouples at the joint interface. A schematic of the experimental set-up is provided in Figure 11, and results of the experiments as a function of tool rotational rate, tool traverse rate, and tool applied load are presented in Figure 12. The results represent separate effects test, meaning that only one of the three parameters was changed at a time, while the other two were held constant.

The advancing edge temperature of the tool was enhanced by increasing the tool rotational rate, while the smallest difference between the advancing and retreating edge of the tool was obtained for a rotational rate of 400 rpm with the tool traverse rate and applied load held constant at 35 ipm and 6 kips, respectively. The advancing edge of the tool was minimally affected by variations in tool traverse rate for a constant tool rotational rate and applied load. Similarly, the smallest temperature gradient across the tool occurred for a traverse rate of 35 ipm, rotational rate of 400 rpm, and applied load of 6 kips. Finally, an increase in the tool applied load from 2 kips to 6 kips resulted in a significant improvement in interfacial temperature, decreasing somewhat for loads above 6 kips due to the limit set on the penetration depth of the tool to avoid

254

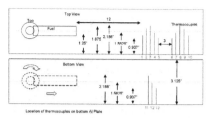

Figure 11. Schematic of the experimental setup, showing the location of the thermocouples placed at the joint interface, and the surrogate SS304 foil also placed at the interface.

impacting a potential foil at the interface. Converse to the other two parameters, the smallest temperature gradient across the tool occurred for a load of 2 kips with a rotational and traverse rate of 400 rpm and 35 ipm, respectively. However, this is caused by inadequate contact of the tool with the work piece surface so that only the pin generated sufficient heat. This artifact was also observed in the surface finish photographs provided in Figure 9, and also results in the significantly lower interface temperatures for this set of conditions.

Observation of these figures clearly shows why the best surface finish and bonding is achieved for a tool rotational rate of 400 rpm, traverse rate of 35 ipm, and applied load of 6 kips. In some cases, variation of the parameters can result in greater interfacial temperature and similar bonding characteristics, but generates the flash and ridges on the surface, pointed out in Figure 9. Thus, careful consideration must be given to the set of parameters that provides the best results for each requirement, namely surface finish and adequate bonding.

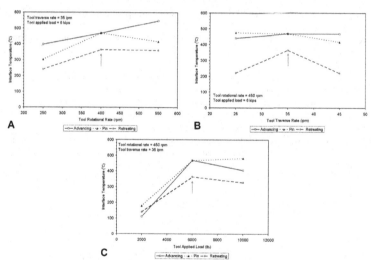

Figure 12. Temperature measured at the joint interface by thermocouples placed under the retreating edge, pin, and advancing edge of the tool. Temperature maps vary as a function of tool rotational rate for tool traverse rate = 35 ipm and applied load = 6 kips in (A), as a function of tool traverse rate for tool rotational rate = 400 rpm and applied load = 6 kips in (B), and as a function of applied load for tool rotational rate = 400 rpm and traverse rate = 35 ipm in (C).

255

Conclusions

This paper shows how a simple modification of the FSW process is now well-positioned for commercial fabrication of new research and test reactor nuclear fuel. Successful fabrication and qualification of this new plate-type nuclear fuel will enable the conversion of reactors currently operating on HEU to LEU, thereby decreasing nuclear proliferation concerns and risks. Solutions to challenges associated with this process, in addition to the novel process itself, offer information for users of the FSW process and as a potential alternative fabrication technique for laminar composite structures.

Acknowledgements

Work was supported by the U.S. Department of Energy, Office of the National Nuclear Security Administration, under U.S. Department of Energy Idaho Operations Office Contract DE-AC07-05ID14517. The FB process has been under continual development since 2004, and much of the progress could not have been made without continual support from the RERTR program and management. Furthermore, the authors would like to acknowledge Mr. Gaven Knighton for his role in the set-up of initial scoping experiments, Mr. Pat Hallinan who has continuously supported experiments, and Mr. Jared Wight who has been influential in the design of hardware for commercial demonstration of the process.

References

1 W. J. Arbegast, "Friction Stir Welding: After A Decade of Development," in Friction Stir Welding and Processing IV, TMS (2007) pp. 3-18.

2 R. S. Mishra and Z. Y. Ma, "Friction stir welding and processing," Mater. Sci. Eng. R, 50 (2005) pp. 1-78.

3 R. Nandan, T. DebRoy, and H.K.D.H. Bhadeshia, "Recent advances in friction-stir welding – Process, weldment structure and properties," Prog. Mat. Sci., 53 (2008) pp. 980-1023.

4 National Nuclear Security Administration (NNSA) website, http://www.nnsa.doe.gov/nuclearnp.htm#1, accessed on January 14, 2008.

5 J. L. Snelgrove, et al., "Development of Very-High-Density Fuels by the RERTR Program," Proceedings of the International Meeting on Reduced Enrichment for Research and Test Reactors, Seoul, Korea (1996).

6 J. Dixon, D. Burkes and P. Medvedev, "Thermal Modeling of a Friction Bonding Process," Proceedings of the COMSOL Conference 2007, Boston, MA (2007) pp. 349-354.

Friction Stir Welding and Processing V
Edited by: Rajiv S. Mishra, Murray W. Mahoney, and Thomas J. Lienert
TMS (The Minerals, Metals & Materials Society), 2009

CORROSION IN 2XXX-T8 ALUMINUM ALLOYS

Christian Widener[1], Tze Jian Lam[1], Dwight Burford[1]

[1]National Institute for Aviation Research Advanced Joining and Processing Laboratory, Wichita State University, Wichita, KS 67260, USA

Keywords: friction stir, corrosion, aluminum lithium, exfoliation, Al 2024, 2219, 2198

Abstract

This experiment investigates an apparent trend in 2XXX-T8 aluminum alloys to possess excellent as-welded exfoliation corrosion resistance in the weld zone compared to the parent material. To better evaluate this trend, friction stir welds were produced in 0.125 (3.2 mm) 2024-T81, 0.080-in. (2 mm) 2219-T87 and 0.153-in. (3.9 mm) 2198-T851 (Al-Cu-Li) material, and then tested in a standard and modified ASTM G34 exfoliation environment. Unlike welding in the –T3 or –T4 tempers, where the weld zone can become anodic to the parent metal and exhibit preferential corrosion, when welded in the –T8 starting temper the weld zone has been found to be relatively cathodic compared to the parent material, and exhibits only mild evidence of corrosion attack.

Introduction

The general aging trends in 2XXX Al-Cu-Mg alloys and 7XXX Al-Zn-Mg-Cu aluminum alloys are well documented and suggest from the outset that there may be similarities in each respective family of aluminum alloys in response to friction stir welding, post-weld artificial aging (PWAA), and subsequently strength and corrosion resistance [1,2]. This does not by any means imply that their behavior will be identical, but only that trends can and should be expected to result.

The 2024-T81 temper has been shown to possess excellent exfoliation corrosion resistance in the as-welded condition following friction stir welding. Compared to 2024-T3 in the as-welded condition, which shows preferential corrosion in the weld zone, 2024-T81 in the as-welded condition appears to be almost immune to exfoliation corrosion in the weld zone, as shown in Figure 1 [3].

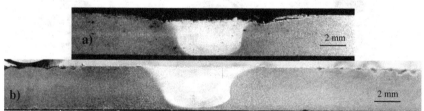

Figure 1: ASTM G34 Exfoliation sample micrograph for a) as-welded 0.125-in. (3.2 mm) thick 2024-T3 and b) as-welded 0.125-in. (3.2 mm) thick 2024-T81 [3]

Similar results have also been reported regarding the corrosion response of both 2195-T8 and 2219-T8. When they were investigated, it was observed that the dissolution of precipitates in the weld nugget and coarsening of the precipitates and precipitate free zones in the thermo-mechanically and heat affected zones did not increase the corrosion susceptibility of the weld zone with respect to the parent material [4,5].

The potential benefit for dissimilar alloy joints between 2XXX-T8 and 7XXX-T7 alloys was also demonstrated by joining 2024-T81 to 7075-T73 followed by a short post weld artificial aging treatment (PWAA) to restore the exfoliation resistance to the joint [6]. Furthermore, the general results and benefits of welding in the -T7 starting temper followed by PWAA in several 7XXX-T7 alloys, specifically 7136-T7, 7055-T74 and 7249-T76, were demonstrated and shown to exhibit similar behavior, although not identical, to that which was reported for 7075-T73 [7,8]. The purpose of the current work, therefore, is to continue to explore the broader trends of these 2XXX-T8 alloys by testing the exfoliation resistance of 0.080-in. (2 mm) thick 2219-T87 (Al-Cu) and 0.153-in. (3.9 mm) thick 2198-T851 (Al-Cu-Li).

Procedure

Welds produced in 0.125 (3.2 mm) 2024-T81 were originally published under a separate investigation of starting tempers in Al 2024. Butt welds were made in 0.125-in. (3.2 mm) thick sheets of 2024-T3 and 2024-T81 parallel to the rolling direction, using the FSW pin tool shown in Figure 2a), which has an approximately 2:1 concave shoulder and a cylindrical threaded probe. The welds were made at 600 rpm and 8 ipm (203 mm/min), under load control with a forging force of 2,150 lbs (9.56 kN) [3].

Additional welds were produced in 0.080-in. (2 mm) 2219-T87 and 0.153-in. (3.9 mm) 2198-T851 (Al-Cu-Li) material for comparison of results with those of 2024-T81. Alloy compositions are shown in Table 1. The tool used for the 2219 and 2198 material is shown in Figure 2 b) and c) respectively. Both tools used a threaded tapered pin tool with three flats; however while the tool in Figure 2b) has a simple scrolled shoulder, the tool in Figure 2c) utilized a patent pending Wiper™ shoulder design [9]. Butt welds were again made parallel to the rolling direction. The welds were made at 600 rpm and 8 ipm (203 mm/min) with 2050 lbs (9.12 kN) forging force for in 2219 and at 900 rpm and 10 ipm (254 mm/min) with 2100 lbs (9.34 kN) forging force.

Figure 2: FSW fixed shoulder pin tools used for the a) 2024 b) 2219 c) 2198

Table 1: Base material compositions

Alloy	% Composition										
	Cu	Li	Mg	Mn	Ag	Ti	V	Zr	Si	Fe	Al
0.125-in (3.2 mm) 2024-T81	3.8 – 4.9		1.2 – 1.8	0.3 – 0.9		<0.15			<0.5	<0.5	Bal.
0.080-in (2 mm) 2219-T87	5.8 – 6.8		<0.02	0.2 – 0.4		0.02 – 0.1	0.05 – 0.15	0.1 – 0.25	<0.2	<0.3	Bal.
0.153-in (3.9 mm) 2198-T851	2.9 – 3.5	0.8 – 1.1	0.25 – 0.80	<0.5	0.1 – 0.5	<0.1		0.04 - 0.18	<0.1	<0.08	Bal.

The standard test method for exfoliation corrosion (EXCO) susceptibility in 2XXX and 7XXX series aluminum alloys is ASTM G34. As a basis for comparison, this test procedure was used for the first stage of the testing. Since a standard exfoliation test has been shown to be overly aggressive for some aluminum lithium alloys (Al-Li), a pre-conditioned solution was also used to evaluate the welds exfoliation resistance. This modified solution was shown by Lee and Lifka [10] to accurately reflect the performance of 2090 (Al-Cu-Li) and 2024 (Al-Cu) in a seacoast atmosphere. Therefore, in the second stage of testing, this modified, less aggressive reagent with a much higher pH (~3.2 as opposed to 0.4) was used.

The standard EXCO ASTM G34 test reagents were prepared according to the following composition: 234 g of NaCl, 50 g of KNO_3 were dissolved in deionized water, and 6.3 mL of concentrated HNO_3. This solution was diluted to 1 L in a beaker and kept at room temperature. The specimens were cleaned with methyl-ethyl-ketone (MEK), labeled with a label maker, and masked with protective tape on the bottom to maintain an un-corroded area for specimen identification. The volume-to-metal surface area ratio was kept in the range of 10 to 30 mL/cm^2. The top FSW surface was kept facing up in a plastic stand to avoid losing exfoliated metal. The dimensions and weight of specimens were recorded.

In the modified exfoliation test, a pre-conditioned solution was prepared based on the standard EXCO test, but with the additional steps of pre-reacting the solution by exposing a piece of 2198-T8 for 24 hours, and heating the solution to 52 °C, as recommended [10]. The experimental setup is shown in Figure 3.

Figure 3: a) Standard EXCO test setup b) Modified EXCO test setup

Results

The same general behavior, which was observed with as-welded 2024-T81 in previous experiments [3,4], was also seen with 2198-T851 and 2219-T87. The weld zone was not attacked, or experienced minimal signs of attack, in an ASTM G34 exfoliation environment. The parent material, on the other hand, experienced varying degrees of attack depending on the alloy type and its degree of exfoliation resistance. An exposed 2198 coupon is shown in Figure 4. Both the weld track and root of the weld experienced a form of cathodic protection from the parent material giving them a pristine appearance even after 96 hours in an aggressive accelerated test environment.

Figure 4: 2198-T851 after standard EXCO test -- a) weld track, b) weld root

Cross-sectional micrographs taken before and after exposure further demonstrate that the weld zone of the 2198-T851 material has not been attacked, reference Figures 5 and 6. Additionally, pre-polished micrographic samples were exposed to a solution more representative of seacoast exposure in order to ensure that the results remained the same in the modified solution, reference Figure 7. After a very light re-polish to reveal the microstructure again, it was observed that the attack was more moderated with the modified solution; however, the bulk anodic/cathodic relationship of the weld zone to the parent material was unaffected. Some exfoliation occurred in the parent material outside of the weld zone, but within the weld nugget and heat-affected zone there were no signs of attack on the cross section.

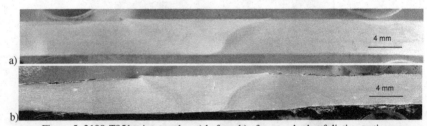

Figure 5: 2198-T851 micrograph – a) before, b) after standard exfoliation testing

Figure 6: 2198-T851 composite micrograph after exfoliation testing

Figure 7: 2198-T851 micrograph exposed to a modified exfoliation solution

This same trend was also observed in 0.080-in. (2 mm) as-welded 2219-T87 samples which were exposed to an ASTM G34 exfoliation environment. After only a short time in the solution (approximately 24 hours) the weld zone is still bright and unattacked while a copper residue had already formed on the parent material, reference Figure 8. After 96 hours, the weld zone still showed very few signs of attack; however, the parent material exhibited typical amounts of exfoliation corrosion. Micrographs of the weld before and after exposure are shown in Figures 9 and 10, respectively. Again, the micrographs reveal that the areas of the weld nugget and HAZ received little to no corrosion attack, while just outside of the weld zone there are obvious signs of exfoliation attack in the parent material.

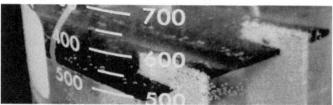

Figure 8: Front and back after standard EXCO test 2219-T8

Figure 9: Macro before standard EXCO test 2219-T8

Figure 10: Macro after standard EXCO test 2219-T8

261

In order to better understand the potential metallurgical contributions to this corrosion behavior, a comparative study of weld nugget microhardness and electrical conductivity was conducted in 2024-T3 versus 2024-T81. The results of a centerline microhardness traverse are shown in Figure 11. While the initial microhardness of the –T81 temper is clearly higher than that of –T3, the resulting microhardness of the nugget material is virtually the same in both alloys. A similar trend is observed with respect to the electrical conductivity values, as shown in Figure 12. Both microhardness and electrical conductivity can be used to characterize weld temper. The fact that welding in either starting temper results in the formation of a weld nugget with similar microhardness and electrical conductivity values suggests that, in this case, both weld nuggets possess essentially the same weld microstructure and effective temper. Based on these results, it can be concluded that the weld nugget is reaching solid solution temperatures and that, in addition to dynamic recrystallization, there is also a resolution of alloying elements occurring in the weld nugget. Therefore the resulting microstructure is more or less comparable to a –T4 temper in this alloy. It is expected that similar results could be obtained for the other two 2XXX-T8 alloys in this investigation. Unfortunately, similar studies of 2219 and 2198 were not possible before publication, since material in the under aged –T3 or –T4 tempers for those alloys was not available.

As a final note, additional PWAA treatments of 2024-T81 at 375°F (191°C) for up to 4 hours were not observed to detrimentally impact the corrosion resistance of the weld zone [3]. Presumably, this is either due to the fact that any microstructural changes would tend to drive the weld nugget temper to match the parent material temper or simply that up to 4 hours exposure is insufficient to dramatically alter the weld zone microstructure, as evidenced by little to no change in the weld zone electrical conductivity over the time period [3]. This finding is the basis for a beneficial combination of 2XXX-T8 and 7XXX-T7 alloys followed by a short duration PWAA treatment to improve the corrosion resistance of the dissimilar alloy joint.

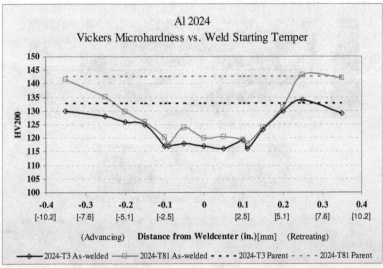

Figure 11: Comparison of as-welded microhardness in the weld zone for 2024-T3 and 2024-T81

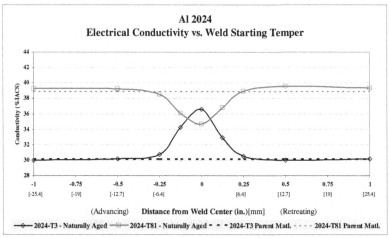

Figure 12: Comparison of electrical conductivity for as-welded 2024-T3 and 2024-T81

Conclusions

It now appears that the excellent exfoliation resistance of the weld zone exhibited in 2024-T81 in the as-welded condition is not just a phenomenon that is restricted to this alloy and temper alone. In a broader sense, it is also displayed by other 2XXX-T8 alloys, namely 2219-T87 and the Al-Cu-Li alloy 2198-T851, and has also been reported for 2195-T8 [4] as well. This exfoliation resistance results from the creation of a weld zone which becomes cathodic to the parent material when these alloys are welded in the –T8 temper and left as-welded. This behavior is believed to be a bulk material phenomenon and to be generally insensitive to weld tool design and material thickness. The general effects of dissolution of precipitates and the formation of a fine grained microstructure and the coarsening of the precipitates and precipitate free zones along the thermo-mechanically and heat affected zones do not increase the corrosion susceptibility of the weld zone with respect to the parent material. This behavior is in stark contrast to what is so commonly displayed by other alloy and temper combinations, where the weld zone becomes anodic to the parent material. Rather, in the case of 2XXX series aluminum alloys, it seems to produce the opposite effect when the parent material is in a –T8 temper.

Future Work

Additional work is needed to better understand the underlying mechanisms which are behind the cathodic response of the weld zone with respect to the parent material. Nevertheless, this work therefore suggests that this broader trend may be expected with the majority of the alloys in this family; however, as results may vary with specific alloys due to the presence of various additional alloying elements, verification of these results for other 2XXX alloys of interest should be performed. Similar results may also be achieved in the closely related –T6 temper. Also, based on these results and previous work, there is a strong potential for enhancing the corrosion resistance of other dissimilar alloy joints of 2XXX-T8 and 7XXX-T7 aluminum alloys. Therefore, dissimilar alloy combinations in addition to 2024-T81 and 7075-T73 should be investigated.

Acknowledgements

The authors would like to thank the State of Kansas and the Federal Aviation Administration for funding this work, the contributions of the students of the Advanced Joining and Processing Lab in the National Institute for Aviation Research at Wichita State University, and Alcan and Lockheed Martin for their contributions of material.

References

1. Starink, M.J., and Jialin, Y., "A Model for Strengthening of Al-Cu-Mg Alloys by S Phase," Proceedings from the Materials Solutions Conference 2003: 1st International Symposium on Metallurgical Modeling for Aluminum Alloys, Pittsburgh, PA, 13-15 October, 2003, pp. 119-126.
2. Starink, M.J., and Li, X.M., "A Model for the Electrical Conductivity of Peak-Aged and Over-aged Al-Zn-Mg-Cu Alloys," Metallurgical and Materials Transactions A, vol. 34A, no. 4, April, 2003, pp. 899-911.
3. Widener, C. A.; Burford, D. A.; Talia, J.E. ; Tweedy, B.M. "Investigation to Restore the Exfoliation Resistance of Friction Stir Welded Aluminum Alloy 2024," Friction Stir Welding and Processing IV, Edited by K.V. Jata, M.W. Mahoney, R.S. Mishra, and T.J. Lienert, TMS (The Minerals, Metals & Materials Society), 2007, pp. 449-458.
4. Wusheng, H. and Meletis, E.I. "Corrosion and Environment-Assisted Cracking Behavior of Friction Stir Welded Al 2195 and Al 2219 alloys," Materials Science Forum, v. 331-337, pt. 3, 2000, pp. 1683-1688.
5. Paglia, C.S. and Buchheit, R.G. "Microstructure, Microchemistry and Environmental Cracking Susceptibility of Friction Stir Welded 2219-T87," Materials Science & Engineering A, v 429, n 1-2, August, 2006, pp. 107-114.
6. Widener, C. A.; Burford, D. A.; Talia, J.E.; Tweedy B.M. "Corrosion in Friction Stir Welded Dissimilar Aluminum Alloy Joints of 2024 and 7075," Friction Stir Welding and Processing IV, Edited by K.V. Jata, M.W. Mahoney, R.S. Mishra, and T.J. Lienert, TMS (The Minerals, Metals & Materials Society), 2007, pp. 459-468.
7. Widener, C.; Kumar, B.; Burford, D.; Talia, J.E. "Evaluation of Post-weld Heat Treatments to Restore the Corrosion Resistance of Friction Stir Welded Aluminum Alloy 7075-T73 vs. 7075-T6," Materials Science Forum, Vols. 539-543 (2007) pp. 3781-3788.
8. Tweedy, B.; Widener, C.; Burford, D. "Fundamental Properties of Friction Stir Welded Al 7136 Including Effects of Post-Weld Artificial Aging," 6th International Friction Stir Welding Symposium, Montreal, Canada, October 10-12, 2006.
9. Burford, D.A., Tweedy, B.M., and Widener, C.A. "The Influence of Shoulder Configuration and Geometric Features on Weld Track Properties," 6th International Friction Stir Welding Symposium, Montreal, Canada, October 10-12, 2006.
10. Lee, S.; Lifka, B.W. "Modification of the EXCO test method for exfoliation susceptibility in 7XXX, 2XXX, and aluminum-lithium alloys," ASTM Special Technical Publication, n. 1134, 1991, p. 1-19.

Friction Stir Welding and Processing V
Edited by: Rajiv S. Mishra, Murray W. Mahoney, and Thomas J. Lienert
TMS (The Minerals, Metals & Materials Society), 2009

CORRELATION BETWEEN ULTRASONIC PHASED ARRAY AND FEEDBACK FORCE ANALYSIS OF FRICTION STIR WELDS

P. Gimenez Britos[1], C.A. Widener[1], J. Brown[1], D.A. Burford[1]

[1]National Institute for Aviation Research, Wichita State University
1845 Fairmount, Wichita, Kansas, 67260, USA

Keyword: Friction stir welding (FSW), ultrasonic phase array, discrete Fourier transformation (DFT), phase space plot

Abstract

Ultrasonic phased array is a powerful non-destructive test (NDT) and is well known for its capability to detect different kinds of FSW indications and defects. A new, process-based NDT technique developed at the South Dakota School of Mines and Technology is the FSW Analysis Software, designed to analyze any specified section of the weld in real time (with a slight computation time delay). With this software, a trained operator or inspector can detect where potential flaws may exist. The purpose of this study is to determine if all of the defects found using ultrasonic phased array inspection can be identified by the software data analysis program. By correlating this software with an ultrasonic phased array inspection, the time and expense associated with 100% inspection of parts could be significantly reduced. The ultimate goal of this research is to support the development of real time quality control to minimize the cost of inspection through statistical process control methods.

Introduction

Friction Stir Welding (FSW) is a relatively new joining technique that relies on localized forging and extrusion around a rotating pin tool to create a solid state joint. FSW is unlike traditional fusion welding processes because the joint is not formed by melting and coalescence, and it does not use any filler material. FSW relies on mechanical forces to stir metal from two or more components into a consolidated joint. When friction stir welding machine is equipped with instruments for welding force feedback, the evaluation of those forces may provide a method for weld quality control. The FSW Analysis Software developed at the South Dakota School of Mines and Technology (SDSM&T) analyzes the feedback forces of the welding process in real time (with a slight computation time delay) in any specified section of the weld. The capability of the software was demonstrated with cylindrical 5651 pin tool design in aluminum 7075-T73 6.3 mm (0.250 in.) thick [1]. The purpose of this current study is to determine the robustness of the software to identify weld defects using a variety of different pin tool designs, and whether or not the results can be correlated with weld defects found using ultrasonic phased array inspection. The ultimate goal of this research is to support the development of real time quality control to minimize the cost of inspection through statistical process control methods.

Procedure

This study investigates the potential for use of the software data analysis package in butt welds using a variety of FSW weld tools in addition to one for which it was originally developed. The methodology covers the use of 5 different tools (Figure 1) to create 10 welds, each with a different combination of processing parameters welding in 6.3 mm (0.250 in.) thick Al 2024-T351. Each weld was repeated three times, for a total of 30 welded plates in order to study the repeatability of the defects and analysis. As shown in Figure 1, the tools vary widely in both pin and shoulder features.

Figure 1. Tool designs

The feedback force capture rate of the FSW machine was set at 10 times the frequency of the spindle in every weld in order to be consistent on the signal quality. This was done to account for every group of welds having different welding parameters. For example, data from a 600 RPM weld would be captured at 100 Hz while data from a 300 RPM weld would be captured at 51.2 Hz. Once the welds were completed, they were analyzed using both the software and ultrasonic phased array. The welds were analyzed by technicians/engineers at both Olympus ND and Cessna Aircraft as part of a round robin NDT investigation. In a later phase of the program the welds will be verified with metallographic inspection and correlated to mechanical test results. Statistical analysis will be developed to evaluate the result of the correlation between the NDT methods and the destructive tests.

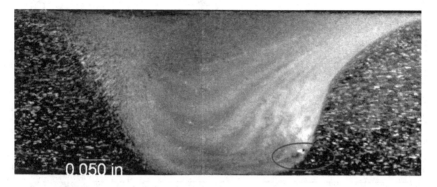

Figure 2, macro picture at 6.7x magnification

Figure 2 shows an example of flaws that can be found in the welds, this picture corresponds to earlier metallographic analysis on a weld with similar welding parameters using the "Wiper™ - large" weld tool, which has a flat tip and tapered probe with three twisted flats and threads.

Results

The round robin test was initiated with the analysis of two replicate sets of plates by Olympus ND and Cessna Aircraft. In this paper only results from one set of duplicate plates are shown because of the full study is yet to be completed. Ultimately, all 30 plates will be analyzed by up to 6 different facilities using, at a minimum, X-ray inspection techniques.

In Figure 3 the ultrasonic phased array analysis performed at Olympus ND is shown. The flaw starts at 222 mm (8.7 in.) from the beginning of the plate, as indicated in the red circle.

267

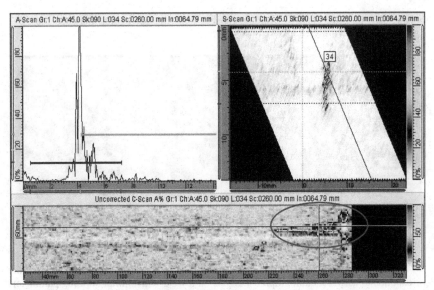

Figure 3, ultrasonic phased array detection of flaws in the weld. Work order
CFSP08502_03 welded with Wiper™ -large pin tool

Figure 4 shows the phased array analysis carried out by Cessna Aircraft
Company. The analysis was completed on work order number CFSP08502_9, which has
the same welding parameters and pin tool as CFSP08502_03. The flaws are located at
195 mm (7.7 in.) from the beginning of the weld and continue to the end of the weld.

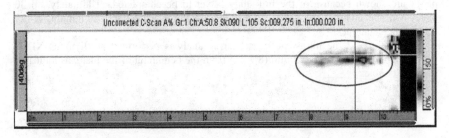

Figure 4, ultrasonic phased array detection of flaws in the weld. Work order
CFSP08502_09 using a Wiper™ -large pin tool

First note that the results from the two facilities, which tested duplicate plates,
produced the same C-Scan results. Further, the analysis performed with the FSW
software analysis package, using a phase space plot (PSP), correlates with the data and
indicates that starting at 215.9 mm (8.5 in.) from the beginning of the weld on

CFSP08502_03 (Figure 5). In this section there are indications of flaws which correlates closely with the results found in both phased array analyses.

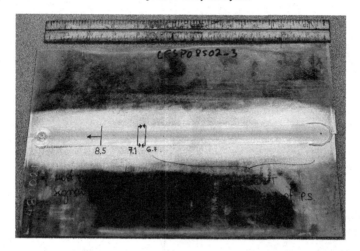

Figure 5, friction stir welded plate with indication of possible flaws according to the FSW Analysis Software

The PSP shown in Figure 6 indicates a possible flaw location at 216 mm from the beginning. This is indicated by a non-uniform shape of the data plot in the lower left graphics window and the high standard deviation displayed at the same plot. Both asymmetry in the plot and the computed high standard deviations have been shown to correlate with weld defects in friction stir welds [2].

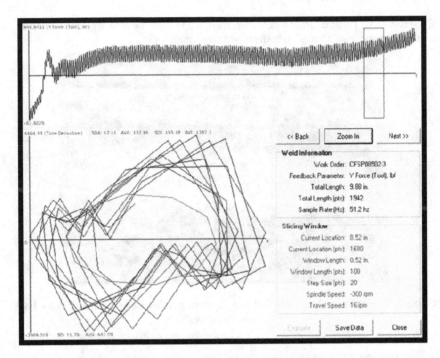

Figure 6, phase space plot for CFSP08502_3

High amplitude signals in the low frequency range, i.e. less than the spindle speed frequency, have been shown to correlate with the presence of weld defects [3]. Figure 7 shows the DFT plot for CFSP08502_3. Looking at the low frequency range, it can be seen that the spindle is the prominent peak located at 5 Hz. The DFT results do not show indication of flaws in any part of the weld based on this technique. It does shows that the amplitude of the low frequency was below 25% of the amplitude of the spindle frequency for the complete weld.

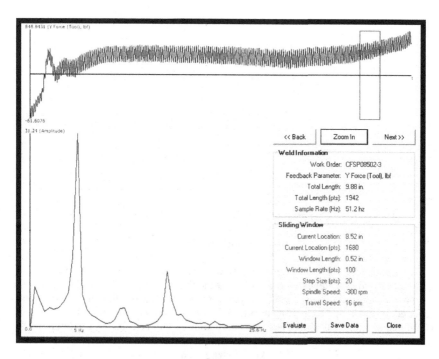

Figure 7, DFT plot for CFSP08502_3

In order to create a quantitative evaluation of the amplitude in the low frequency range, four levels of the spindle frequency amplitude were set (25, 50, 75, and 100%). The results are summarize in Figure 8 where a complete software analysis for the plate CFSP08502_3 is presented superimposed on a schematic of the plate.. For discrete Fourier transformation (DFT) analysis, values listed represent percent of spindle frequency amplitude. For phase space plot (PSP) standard deviation (SD) and distorted orbits are shown by location on plate. Further investigation will analyze if there is a relationship between the levels of the low frequency range and sizes of the flaws.

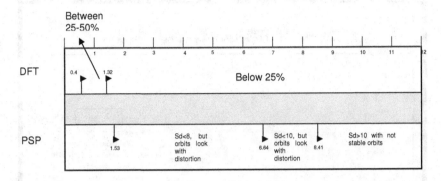

Figure 8. Complete software analysis for the plate CFSP08502_3.

Conclusions

In this ongoing project, preliminary work shows that there is a strong correlation between phased array analyses from different facilities. A promising correlation between the phased array and the software analysis can also be seen but further investigation is required in order to better evaluate the reliability of the correlation. Also, a quality control correlation analysis for lap weld configurations will be conducted in future work.

Acknowledgments

The authors will like to thank Enkhsaikhan Boldsaikhan for the training on FSW Analysis Tool software and Jason Grube, Galen Werth, Vien Nguyen, and Vishwanath Ananthanarayanan from the Advanced Joining and Processing Lab in the National Institute for Aviation Research. The authors also will like to thank Cessna Aircraft and Olympus ND for the analyses done.

References

1. E. Boldsaikhan. E. Corwin, A. Logar, and W. Arbegast, "Neural network evaluation of weld quality using FSW feedback data". *6th International FSW symposium, Montreal*, 2006.

2. E. Boldsaikhan E. Corwin, A. Logar, J. McGough, and W. Arbegast, "Phase space analysis of friction stir weld quality". *Friction stir welding and processing IV*, 2007, p 101-11.

3. M. Toshio, Master of Science graduating thesis, South Dakota School of Mine and Technology, Rapid City, SD. 2004.

Friction Stir Welding and Processing V
Edited by: Rajiv S. Mishra, Murray W. Mahoney, and Thomas J. Lienert
TMS (The Minerals, Metals & Materials Society), 2009

CORROSION AND FATIGUE EVALUATION OF SWEPT FRICTION STIR SPOT WELDING THROUGH SEALANTS AND SURFACE TREATMENTS

Jeremy Brown[1], Christian Widener[1], Dwight Burford[1],
Walter Horn[2], George Talia[3], Bryan Tweedy[4]

[1]National Institute for Aviation Research Advanced Joining and Processing Laboratory, Wichita State University, Wichita, KS 67260, USA
[2]Department of Aerospace Engineering, Wichita State University, Wichita, KS 67260, USA
[3]Department of Mechanical Engineering, Wichita State University, Wichita, KS 67260, USA
[4]H.F. Webster Engineering and Professional Services, Rapid City, SD 57702 USA

Keywords: friction stir, FSSW, alodine, anodized, 2024, sealant

Abstract

This experiment investigates the capability of welding though sealants and surface treatments with swept Friction Stir Spot Welding in thin gauge 2024-T3 aluminum alloy. The aluminum sheets have a sealant applied and are pre-treated with various surface coatings. The uncured sealants were applied to the faying surface of the test coupons shortly before joining. The results are also compared to bare sheets in the untreated condition. Corrosion testing was performed through alternate immersion in a 3.5% salt solution. The samples were evaluated through metallography and testing of residual strength. Fatigue testing was performed per the NASM 1312-21 specification. S-N data was collected for 5 load levels for each sample type. Riveted data has also been collected using this method. Work in this area is important to support increased implementation of FSSW in production applications as a replacement for other discrete fastening methods, like riveting and resistance spot welding.

Introduction

The plunge type of Friction Stir Spot Welding (FSSW) was the first Friction Stir Welding (FSW) process to utilize the concept of spot welding [1]. In this type of FSSW, a FSW pin tool is plunged into the material while rotating. The faying surface between the two sheet materials in the immediate vicinity of the tool is stirred and then the tool is then retracted, leaving behind a consolidated joint. This process is shown in Figure 1. Refill FSSW is a similar type of FSSW. This process uses a tool with a pin and shoulder which can move independently of each other. This creates a nearly flush surface at the spot top surface by eliminating the exit hole [2].

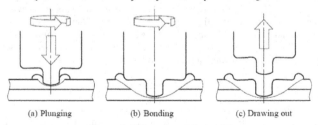

(a) Plunging   (b) Bonding   (c) Drawing out

Figure 1. Plunge Spot Motion [3]

Other types of FSSW could include stitch FSSW, developed by GKSS, swing FSSW, developed by Hitachi [4], and swept FSSW, which was called a Squircle™ by TWI [5]. These processes begin with a simple plunge. The tool is then translated or rotated in the material. This creates a larger weld zone than a simple plunge and retract spot. The increase in shear area typically results in an increase in spot strength.

Swept FSSW has been under significant development at Wichita State University (WSU) since 2005. WSU's pattern was dubbed the Octaspot™, shown in Figure 2, and is based closely on TWI's pattern. An advantage of swept spots over plunge spots is the increased strength primarily due to the resulting increased shear area. The flow of material during the plunge and dwell of typical spots can create an upturn of the sheet interface. The thinning of the effective sheet thickness typically correlates with a decrease in strength, or at the least a loss of peel strength. The sweeping pattern of swept FSSW may consume this interface upturn leaving a nearly straight joint.

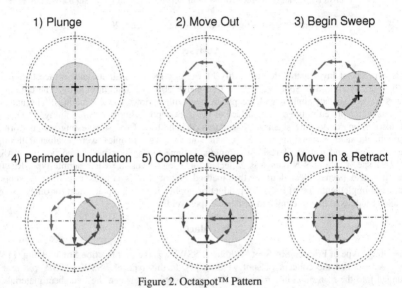

Figure 2. Octaspot™ Pattern

Sealants and surface treatments are used in order to protect against corrosion. It has been shown that swept FSSW can maintain good strengths when welding through a surface treated sheet onto a bare sheet [6] and through two surface treated sheets [7]. These strengths can also be maintained when welding through a sealant [8]. While good static strengths have been shown, little has been shown on the effectiveness of FSSW to maintain the corrosion resistance. Information on the fatigue characteristics of FSSW is also lacking. The goals of this project were to determine if FSSW can maintain the corrosion resistance of the sealant and surface treatments and to determine the fatigue characteristics of swept FSSW.

Approach

This experiment dealt exclusively with swept friction stir spot welding through 1 mm (0.040 inch) thick aluminum alloy 2024-T3 sheets. The material was untreated (bare), aluminum clad, Alodine conversion coated, or chromic acid anodized (CAA). The Alodine and CAA treatments were done to bare sheets. The corrosion testing evaluated two sealants. The first sealant was the PRC-DeSoto PR-1432 GP. The second sealant was the Pelseal PLV 6032. The fatigue portion of this experiment used only the PR-1432 GP sealant.

Figure 3. Psi™ Pin Tool

The welds were made using a fixed pin tool, called a Psi™ tool. It is shown in Figure 3. The Psi™ tool has a concave shoulder, and the pin has three flats and two vertical flutes. All welds for this experiment were made at 1500 RPM and 5.93 mm/s (14.0 IPM). The Octaspot™ travel diameter was 4.06 mm (0.160 inch). The advancing side was placed on the outside of the Octaspot™ as it has been show to give slightly higher strengths in that configuration [9]. The forge load was 6.01 kN (1350 lbf), and the tilt angle was 0.5 degrees. These parameters were selected based on previous experiments [7].

The PR-1432 GP sealant was applied by roller to a thickness between 0.06 mm and 0.09 mm (0.0025 inch to 0.0035 inch). The thickness was measure by use of a wet film thickness gauge. Welding took place immediately after the sealant application. The PLV 6032 was brushed on to an undetermined thickness. The sheets were laid together with a small weight on top which pushed out excessive sealant. The sealant was allowed to partially cure for 24 hour prior to welding. The surface treatments and sealants were oriented, as shown in Figure 4, with the same surface treatment on both sheets and the sealant at the faying surface.

Figure 4. Sealant and Surface Treatment Placement

The corrosion coupons were made per the NASM 1312-4 specification as shown in Figure 5. The coupons were place in an alternate immersion tank for 240 hours per ASTM G44. The samples were immersed in a 3.5% NaCl solution for 10 minutes and then left to dry for 50 minutes every hour. All surface treatments were tested with the PR-1432 GP sealant. Coupons were also tested with bare sheets and PLV 6032 as well as bare sheets without sealant. Five of each sealant/surface treatment combination were corroded. Four were pulled for residual strength. The remaining coupon was cross sectioned for macroscopic inspection.

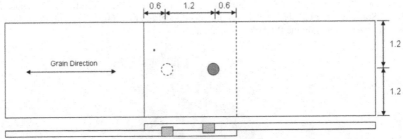

Figure 5. NASM 1312-4 Coupon

Fatigue testing was done per the NASM 1312-21 specification. The coupons were 100% load transfer coupons from that specification, as shown in Figure 6. The coupons were guided with the restraint fixture shown in Figure 7. Each surface treatment was tested with PR-1432 GP. PLV 6032 was not used during the fatigue experimentation. Bare sheets were also tested without sealant. Three of each sealant/surface treatment combination were tested at five load levels. All coupons were fatigued at a load ratio of 0.1. Coupons were tested at a frequency of 40 Hz. If the coupons survived to 5 million cycles, the frequency was increased to 80 Hz. If coupons reached 11 million cycles without failure, their testing was stopped.

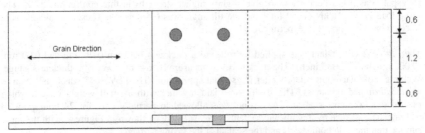

Figure 6. NASM 1312-21 100% Load Transfer Coupon

Figure 7. Fatigue Restraint Fixture

276

Corrosion Results

Both sealants tested were effective against preventing crevice corrosion. The bare material with no sealant, shown on the left in Figure 8, has apparent signs of solution ingress and crevice corrosion. The bare material with PLV 6032 and the bare material with PR-1432 GP are shown in the middle and on the right of Figure 8, respectively. The PLV 6032 occasionally allowed small amounts of solution to penetrate to the faying surface and light crevice corrosion. The PR-1432 GP allowed no ingress or crevice corrosion.

Figure 8. Faying Surfaces Post Corrosion

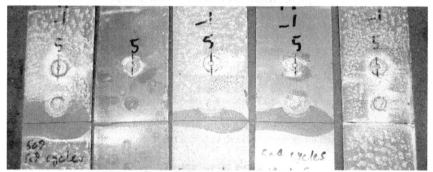

Figure 9. Coupons Post Corrosion

The surface treated coupons faired better than the untreated bare coupons in the corrosive environment. Figure 9 shows coupons after corrosion. They are, from left to right, bare with PR-1432 GP, clad with PR-1432 GP, CAA with PR-1432 GP, Alodine with PR-1432 GP, and bare with PLV 6032. The Alodine surface faired best. It was followed by clad, which tarnished, CAA, and then the bare surfaces. There was attack at the spot where the surface treatments had been displaced by the tool. There was also attack on the anvil side of the spot. It appeared to occur in the HAZ. The top and bottom surface attacks are shown for the CAA material in Figure 10.

Figure 10. CAA Spot Surface Post Corrosion

The cross sections in Figures 11 through 16 show very little sign of attack. The bare coupons show indications that some surface attack occurred. However, the other cross sections do not show evidence of attack.

Figure 11. Bare without Sealant Cross Section

Figure 12. Bare with PR-1432 GP Cross Section

The weld parameters were not optimized for welding through the partially cure PLV 6032. This resulted in welds which were not fully consolidated. The cross section of the bare material and PLV 6032 in Figure 12 shows a wormhole in the bottom right of the nugget.

The PLV 6032 also had negligible adhesive properties. It averaged only 261 N with a standard deviation of 94 N in coupons joined with sealant only. The PR-1432 had an average strength of 5,962 N with a standard deviation of 1,017 N in sealant only coupons.

The wormhole and the low adhesive strength of the PLV 6032 resulted in the bare coupons with PLV 6032 having overall lower strengths both before and after corrosion.

Figure 13. Bare with PLV 6032 Cross Section

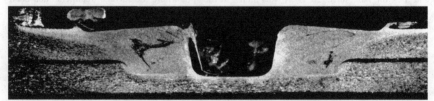

Figure 14. Clad with PR-1432 GP Cross Section

278

Figure 15. CAA with PR-1432 GP Cross Section

Figure 16. Alodine with PR-1432 GP Cross Section

The tensile results for the pristine pre-corrosion coupons and post-corrosion samples are shown in Figure 17. Only the bare coupons with PR-1432 GP and without sealant had any significant loss in strength, demonstrating that the sealants were effective in preventing crevice corrosion in these tests The clad and Alodine coupons actually averaged slightly higher strengths post corrosion; however, this observation could be attributable to the low number of coupons tested. It was also noted that the standard deviation of each data set was higher for the post corrosion coupons than the pristine coupons.

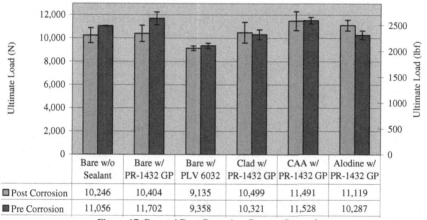

	Bare w/o Sealant	Bare w/ PR-1432 GP	Bare w/ PLV 6032	Clad w/ PR-1432 GP	CAA w/ PR-1432 GP	Alodine w/ PR-1432 GP
Post Corrosion	10,246	10,404	9,135	10,499	11,491	11,119
Pre Corrosion	11,056	11,702	9,358	10,321	11,528	10,287

Figure 17. Pre and Post Corrosion Coupon Strengths

Fatigue Results

The riveted coupons in this experiment were NAS 1097 AD 4 flush countersunk rivets. The coupons with rivets had the same spacing and dimensions as the FSSW coupons. The riveted data only covers mid to lower stress levels. The coupons were made without sealant. The fatigue results are shown in Figure 18 and 19. Table I gives the R^2 values for the trendlines in Figure 19 showing that they are representative of the data.

279

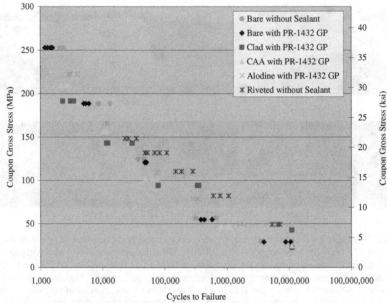

Figure 18. Fatigue Results

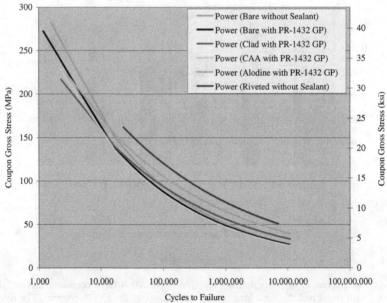

Figure 19. Trendlines of Fatigue Results

Table I. Coefficient of Determination Values for Trendlines of Fatigue Results

	Bare w/o Sealant	Bare w/ PR-1432 GP	Bare w/ PLV 6032	Clad w/ PR-1432 GP	CAA w/ PR-1432 GP	Alodine w/ PR-1432 GP
R^2	0.983	0.988	0.911	0.996	0.964	0.972

The FSSW coupons tended to be just slightly below the fatigue properties of the riveted coupons, as shown in Figures 18 and 19. The data also suggests that sealants and surface treatments do not have a detrimental affect on the fatigue performance of swept FSSW coupons for these conditions, since the grouping is fairly close for all of the FSSW coupons.

The process parameters for the FSSW coupons were optimized for tensile strength in CAA material through the PR-1432 GP sealant and not specifically for fatigue. Parameters were chosen for fatigue testing based on the best cross sectional appearance from the six strongest parameters from a previous tool and parameters selection study [7].

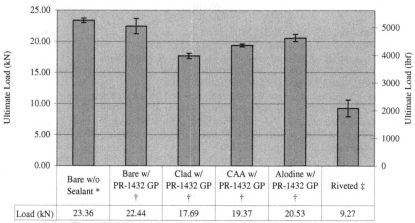

| Load (kN) | 23.36 | 22.44 | 17.69 | 19.37 | 20.53 | 9.27 |

Figure 20. Ultimate Load of Guided NASM 1312-21 Coupons
*Six Data Points †Five Data Point ‡Two Data Points

Figure 20 shows the results of ultimate tensile strength testing. The coupons were NASM 1312-21 coupons tested in guided lap shear. The spacing for the rivets and the FSSW were the same. The rivets were the NAS 1097 AD4 countersunk rivets. The results show that FSSW is much stronger than rivets. More work is needed to understand the fatigue performance of FSSW as compared to rivets. Additional testing is underway at Wichita State to compare FSSW to rivets in fatigue and to understand fatigue crack initiation in FSSW.

Conclusions

Both sealants were successful in preventing ingress of solution to the faying surface in this testing. The PLV 6032 only had slight penetration of the solution. The PR-1432 GP completely prevented solution from entering the faying surface.

The surface treatments also provided protection against corrosion. The Alodine and clad faired much better than the bare surfaces. The CAA surfaces also did better than bare surfaces at

281

preventing a corrosion attack. The exposed top surface area of the FSSW, where the tool displaced the surface treatments, would likely need to be treated after welding since this untreated area was the most vulnerable to corrosion attack. Attack was also noticeable on the anvil side of the coupons in the HAZ beneath the stir zone, and would also need an additional protection scheme applied after welding, such as primer and paint, etc.

The coupons with surface treatments were effective in limiting the decrease in strength from corrosion testing. The CAA had only a small decrease in average strength while the clad and Alodine showed higher strengths in the corroded coupons. There were not enough points, however for this difference to be statistically relevant.

In this experiment, the results indicate that the sealants and surface treatments have little influence on the fatigue characteristics of FSSW. Also, while the FSSW fatigue results fell slightly below the riveted coupons, additional work is already underway at Wichita State University to better understand the fatigue performance of these joints.

References

1. T. Iwashita, U.S. Patent 6,601,751
2. C. Schilling and J. dos Santos, U.S. Patent 6,722,556 B2
3. R. Sakano et al., "Development of Spot FSW Robot System for Automobile Body Members," *Proceedings of the Third International Symposium of Friction Stir Welding* (Kobe, Japan), TWI, Sept 27-28, 2001
4. Pan, T., "Friction Stir Spot Welding (FSSW) – A Literature Review," *Proceedings of the 2007 SAE World Congress*, (Detroit, MI, USA), April 16-19, 2007
5. A.C. Addison and A.J. Robelou, "Friction Stir Spot Welding: Principle Parameters and Their Effects," Proceedings of the 5th International Symposium on Friction Stir Welding (Metz, France), TWI, Sept 14-16, 2004
6. B. Tweedy, C. Widener, and D. Burford, "The Effect of Surface Treatments on the Faying Surface of Friction Stir Spot Welds," *Friction Stir Welding and Processing. TMS 2007 Annual Meeting*, (Orlando, FL, USA), Feb 25 – March 1, 2007.
7. J. Brown et al, "Evaluation of Swept Friction Stir Spot Welding through Sealants and Surface Treatments," *8th International Conference on Trends in Welding Research*, (Pine Mountain, GA, USA), ASM International, June 1-6, 2008.
8. T. Li et al. "Friction Stir Lap Joining of Al 7075 with Sealant," *Friction Stir Welding and Processing. TMS 2007 Annual Meeting*, (Orlando, FL, USA), Feb 25 – March 1, 2007.
9. B. Tweedy et al, "Factors Affecting the Properties of Swept Friction Stir Spot Welds," *Proceedings of the 2007 SAE World Congress*, (Detroit, MI, USA), April 14-17, 2008

FRICTION STIR WELDING AND PROCESSING V

Session Session VI

Session Chair

Anthony P. Reynolds

Friction Stir Welding and Processing V
Edited by: Rajiv S. Mishra, Murray W. Mahoney, and Thomas J. Lienert
TMS (The Minerals, Metals & Materials Society), 2009

MONOTONIC AND CYCLIC DEFORMATION BEHAVIOR OF FRICTION STIR WELDED (FSW) MG/MG- AND AL/MG-JOINTS

Otmar Klag, Guntram Wagner, Dietmar Eifler

Institute of Materials Science and Engineering;
University of Kaiserslautern; P.O. Box 3049, 67653 Kaiserslautern, Germany

Keywords: Friction Stir Welding (FSW), AZ91, AA5454

Abstract

In the automotive and aircraft industry as well as for railway vehicles weight reduction is a predominant aspect for innovative products. But for the integration of light weight metals in complex mechanical components appropriate joining technologies are necessary. By using friction stir welding (FSW) it is possible to join light weight metals in solid state [1, 2]. However there are still many gaps in the knowledge about the resulting microstructure and the monotonic and the cyclic deformation behavior of FSW joints.
At the Institute of Materials Science and Engineering (WKK) at the University of Kaiserslautern Mg/Mg- and Mg/Al-joints were realized by FSW. Besides others AZ91/AZ91- and AZ91/AlMg3Mn-joints were investigated concerning their monotonic and cyclic deformation behavior. Light microscopy was used to analyze the microstructure in the welding zone. Furthermore by non-destructive testing methods it could be impressively demonstrated that for die cast magnesium alloys the friction stir welding process leads to a significant reduction of the pore portion in the welding zone.

Welding Equipment

The WKK is using a conventional milling machine from the type Deckel Maho DMU80T to realize friction stir welds (Figure 1).

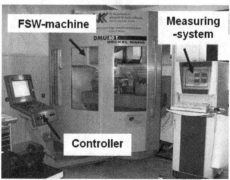

Figure 1. FSW-Equipment at the Institute of Materials Science and Engineering at the University of Kaiserslautern, Germany

The milling machine is proved to perform friction stir welds of light weight material sheets up to a thickness of 10 mm. The maximum dimension of the sheets is 800 x 600 mm^2. The integrated measurement equipment enables to record temperatures in the welding zone by thermocouples and welding forces in x-, y- and z-direction online during the FSW process. By a force control system the welding force through the joining process is stabilized. To perform friction stir welds in a repeatable and constant quality a special clamping unit was developed at the WKK (Figure 2).

Figure 2. Clamping on the FSW-system developed at the WKK

The welding experiments were carried out by using a standard FSW-welding tool. The welding tool was made of hot work tool steel with high temperature strength and high wear resistance. With the aim to optimize the welding process the temperature in the welding zone was measured by minimum three thermocouples equispaced implemented along the seam [3]. Additionally, the surface temperature was recorded with an infrared camera above the machine spindle (Figure 3).

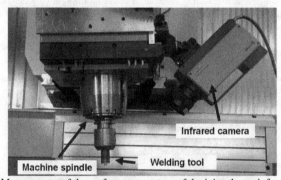

Figure 3. Measurement of the surface temperature of the joints by an infrared camera

In Figure 3 the characteristic courses of the welding force and the temperature at three measuring points in the seam for a high quality AZ91/AA5454-FSW-joint are shown. It clearly can be seen that the integrated force control system of the FSW-machine ensures a constant welding force, in this case of 18 kN, during the whole welding process. The temperatures in the butt weld increase to a maximum of nearly 400°C for each measuring point and are nearly identical over the whole seam weld. The surface temperature rapidly increases up to 280°C during the first 10 mm of the seam. After this, the process is stable.

286

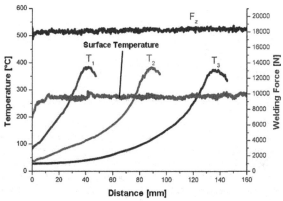

Figure 4. Welding force and welding temperatures of an AZ91/AA5454-FSW-joint

Materials

The magnesium die cast alloy AZ91 was used for the investigations. Furthermore dissimilar joints out of AZ91 and AA5454 were realized. The chemical composition of these alloys is given in Table I.

Table I. Chemical composition of the investigated light weight materials

(wt-%)	Al	Mg	Mn	Fe	Si	Zn	Cr	Cu
AA5454	bal.	2.85	0.84	0.31	0.13	0.01	0.08	0.03
AZ91	8.75	bal.	0.28	<0.01	0.04	0.65	-	<0.01

The microstructure of AA5454 consists of an intermetallic Al-Mg-phase (β-phase) embedded in a solid solution aluminum α-phase (Figure 5a)). Both phases are elongated in longitudinal direction due to the rolling process. Figure 5b) gives an overview over the typical cast microstructure of AZ91. The high aluminum content of nearly 8% leads to a high strength β-phase at the grain boundaries of the more ductile α-phase with only a low fraction of Al [4].

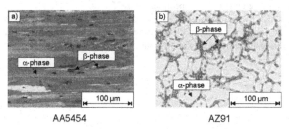

AA5454   AZ91

Figure 5. Micrographs of the investigated alloys AA5454 a) and AZ91 (IfW, TU Freiberg) b)

Some mechanical properties of the investigated alloys are shown in Table II. The aluminum alloy was investigated in condition H22 that means rolled and annealed to a hardness of 90 HV10. Because of its improved ductility in this condition, AA5454 is often used for cupping components in automotive industry. Due to his hexagonal lattice the ductility of magnesium is significantly lower then of the Al-alloy.

Table II. Mechanical properties of the investigated materials

	$R_{p0.2}$ [MPa]	R_m [MPa]
AA5454 (H22)	220	300
AZ91	120	160

Specimens

The metal sheets with the dimensions of $125 \cong 300 \cong 7.5$ mm³ were butt welded by using standard tool geometry with a shoulder diameter of approximately 25 mm and a M8 pin. In systematic investigations the welding process was optimized with regard to attain a high tensile strengths of the FSW joints. The tests show that a tool rotation in a range between 250 and 1200 rpm and a welding velocity between 10 and 100 mm/min are best suited to produce pore-free and high strength Mg/Mg and Mg/Al-welds. After welding, the sheets were cut using an abrasive water jet to avoid any thermal influence of the preparation procedure to the microstructure of the specimens and then polished for the monotonic and cyclic investigations according to the details in Figure 6.

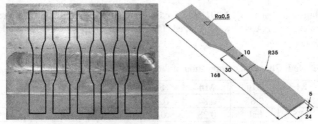

Figure 6. Extraction position and geometry of the specimens

Experimental Setup

To receive information exclusively from the welding zone, the gauge length for the monotonic and cyclic investigations was chosen correspondingly to the diameter of the shoulder of the welding tools. The gauge length of the specimens was 25 mm.

In order to characterize the fatigue behavior of the friction stir welded joints actually axial stress-controlled load increase tests for Mg/Mg-joints were performed with a load ratio of $R \approx 0$ at ambient temperature using a frequency of 5 Hz and triangular waveforms [5-7].

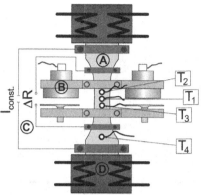

Figure 7. Experimental setup for fatigue tests

The experimental setup for the fatigue tests is shown in Figure 7. The polished specimen (A) was fixed in thermostatic controlled clampings (D) of the servo hydraulic testing equipment. To record the stress-strain-hysteresis two parallel plate condensers (B) were used. Four thermocouples applied on the specimen surface allow highly accurate temperature measurements. The local temperatures T_1, T_2 and T_3 were measured in the gauge length, T_1 at the weld nugget in the middle of the joint, T_2 at the retreating side and T_3 at the advancing side of the joint. The cross-sectional area at the measuring point T_4 is significantly larger in comparison to the gauge length, where T_1, T_2 and T_3 are measured. Consequently the deformation in this position of the specimen is completely elastic and the temperature T_4 can be used as a reference value to evaluate the temperature increase due to cyclic plastic deformation at the positions T_1 to T_3. Using the following equitation (*Eq. 1*) thermal influences from the environment during the fatigue test on the measured values can be considered:

$$\Delta T_n = (T_n - T_4) \ with \ n = 1,2,3 \tag{1}$$

The change in specimen temperature is caused by energy dissipation as a consequence of plastic deformation in the gauge length.

For the electrical resistance measurements a DC power supply was fixed at both specimen shafts (C). The change in the electrical resistance ΔR was measured between the gauge length and the shafts. The measured electrical resistance depends on the geometric values length L and cross section area A as well as on the resistivity ρ^* (*Eq. 2*). This value is strictly related to the defect density of the material like dislocazion density and structure as well as pores and microcracks [5].

$$\Delta R = \rho^* \cdot L / A \tag{2}$$

Results

Figure 8 shows a comparison of the tensile strength of the parent material AZ91 and a typical friction stir welded AZ91/AZ91-joint. The strength of the joint is nearly on the same level as the parent material. This demonstrates impressively the efficiency of the friction stir welding process, which really does not cause a considerable decrease of the strength compared to the parent material.

289

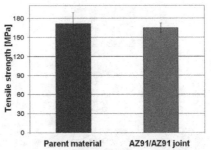

Figure8. Comparison of the tensile shear strengths

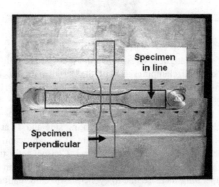

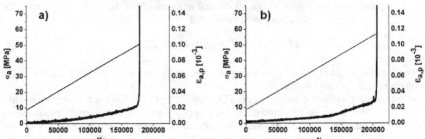

Figure 9. Continuous load increase test of an AZ91/AZ91-FSW-joint
a) specimen orientation perpendicular
b) specimen orientation in line

Continuous load increase tests (LIT) were performed to estimate the influence of the specimen orientation on the numbers of cycles to failure. For the AZ91/AZ91-FSW-joints the LIT were started at 8 MPa. The amplitude was continuously increased with a rate of $d\sigma_a/dt = 2.5 \cong° 10^{-3}$ MPa/s until specimen failure. The LIT of AZ91/AZ91-FSW-perpendicular specimen orientation yields to a number of cycles to failure of $175 \cong° 10^5$ (Figure 9a). The failure occurs in this case in the parent material outside the joining area. A LIT of sampling in line of the welding with only the microstructure caused by the FSW-process leads to $22 \cong° 10^5$ cycles to failure (Figure 9b). This result clearly shows that the friction stir welding process improves the cyclic properties of the AZ91. This behavior can be traced back one the fine grained microstructure

290

which was observed in the weld nugget (Figure 10). In comparison to the parent material the grain size is reduced about 85% from 424 μm to 63 μm in average.

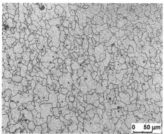

Figure 10. Fine-grained microstructure of the weld nugget of an AZ91/AZ91-FSW-joint (IfW, TU Freiberg)

Further more the significant reduction of the pore fraction caused by the FSW-process improves the fatigue behavior of the samples. This reduction can be clearly verified with a non-destructive ultrasound investigation of friction stir welded die cast sheets of Mg-alloys with a high pore density (Figure 11). In the weld area a clear reduction of the pores can be detected in comparison to the parent material. The only visible defect within the welding zone is the end hole of the weld. Hence it is convincingly demonstrated that the FSW process is appropriate to reduce defects in die cast Mg-alloys.

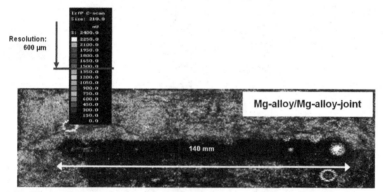

Figure 11. Ultrasonic inspection of an Mg-alloy-FSW - die cast-joint (IZFP, Saarbrücken)

With appropriate welding parameters and welding setup it is possible to produce hybrid joints between AZ91 and rolled AA5454 wrought aluminum alloy. **Figure 12** shows a light micrograph of a joint with the magnesium on the left and the aluminum on the right side. The micro-hardness mapping beneath demonstrates the challenges associated in joining this material combination. The map shows on the left side the typical hardness pattern of a cast material with pronounced hardness differences. On the right side a very homogeneous hardness distribution typical for AA5454 appears. Between the two sections a nugget area with considerable higher hardness values up to over 2000 HM has developed as a result of intermetallic phases (Figure 12). To avoid this effect the temperature distribution in the joining area has to be optimized in the next investigations.

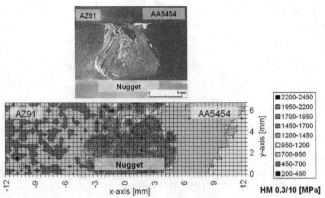

Figure 12. Micrograph and micro-hardness profile of an AZ91/AA5454 FSW joint

The following diagram (Figure 13) shows the influence of the pin position in relation to the butt of the weld and the achievable tensile strengths. It can clearly be seen that the pin directly positioned in the but leads to the best results with tensile strengths of about 60 MPa.

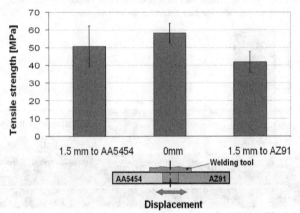

Figure 13. Influence of the pin position on the tensile strength of a AZ91/AZ91-joint

292

Conclusions

The tensile strength of friction stir welded AZ91/AZ91-joints achieves about 95% of the tensile strength of the parent material. LIT of AZ91-joints in line to the welding seam show that the FSW-process leads to a lower failure rate and hence to a higher number of cycles to failure. This can be traced back to a lower pore density in the welding area and a fine-grained microstructure in the welding zone in comparison to the parent material. This effect of a significant pore reduction can also be clearly demonstrated by using non-destructive ultrasound test methods. Furthermore it was shown, that the FSW process is an appropriate method to produce reliable Mg/Al-hybrid-joints.

Acknowledgement

The investigations are supported by the DFG Priority Program 1168: "Erweiterung der Einsatzgrenzen von Magnesiumlegierungen".
Special thanks to the IZFP, Saarbrücken, Germany and the IfW, TU Freiberg, Germany for the support of our investigations.

References

[1]   S. W. Kallee, J. M. Kell, W.M. Thomas, C. S. Wiesner, *Development and Implementation of Innovative Joining Processes in the Automotive Industry*. DVS Annual Welding Conference, Essen, (2005)

[2]   M. R. Johnsen, *Friction Stir Welding Takes Off at Boeing*. *Welding Journal*. 78 (1999) 35-39

[3]   G. Wagner, M. Gutensohn, D. Eifler, *Monotonic Properties and Cyclic Deformation Behavior of Friction Stir Welded Mg/Mg-, Mg/Al- and Al/Al-Joints*. Magnesium, Proceedings of the 7th International Conference on Magnesium Alloys and Their Applications, (2006) 1092-1098

[4]   H. Westengen, T. K. Aune, *Magnesium Casting Alloys*. Magnesium Technology, Springer, (2006)

[5]   F. Walther, D. Eifler, *Cyclic deformation behavior of steels and light-metal alloys*. Mat. Sci. Eng. A, 468-470 (2007) 259-266

[6]   P. Starke, F. Walther, D. Eifler: *Hysteresis, temperature and resistance measurements for the characterization of the fatigue behavior of metals*. ISCS 2007, 5th Int. Conf. Structural Integrity of Welded Structures - Testing & Risk Assessment in the Development of Advanced Materials and Joints, Timisora, Romania, www.ndt.net, Issue 2008-04 (2008) 1-9

[7]   G. Wagner, M. Gutensohn, O. Klag, D. Eifler: *Microstructure and Deformation Behavior of Friction Stir Welded Light Metals*. 7th International Symposium on Friction Stir Welding, TWI Ltd (2008), CD-Edition, ISBN 13-978-1-903761-06-9

Friction Stir Welding and Processing V
Edited by: Rajiv S. Mishra, Murray W. Mahoney, and Thomas J. Lienert
TMS (The Minerals, Metals & Materials Society), 2009

DISSIMILAR FRICTION STIR WELDING OF AA2024-T3 AND AA7075-T6 ALUMINIUM ALLOYS

E. Aldanondo, A.A.M da Silva, P.Alvarez, A. Lizarralde, A. Echeverria

[1]Asociación Centro de Investigación en Tecnologías de Unión LORTEK;
Barrio La Granja, s/n; Ordizia, Guipúzcoa 20240, Spain

Keywords: Friction Stir Welding, Aluminium, Microstructure

Abstract

The scope of this investigation is to evaluate the effect of joining parameters on the mechanical and microstructural properties of dissimilar aluminium alloys (3 mm thick AA2024-T3 and AA7075-T6 sheets) joints produced by friction stir welding. The influence of the location of the base materials under different conditions during welding on the material mixing pattern has also been investigated. Microstructural features have been analysed; while mechanical performance has been investigated in terms of hardness and tensile testing. In both welding configurations no onion ring formation has been observed; the boundary between both base materials at the stir zone is clearly delineated, i.e., no material mixing is observed. Microstructural observation has revealed the development of a recrystallised fine-grained stir zone, with two different grain sizes resulting from the two different base materials. Failure of tensile specimens occurs at the HAZ of the retreating side (AA2024-T3) in one condition; while in the other condition failure has occurred at the stir zone.

Introduction

Friction Stir Welding (FSW) unlike fusion welding is a solid-state joining process whereby no bulk melting of the base material occurs during joining. In comparison to fusion welds there is no evidence of an as cast structure in the join zone. Other advantages of FSW are less distortion, no need of cleaning operations of the surface oxide prior to the welding process as well as no presence of defects arising from melting and solidification during conventional fusion welding processes. The process was developed and then patented in 1991 by The Welding Institute (TWI, Cambridge) [1]. The workpieces are joined by means of frictional heating and plastic deformation typically at temperatures below that of the absolute melting temperature of the alloys being joined. This is brought about by the interaction of a non consumable and rotating tool with the interfacing surfaces that is plunged into and then traversed along the interface between the workpieces [2]. The actual mechanism of weld formation is most nearly described as a combination of *in-situ* extrusion combined with forging [3].

High strength aluminium alloys (2xxx- and 7xxx-series alloys) that are usually employed in the aeronautic industry are normally difficult to fusion weld since dendritic structure is formed in the fusion zone when TIG and laser welding are used leading to a drastic reduction of the mechanical behaviour (static and dynamic properties) [4]. As a solid-state process, FSW is capable of producing high-quality defect free welds (no solidification defects and/or brittle intermetallic phases) when optimised parameters are used [2-11].

The ability to join dissimilar alloys and/or materials can have enormous potential for applications particularly in the aeronautic industry. Successful application of FSW to the alloys investigated in this work would allow for new structural and design possibilities as well as to contribute to weight reduction thereby increasing the efficiency and reducing the cost of production. According to Zettler et al [2], FSW of dissimilar alloy combinations raises complex issues such as the heat produced by the process which will more readily flow in the material with the larger thermal conductivity. As a consequence this can lead to incomplete bonding through too rapid a heat loss from the higher conductivity alloy, or may cause excessive heating of the material possessing the lower thermal conductivity.

Another important feature is the degree of material mixing as well as the microstructure distribution which can have a great effect on the final mechanical behaviour of the FSW joint. According to Sato et al [7] investigating the microstructural features of dissimilar FSW between AA2024 and AA7075 aluminium alloys, the heat input plays an important role in material mixing, i.e., high heat inputs have lead to a stir zone with onion ring patterns consisting of AA2024 and AA7075 bands, including a gradual change in hardness and chemical composition in regions between these bands; while low heat inputs have produced a stir zone clearly divided into AA2024 and AA7075 regions by a zigzag feature.

The objective of the present study is to investigate the microstructural features and the mechanical properties in terms of hardness and tensile behaviour of dissimilar FSW joints of AA2024-T3 and AA7075-T6 high strength aluminium alloys. Material mixing is also going to be investigated by analysing microstructure in two different directions (top and transversal) according to the fixed location of both materials (AA7075 in the advancing side and AA2024 in the retreating side and vice versa).

Experimental Procedure

Materials

Bare aluminium alloys AA2024-T3 and 7075-T6 of 3 mm thickness have been used to produce dissimilar FSW butt joints. The nominal composition in weight percent (major alloying additions) of both alloys is presented in Table I. Table II presents the tensile properties (tensile strength, R_m; yield strength, $R_{p0.2}$ and elongation) of both alloys. The AA2024-T3 aluminium alloy microstructure consists of elongated grains (in the rolling direction) with a random distribution of constituent precipitates (small black particles); while the AA7075-T6 aluminium alloy microstructure has elongated grains with constituent precipitates (black particles) randomly distributed across the aluminium grains. Figure 1 presents the 3D schematic representation of the microstructure of base materials.

Table I. Nominal chemical composition of AA2024-T3 and AA7075-T6 aluminium alloys used in this investigation.

	Cu	Mg	Cr	Zn	Mn	Si	Fe	Al
AA2024-T3	4.46	1.42	0.001	0.048	0.631	0.042	0.112	balance
AA7075-T6	1.70	2.43	0.20	5.64	0.08	0.04	0.11	balance

Table II. Tensile properties of AA2024-T3 and AA7075-T6 aluminium alloys used in this investigation (LT direction).

	R_m (MPa)	$R_{p0.2}$ (MPa)	Elongation (%)
AA2024-T3	458	305	18
AA7075-T6	565	491	13

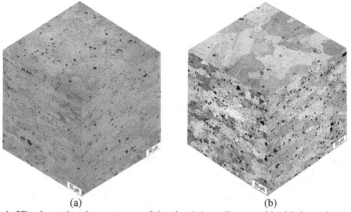

| (a) | (b) |

Figure 1. 3D schematic microstructure of the aluminium alloys used in this investigation. (a) AA7075-T6. (b) AA2024-T3.

Welding Procedure

FSW joints have been performed at LORTEK (Ordizia, Spain) using a MTS ISTIR PDS FSW machine capable of applying axial forces up to 10 t and maximum rotational speed of 3000 rpm. Welds have been produced in force control with a tilt angle of 3°. The Advanced MTS application software and the TestStarTM controller enable continuous data acquisition and management of the key FSW process parameters. The FSW tool system comprised a threaded pin (4 mm diameter and 2.85 mm length) and flat shoulder (12 mm diameter). The tool has been machined from M42 high speed steel (1.08 wt pct C, 3.85 wt pct Cr, 8 wt pct Co, 9.5 wt pct Mo, 1.5 wt pct W and 1.2 wt pct V), heat treated to a hardness of 52-56 HR$_c$.

The rolling direction of the base materials sheets was in parallel with the joining direction. Two different weld configurations have been used placing both aluminium alloys at either the advancing or the retreating sides (AS and RS, respectively). When the AA2024-T3 was placed in the AS the weld has been designated as 2024T3/7075T6; while the opposite configuration (AA7075-T6 placed in the AS) the weld has been expressed as 7075T6/2024T3. Tool rotational speed and travel speed have been held constant at 1000 rpm and 254 mm/min, respectively; while force has been varied according to the configuration of the weld, see Table III.

Table III. Dissimilar FSW joining parameters of AA2024-T3 and AA7075-T6 aluminium alloys used in this investigation.

	Rotational speed (min^{-1})	Travel speed (mm/min)	Force (kN)
2024T3/7075T6	1000	254	16.5
7075T6/2024T3	1000	254	12

Microstructural Characterisation and Mechanical Tests

An Olympus GX51 light optical microscope (OM) has been used to investigate the microstructural features and to assess the quality of the FSW joints. The samples for evaluating the microstructure of the FSW welds have been taken from two different directions: transversal (cross-section) and top. The specimens for metallographic examination have been carefully prepared according to standard metallographic procedures and etched using Keller´s solution (2

297

ml HF, 5 ml HNO$_3$, 3 ml HCl and 190 ml distilled water) for microscopic examination. In all micrographs the advancing side is always placed in the left hand side of the picture.

Vickers microhardness profiles have been performed in a Struers Duramin computer controlled hardness testing machine using 1.96 N load (200 gf) and 0.5 mm spacing between the indentations. Three different sets of microhardness measurements have been performed at each FSW condition: top (1.0 mm above mid-thickness line), mid-thickness and bottom (1.0 mm below mid-thickness line). The total length investigated for every single row was approximately 25 mm, thus enabling a comparison between microhardness values of the base material, heat affected zone, thermo-mechanically affected zone and stir zone regions. Microhardness measurement has been performed after 50 days of natural aging.

Static tensile tests have been carried out on 3 mm-thick FSW joints according to the EN 895 Standard [12] (gauge length Lo, 50 mm; and width, 25 mm). The specimens have been extracted transversely to the welding direction, which is coincident with the rolling direction of the plates. The equipment utilized for performing the tensile tests was an electro-hydraulic testing machine Hoytom TN-MD powered by a Hoywin controller (300 kN load capacity) with a displacement rate of 1.6 mm/min. Tensile testing has been performed after 50 days of natural aging.

Results and Discussion

Microstructural Characterisation

Microstructural characterisation of both FSW conditions in transversal (cross-section) and top (top-section) directions is presented in Figures 2 to 6. In the cross-sections shown in Figures 2 and 3 one can observe that all joints have no defects (lack of penetration, porosity, etc.). Almost negligible effect of the alloy location either in the retreating or advancing sides has been observed. The stir zones in both cases are very similar in size as well as in the mixture pattern of both alloys. Typical microstructural features of FSW welds such as the stir zone (SZ), thermo-mechanically affected zone (TMAZ) and heat affected zone (HAZ) have been identified. The HAZ is characterised by being influenced only by the thermal history. The HAZ microstructure on both configurations and materials is very similar to that of the base material, i.e., no grain growth has occurred. The TMAZ region experiences both heating and plastic deformation during FSW process being characterised by a deformed structure induced by the mechanical stirring of the tool (Figures 2a and 3a). The SZ microstructure of both materials consists of recrystallised and fine homogeneous equiaxed grains (Figure 4). The transition between TMAZ (elongated and deformed grain region) and SZ (fine equiaxed grain region) in the advancing side is abrupt showing a very narrow boundary (Figures 2a and 3a); while a region with progressive and smooth distribution of different grain sizes is visible in the retreating side (Figures 2c and 3c).

The SZ is formed by two regions consisting of recrystallised fine grains of both 2024-T3 and 7075-T6 alloys. The grain size is approximately 3-5 μm in the 7075-T6 region and 5-7 μm in the 2024-T3 region as shown in Figures 4a and 4b, respectively. The boundary separating the two regions is clearly visible in Figures 2b and 3b. Such boundary is very narrow suggesting that there would not be any significant diffusion phenomena between both alloys. No onion ring formation has been observed in any of the analysed micrographs suggesting a lack of intermixing between both alloys that would result in a banded microstructure. However, some evidence of intermixing initiation can be seen in the bottom of the SZ as some small regions of the alloy located in the advancing side are pushed into the retreating side. That is not the case for the top part of the SZ where a narrow and continuous boundary separates both alloys.

298

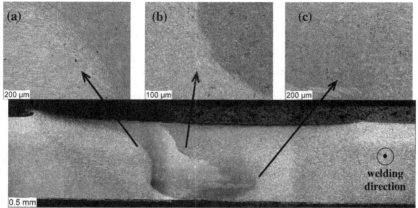

Figure 2. OM macrograph of the cross-section of condition 7075T6/2024T3. (a) TMAZ-SZ boundary, AS. (b) SZ-SZ interface between 7075-T6 and 2024-T3. (c) TMAZ-SZ boundary, RS and detail view of the interface between 7075-T6 and 2024-T3.

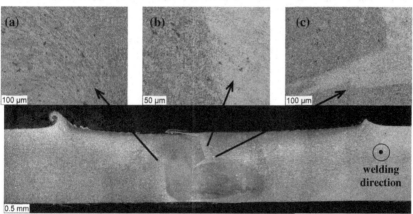

Figure 3. OM macrograph of the cross-section of condition 2024T3/7075T6. (a) TMAZ-SZ boundary, AS. (b) and (c) SZ-SZ interface between 7075-T6 and 2024-T3.

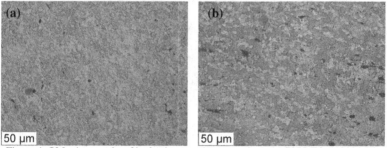

Figure 4. OM micrographs of both stir zones (cross-section). (a) 7075-T6. (b)2024-T3.

The mixture pattern of 2024-T3 and 7075-T6 alloys in dissimilar FSW has been discussed before under different welding parameters. Sato et al [7] distinguished between high and low heat-input welds. In high heat input welds (weld pitch of 0.012 mm/rev) a banded structure showing an onion ring pattern was obtained; while in low heat-input welds (weld pitch of 0.684 mm/rev) 2024-T3 and 7075-T6 alloys were separated by a zigzag boundary (no banded structure was identified). The FSW joints of the present investigation have been carried out with a weld pitch of 0.254 mm/rev which would lie between these "extreme conditions" used by Sato et el; therefore, as expected, the obtained microstructure has not shown a pure banded or zigzag feature but a mixture of these two features.

Top section images of the 7075T6/2024T3 and 2024T3/7075T6 welds are shown in Figures 5 and 6, respectively. Once again no effect of the material location on the obtained microstructure has been observed. The SZ is formed by two regions comprising recrystallised fine grains of 2024-T3 and 7075-T6 alloys. The boundary between these two regions is a continuous and straight line parallel to the welding direction. However this boundary is wider than the boundary observed in the cross sections (Figures 2 and 3) showing a transition region between both alloys (Figures 5b and 6b) that is indicative of further mixing. The SZ is surrounded by the TMAZ in both advancing and retreating sides where the typical highly deformed structure induced by the mechanical stirring of the tool can be observed (Figures 5a and 6c, 7075-T6; Figures 5c and 6a, 2024-T3). The transition between the TMAZ and the SZ is abrupt in the advancing side and more progressive in the retreating side.

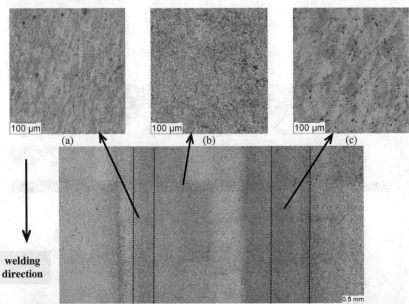

Figure 5. OM macrograph of the top-section of condition 7075T6/2024T3. (a) TMAZ, AS. (b) SZ-SZ interface between 7075-T6 and 2024-T3 (c) TMAZ, RS.

300

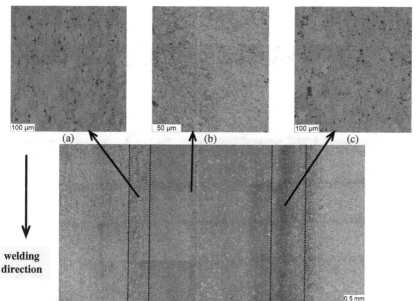

Figure 6. OM macrograph of the top-section of condition 2024T3/7075T6. (a) TMAZ, AS. (b) SZ-SZ interface between 7075-T6 and 2024-T3 (c) TMAZ, RS.

Hardness Testing

Microhardness profiles of FSW cross-sections of both configurations performed at mid-thickness line as well as at 1 mm above and below the mid-thickness line (see Figure 7) are presented in Figure 8. Microhardness of base materials, AA2024-T3 and AA7075-T6, is approximately 137 $HV_{0.2}$ and 172 $HV_{0.2}$ respectively. Minimum hardness values have been found in both configurations (2024T3/7075T6, Figure 8a and 7075T6/2024T3, Figure 8b) at the HAZ regions of the retreating side with values varying from 122 $HV_{0.2}$ to 130 $HV_{0.2}$ depending on the welding configuration. According to several investigations [7, 9, 10], this softening effect is related to the dissolution and/or coarsening of the precipitates at this particular region. Furthermore, there has been found a variation on the hardness distribution according to the location of the material – advancing or retreating side. For instance, when the AA7075-T6 base material was located at the advancing side the minimum hardness value was approximately 130 $HV_{0.2}$; while when located at the retreating side was approximately 122 $HV_{0.2}$. According to Zadpoor et al [9], such behaviour results from different deformations at both sides. The material at the advancing side is more largely influenced by the vortex velocity field since it has different direction on the retreating side but the same direction on the advancing side. Hence, the shear strain is more severe on the advancing side than in the retreating side resulting in slightly higher hardness values.

One can clearly observe that the microhardness distribution is quite heterogeneous, particularly at the stir zone. It should be noted that in the case of mid-thickness and 1 mm below mid-thickness lines the stir zone has approximately the size of the pin diameter; while at 1 mm above the mid-thickness line is relatively larger. This hardness distribution has been attributed to the absence of material mixing between both aluminium alloys in which the interface between both stir zones is clearly defined (Figure 7). Therefore consecutive measurements that have been taken inside the SZ but in different base materials close to the interface can differ substantially.

Figure 7. Schematic representation of the microhardness profiles performed at the FSW cross-sections joints (7075T6/2024T3).

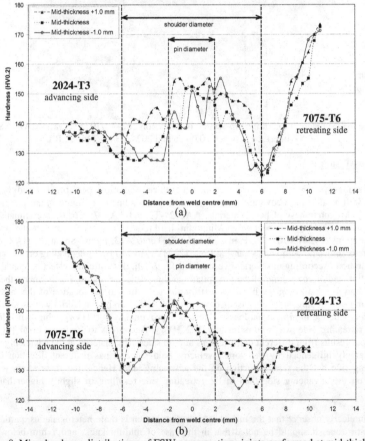

Figure 8. Microhardness distributions of FSW cross-sections joints performed at mid-thickness as well as at 1 mm below and above the mid-thickness line. (a) 2024T3/7075T6. (b) 7075T6/2024T3.

<u>Tensile Testing</u>

The static behaviour of dissimilar FSW joints in both configurations as well as the fracture location of the tensile specimens is summarised in Tables IV and V. Maximum tensile performance has been obtained in condition 7075T6/2024T3 with tensile and yield strength (Rm and $Rp_{0.2}$, respectively) of 443 MPa and 229 MPa as well as elongation of 7.6 %. The weld efficiency in terms of tensile strength in conditions 2024T3/7075T6 and 7075T6/2024T3 has reached 93 % and 96 %, respectively. As stated before, there has been found an absence of material mixing at the stir zone on both conditions; however such feature has not influenced the final mechanical properties of condition 7075T6/2024T3 since the failure has occurred in the HAZ of the retreating side. Nevertheless, in condition 2024T3/7075T6 failure has occurred at the stir zone; therefore, further investigation is needed to fully identify the influence of material mixing as well as the possible presence of small volumetric defects and/or lack of penetration (not observed using OM) on the tensile behaviour of this condition.

Table IV. Tensile results of the FSW cross-sections joints (condition 2024T3/7075T6).

	Rm (MPa)	$Rp_{0.2}$ (MPa)	Elongation (%)	Fracture location
2024T3/7075T6 - 1	419	254	4.7	SZ
2024T3/7075T6 - 2	444	268	7.2	SZ
2024T3/7075T6 – 3	424	276	3.4	SZ
Average ± STD	429 ± 13.2	266 ± 11.1	5.1 ± 1.9	-

Table V. Tensile results of the FSW cross-sections joints (condition 7075T6/2024T3).

	Rm (MPa)	$Rp_{0.2}$ (MPa)	Elongation (%)	Fracture location
7075T6/2024T3 - 1	438	269	7.1	HAZ 2024
7075T6/2024T3 - 2	447	224	8.0	HAZ 2024
7075T6/2024T3 – 3	445	253	7.8	HAZ 2024
Average ± STD	443 ± 4.7	229 ± 22.8	7.6 ± 0.5	-

Conclusions

Mechanical properties and material mixing pattern in dissimilar AA2024-T3 and AA7075-T6 FSW joints have been investigated. Welds have been performed at fixed joining conditions (weld pitch of 0.254 mm/rev) but varying the materials located at the retreating and advancing sides. The following conclusions can be drawn:

1. Comparable microstructural and mechanical results have been obtained irrespectively of material location, although 7075T6/2024T3 configuration yielded slightly better mechanical properties with reduced process force requirements;

2. Typical microstructural features of FSW welds such as SZ, TMAZ and HAZ regions have been identified. A sharp transition from TMAZ to SZ has been observed in the advancing side; while in the retreating side such transition is more gradual. The SZ region is characterized by a clearly unmixed narrow boundary at the top section, whereas some evidences of intermixing have been observed at the bottom. No onion rings or banded structures have been observed.

3. The minimum hardness value of naturally aged samples has been found in the HAZ at the retreating side (about 88 % and 71 % of 2024-T3 and 7075-T6 base materials, respectively).

4. The weld efficiency in terms of tensile strength for 7075T6/2024T3 condition is approximately 96 %; while for 2024T3/7075T6 condition is approximately 93 %. Fracture of 7075T6/2024T3 specimens has occurred in the retreating side HAZ; while in the case of 2024T3/7075T6 samples fracture has taken place in the stir zone (although no evidence of volumetric defects or lack of penetration has been found).

5. Further investigation is needed to fully identify the micromechanisms of fracture as well as the influence of material mixing on the mechanical performance of the dissimilar FSW joints.

References

1. W. M. Thomas et al., International Patent Application No. PCT/GB92/02203 and GB Patent Application No. 9125978.9, 1991

2. R. Zettler et al., "Dissimilar Al to Mg Alloy Friction Stir Welds," *Advanced Engineering Materials*, 8 (5) (2006), 415-421.

3. L. Cederqvist, and A.P. Reynolds, "Factors Affecting the Properties of Friction Stir Welded Aluminum Lap Joints," *Welding Journal*, 80 (2001), 281s-287s.

4. P. Cavaliere et al., "Mechanical and Microstructural Behaviour of 2024-7075 Aluminium Alloy Sheets Joined by Friction Stir Welding," *International Journal of Machine Tools & Manufacture*, 46 (2006), 588-594.

5. Y.J. Chao, Y. Wang, and K.W. Miller, "Effect of Friction Stir Welding on Dynamic Properties of AA2024-T3 and AA7075-T7351," *Welding Journal*, 80 (2001), 196s-200s.

6. P. Cavaliere, E. Cerri, and A. Squillace, "Mechanical Response of 2024-7075 Aluminium Alloys Joined by Friction Stir Welding," *Journal of Materials Science*, 40 (2005), 3669-3676.

7. Y.S. Sato, Y. Kurihara, and H. Kokawa, "Microstructural Characteristics of Dissimilar Butt Friction Stir Welds of AA7075 and AA2024," (Paper presented at the 6th International Symposium on Friction Stir Welding, Toronto, Canada, 10-13 October 2006).

8. S.A. Khodir, and T. Shibayanagi, "Friction Stir Welding of Dissimilar AA2004 and AA7075 Aluminum Alloys," *Materials Science and Engineering B*, 148 (2008), 82-87.

9. A.A. Zadpoor et al., "Mechanical Properties and Microstructure of Friction Stir Welded Tailor-Made Blanks," *Materials Science and Engineering A*, 494 (2008), 281-290.

10. R.S. Mishra, and Z.Y. Ma, "Friction Stir Welding and Processing," *Materials Science and Engineering R*, 50 (2005), 1-78.

11. M.W. Mahoney. "Mechanical Properties of Friction Stir Welded Aluminum Alloys," Friction Stir Welding and Processing, ed. R.S. Mishra and M.W. Mahoney (Materials Park, OH: ASM International, 2007).

12. DIN EN 895. "Destructive Testing of Welds in Metallic Materials – Transverse Tensile test," (Brussels, Belgium: CEN European Committee for Standardization, 1995).

Friction Stir Welding and Processing V
Edited by: Rajiv S. Mishra, Murray W. Mahoney, and Thomas J. Lienert
TMS (The Minerals, Metals & Materials Society), 2009

FATIGUE BEHAVIOR OF FRICTION STIR WELDED (FSW) ALUMINUM JOINTS

Markus Gutensohn[1], Guntram Wagner[1], Masahiro Endo[2], Dietmar Eifler[1],

Institute of Materials Science and Engineering, University of Kaiserslautern,
P.O. Box 3049, 67653 Kaiserslautern, Germany[1],
Department of Mechanical Engineering, Fukuoka University,
8-19-1 Nankuma, Fukuoka, Japan[2],

Keywords: Friction Stir Welding (FSW), Aluminum, Fatigue Behavior

Abstract

The application of friction stir welded parts in monotonic and cyclic loaded components requires a detailed knowledge of their deformation behavior. In the present work the monotonic and cyclic deformation behavior of aluminum AA5454 FSW-joints are discussed. The yield strength of friction stir welded joints reaches up to 90% of the value of the parent material. In fatigue tests high-resolution plastic strain amplitude, temperature and electrical resistance measurements were performed to describe the cyclic deformation behavior of the joints. With FSW-joints welded at different welding forces single step tests were carried out. Higher axial welding forces lead to an increase of the softened volume in the welding zone and consequently the initial plastic deformation during the first cycles increases. The cyclic deformation behavior is characterized by initial cyclic hardening followed by a saturation state with a plastic strain amplitude nearby zero until macro crack growth. Besides the mechanical hysteresis measurements the change in temperature and the change in electrical resistance can be used to describe the fatigue behavior of friction stir welded aluminum joints in more detail.

Introduction

Friction stir welding (FSW) is a solid state joining technique, which was developed as an improvement of the traditional friction welding process [1]. It is well suited to join structural materials like aluminum or magnesium alloys [2]. A friction stir weld is produced by plunging a rotating steel tool, consisting of a shoulder and a profiled pin into the components which have to be welded. The shoulder grinds on the surface of the joining partners and the pin deforms the bulk in the range of the joint line. The friction between the rotating tool and the joining partners generates heat with $T_{welding} < T_{melting}$ [3], that leads locally to a more ductile condition of the materials to be welded. In this material condition the rotating welding tool can be easily be moved along the weld seam.

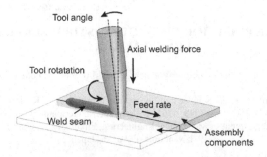

Figure 1. Friction Stir Welding Process, schematically

The most important process parameters, which govern the generated temperature in the welding zone, are the welding feed, the rotational speed, the tool angle and the axial welding force (figure 1) [4-6]. Furthermore the geometry of the welding tool has an impact on the welding temperature [7,8]. All these factors influence the resulting microstructure in the weld seam and consequently the mechanical properties of the joints [9].

Welding equipment, used material and specimen geometry

The friction stir welds were produced by using a conventional Deckel Maho four axis milling machine, as shown in figure 2. The machine was equipped with a force measuring system integrated in a pneumatic clamping unit. Besides the process control the force measuring system was used to realize a force controlled welding process. Additionally the welding temperature was measured by thermocouples, which were directly inserted in the joint line before the welding process was started. The investigated FSW-joints were produced by using a standard welding tool with a shoulder diameter of approximately 15 mm and a M3.5 pin.

Figure 2. Used welding equipment

The investigations were carried out with the wrought aluminum alloy AA5454. Its chemical composition is given in Table I. Besides the corrosion behavior the manganese and chromium fraction particularly improve the monotonic and cyclic strength, because both elements together with the iron form very stable intermetallic precipitations.

Table I Chemical composition of AA5454

(wt-%)	Mg	Mn	Fe	Si	Zn	Ti	Cr	Cu
AA5454	2.85	0.84	0.31	0.13	0.01	0.01	0.08	0.03

The microstructure of the used aluminum sheets is shown in the light micrograph in figure 3. It consists of a solid solution aluminum phase (α-phase) and intermetallic precipitations rich in iron and manganese. Both phases are elongated in longitudinal direction due to the rolling process.

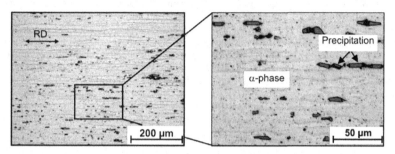

Figure 3. Light micrograph of the base material, ↔ rolling direction

The investigated aluminum alloy AA5454 was used in condition H22, that means work hardened by cold rolling followed by an annealing process. The annealing process is configured to adjust a hardness condition of ¼ hard. In figure 4 selected mechanical properties of the parent material are given and a characteristic σ_n-ε_t-curve is shown. The σ_n-ε_t-curve is characterized by serrations as a result of the Portevin-Le Chatelier effect (PLC). The interaction between diffusing atoms, like e.g. Mg, and moving dislocations leads at a defined ratio of temperature and deformation rate to a serrated material flow as a critical deformation is exceeded.

		AA5454 (H22)
$R_{p0.2}$	[MPa]	220
R_m	[MPa]	295
E	[GPa]	70.6
A	[%]	7.2
HB 5	[ ]	84.4

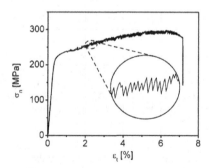

Figure 4. Selected mechanical properties and σ_n-ε_t-curve of AA5454 (H22)

In the present work, the aluminum sheets with the dimensions $125 \cong 300 \cong 3.5$ mm^3 were butt welded perpendicular to the rolling direction along the longer side of the sheets. The used process parameters with a tool rotation of 1500 1/min and a feed rate of 200 mm/min in combination with axial welding forces of 5 kN, 6.5 kN and 8 kN are well suited to produce

FSW-joints free of pores with high monotonic and cyclic strength. After welding, the sheets were cut, flat tensile and fatigue specimens were milled perpendicular to the weld seam, and finally the specimens were polished to achieve a surface roughness of $R_z = 0.5$ μm.

Experimental procedures

Two-dimensional martens hardness measurements allow to quantify the local hardness gradients resulting from the welding process with high precision. Besides monotonic tensile tests, axial stress-controlled fatigue tests were performed at a load ratio of $R = 0$ at ambient temperature using a frequency of $f = 5$ Hz and triangular waveforms [10-12]. In figure 5 the experimental setup for the fatigue tests is shown. The polished specimen (mark (A)) is fixed in the thermostatically controlled grips (mark (D)). To receive information about the deformation behavior of the welding zone itself, the clamping of the extensometer was positioned with a displacement of 15 mm, corresponding to the diameter of the shoulder of the welding tool (mark (B)).

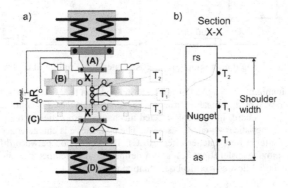

Figure 5. Experimental setup for fatigue tests a)
and zoomed section X-X with positions of thermocouples b)

Furthermore, high-precision temperature and electrical resistance measurements were performed to describe the cyclic deformation behavior in detail [9-12] (figure 5 a) and b)). Therefore four thermocouples were fixed at the specimen surface, three in the gauge length (T_1-T_3) and one at the shaft of the specimen (T_4). At the measuring point T_4 the cross section is wide enough, that no plastic deformation occurs. With the equation

$$\Delta T_n = (T_n - T_4) \text{ with } n = 1, 2, 3 \qquad (1)$$

thermal influences form the environment on the temperatures measured in the gauge length can be eliminated. The thermocouple at the measurement position T_1 is located at the weld nugget, T_2 at the retreating side and T_3 at the advancing side of the joint, shown in figure 5 b). This configuration allows to describe locally different plastic deformations related to the local microstructure. The electrical resistance depends on the geometrical values length L and the cross sectional area A and in particular on the resistivity ρ^*.

$$R = \rho^* \cdot L/A \qquad (2)$$

As the value ρ* depends amongst others on lattice imperfections like dislocations, pores and microcracks, with the value R deformation-induced changes in the microstructure of a metallic material can be detected. For the electrical resistance measurements a DC power supply was fixed at both specimen shafts (mark (C)). The change in resistance was measured outside the gauge length nearby the extensometer clamping.

Results

Characterization of the welding process
During the force controlled welding process the welding temperature in the joint line T_{Weld} was measured by the inserted thermocouples. Depending on the axial welding force, especially the cooling can be different, as shown in figure 6.

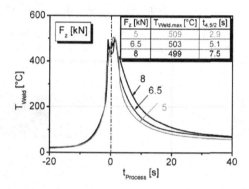

Figure 6. Curves of the welding temperature for different axial welding forces F_z

Independent of the axial welding force, the maximum welding temperature in the joint line is about 500 °C. Besides the maximum temperatures, with the thermometric measuring method the temperature increase and decrease can be described. Especially the cooling time differs depending on the welding force F_z. In the following, the cooling time from 450 °C to 200 °C is named $t_{4.5/2}$. For the highest welding force of 8 kN it takes $t_{4.5/2} = 7.5$ s for the cooling as a result of a wider shoulder contact area in contrast to the lower welding force with $F_z = 5$ kN and a resulting $t_{4.5/2} = 2.9$ s.

Microstructure and martens hardness
The different cooling times $t_{4.5/2}$, influence the microstructure in the welding zone (figure 7). The lower welding force with the shorter cooling time leads to a smaller weld nugget compared to the nugget resulting from the higher welding force. As a result of the elevated temperatures and high plastic deformation in the area of the weld nugget dynamic recrystallization occurs. The joint welded with $F_z = 5$ kN has a nugget with a cross sectional area of $A_{Nugget} = 13.7$ mm^2, the joint welded with $F_z = 8$ kN has a nugget with $A_{Nugget} = 17.2$ mm^2.

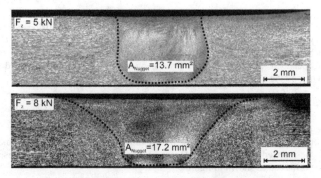

Figure 7. Microstructure of FSW-joints welded with different axial welding forces F_z

Besides the microstructure, the hardness distribution in the area of the weld seam can be used to show the influence of the axial welding force. In figure 8 the martens hardness profiles of the welding zone for axial welding forces of $F_z = 5$ kN and $F_z = 8$ kN are shown.

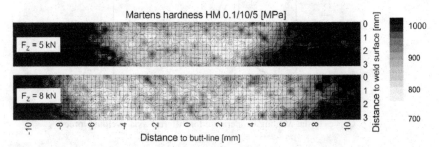

Figure 8. Martens hardness profiles of FSW-joints welded with different axial welding forces F_z

As a result of the thermo-mechanical interaction between the rotating welding tool and the joining material the joints show a softened area in the middle of the cross section. In both profiles, the softened area is not restricted to the weld nugget. A certain broadening is caused by the frictional heat input of the grinding shoulder. The joint welded with 8 kN is characterized by a more pronounced softened area compared to 5 kN. This difference is caused by the different heat input as a consequence of the chosen axial welding forces.

Monotonic behavior
In figure 9 selected monotonic properties of FSW-joints welded with the welding forces 5, 6.5 and 8 kN are plotted. With increasing welding force, the yield strength $R_{p0.2}$ decreases and the elongation after fracture A increases more than 50% due to the pronounced softening effects in the TMAZ. No influence of the welding force on the tensile strength R_m was detected. The continuous serrations in σ_n in the course of the σ_n-ε_t-curves are caused by the Portevin-Le Chatelier (PLC) effect.

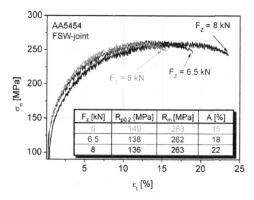

Figure 9. Monotonic stress-strain curves of FSW-joints welded with different axial welding forces

Cyclic deformation behavior

For the microstructure-related fatigue assessment of FSW-joints, constant amplitude tests were performed. In figure 10 the cyclic $\varepsilon_{a,p}$-N-, ΔT-N-, ΔR-N- and $\varepsilon_{m,p}$-N-curves at a stress amplitude $\sigma_a = 120$ MPa are plotted for a FSW-joint welded with $F_z = 6.5$ kN.

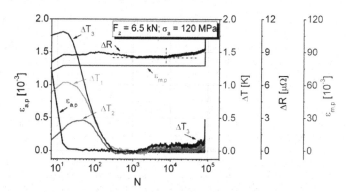

Figure 10. Cyclic deformation, temperature, electrical resistance and creep curves for a FSW-joint welded with $F_z = 6.5$ kN

The cyclic deformation behavior of the welding zone is characterized by pronounced cyclic hardening within the first 20 cycles, followed by a saturation state with a plastic strain amplitude $\varepsilon_{a,p}$ nearby zero until macro-crack growth. Further-more, the initial cyclic plastic deformation causes an initial change in the temperature ΔT_1, ΔT_2 and ΔT_3. The signal of the temperature is a little bit shifted compared to the mechanical values because of thermal conduction. The different development of the local temperatures can be taken as a very sensitive indicator, that the cyclic deformation behavior differs locally in the welding zone. Especially the development of $\Delta T3$ for $N > 3 \cdot 10^3$ cycles can be regarded as an indicator of a proceeding fatigue damage at the advancing side, where finally the fatigue failure occurs. As a consequence of the accumulated plastic deformation at the beginning of the test, the change in electrical resistance reaches values of about 9 μΩ. For a better understanding of the cyclic deformation behavior the cyclic creep curve is plotted, too. In the first cycles the plastic deformation is very distinctive and consequently $\varepsilon_{m,p}$ increases. The change in electrical resistance at test beginning is in good accordance to the change in specimens geometry. But the secondary increase of ΔR isn't related to a change of $\varepsilon_{m,p}$. For $N > 8 \cdot 10^3$ cycles ΔR increases due to deformation-induced changes in microstructure as well as micro-crack formation and growth in the gauge length.

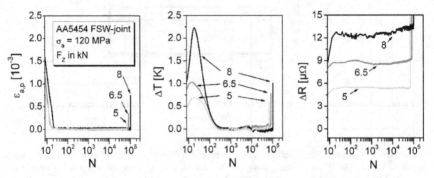

Figure 11. Cyclic $\varepsilon_{a,p}$-N-, ΔT-N-and ΔR-N-curves for FSW-joints welded with different F_z

To evaluate the influence of the welding force on the cyclic deformation behavior of FSW-joints, in figure 11 $\varepsilon_{a,p}$-N-, ΔT-N- and ΔR-N-curves are plotted for the stress amplitude $\sigma_a = 120$ MPa and axial welding forces of 5, 6.5 and 8 kN. The initial plastic deformation increases from $\varepsilon_{a,p} = 1.0 \cdot 10^{-3}$ to $\varepsilon_{a,p} = 1.6 \cdot 10^{-3}$ with increasing welding forces. Although plastic deformation at the beginning of the tests is more pronounced, increasing welding forces result in a slight but systematic increase of the number of cycles to failure, from $6.7 \cdot 10^4$ for $F_z = 5$ kN to $10.4 \cdot 10^4$ cycles for $F_z = 8$ kN. The increase of the softened area in the welding zone for higher axial welding forces leads to an increase of the plastic deformation, but the wider weld nugget leads to a more pronounced cyclic hardening, and consequently the lifetime increases. It can be noticed, that all three measured values $\varepsilon_{a,p}$, ΔT and ΔR are well suited to characterize the fatigue behavior of friction stir welded aluminum joints in detail.

In figure 12 the measured mechanical, thermal and electrical data in the very early fatigue state $N = 10$ cycles and furthermore the lifetime is plotted as a function of the axial welding force F_z. This diagram impressively shows, that besides the mechanical values, the thermal and the electrical values can be used to estimate the lifetime on the basis of data measured after $N = 10$ cycles.

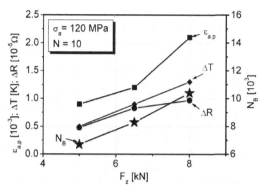

Figure 12. Measured mechanical, thermal and electrical data and lifetime as a function of F_z

Conclusions

The axial welding force is a very important process parameter which influences the microstructure of the welding zone and consequently the deformation behavior of the joints. Besides the hardness and the monotonic behavior mainly the cyclic deformation behavior of FSW-joints is influenced by F_z. Different monotonic properties and changes in plastic deformation capability of the joints were measured in dependence of the welding force. For the fatigue assessment high-precision plastic strain amplitude, temperature and electrical resistance measurements were performed. With higher welding forces, the plastic deformation at the beginning of the tests increases as a result of the size of the softened area, but despite of this the fatigue life of the joints increases with higher welding forces. This can be explained with the homogeneous hardness distribution in the gauge length and the larger weld nugget, that provides a good cyclic deformation behavior by its fine-grained microstructure. The local change in temperature and especially the change in electrical resistance are very powerful tools to describe and evaluate the actual fatigue state of FSW-joints in detail.

References

[1]   M. Thomas, E.D. Nicholas, J.C. Needham, M.G. Murch, P. Temple-Smith: *Improvements to Friction Welding*, in: TWI (Ed.), Vol. EU Patent 0 615 480 B1, GB, 1991.

[2]   S. Kallee, Wayne, M. Thomas, E.D. Nicholas: *Friction Stir Welding of Lightweight Materials*, Magnesium Alloys and their Applications, K.U. Kainer (Ed.), Wiley-VCH-Verlag (2006) 173-190.

[3]   S. Benavides, Y. Lia, L.E. Murra, D. Browna, J.C. McClurea: *Low-temperature friction-stir welding of 2024 aluminum* Scripta Materialia 41 (8) (1999) 809-815

[4]   M. Ericsson, R. Sandstrom: *Influence of welding speed on the fatigue of friction stir welds, and comparison with MIG and TIG*, International Journal of Fatigue 25 (12) (2003) 1379-1387.

[5]   M.N. James, D.G. Hattingh, G.R. Bradley: *Weld tool travel speed effects on fatigue life of friction stir welds in 5083 aluminium*, International Journal of Fatigue 25 (12) (2003) 1389-1398.

[6]   J.E. Mitchell, G.E. Cook, A.M. Strauss: *Force sensing in friction stir welding*, 6th International trends in welding research conference proceedings, Pine mountain, ASM International (2002).

[7]   K. Kumar, S.V. Kailas: *The role of friction stir welding tool on material flow and weld formation*, Materials science & engineering A 485 (2008) 367-374.

[8]   L. Cederqvist, A.P. Reynolds: *Factors affecting the properties of friction stir welded aluminum lap joints*, Welding Journal 80 (2001) 281-287.

[9]   Y.S. Sato, H. Kokawa: *Microstructural Factors Governing Mechanical Properties in Friction Stir Welds*, Key Engineering Materials (2007) 1493-1496.

[10]   M. Gutensohn, G. Wagner, F. Walther, D. Eifler: *Cyclic Deformation Behavior of Friction Stir Welded Alumnium Joints*. Aluminium Alloys – Their Physical and Mechanical Properties, Proceedings of the 11th Internat. Conference on Alumnium Alloys. Wiley-VCH Verlag, Weinheim (Germany), (2008), 1998-2004

[11]   G. Wagner, M. Gutensohn, O. Klag, D. Eifler: *Microstructure and Deformation Behavior of Friction Stir Welded Light Metals*. 7th International Symposium on Friction Stir Welding, TWI Ltd (2008), CD-Edition, ISBN 13-978-1-903761-06-9

[12]   P. Starke, F. Walther, D. Eifler: *Hysteresis, temperature and resistance measurements for the characterization of the fatigue behavior of metals*. ISCS 2007, 5th Int. Conf. Structural Integrity of Welded Structures - Testing & Risk Assessment in the Development of Advanced Materials and Joints, Timisora, Romania, www.ndt.net, Issue 2008-04 (2008) 1-9

Friction Stir Welding and Processing V
Edited by: Rajiv S. Mishra, Murray W. Mahoney, and Thomas J. Lienert
TMS (The Minerals, Metals & Materials Society), 2009

Friction Stir Welding of Sc-Modified Al-Zn-Mg-Cu Alloy Extrusions

C. Hamilton[1], S. Dymek[2], O. Senkov[3]

[1]Department of Mechanical and Manufacturing Engineering
Miami University, Oxford, OH

[2]Faculty of Metals Engineering and Industrial Computer Science
AGH University of Science and Technology, Kraków, Poland

[3]UES, Inc., 4401 Dayton-Xenia Rd., Dayton, OH 45432-1894, USA

Keywords: friction stir welding, aluminum, scandium, mechanical properties

Abstract

Small additions of scandium to Al-Zn-Mg-Cu 7000 series alloys can significantly improve mechanical properties and augment the strength retention at low and elevated temperatures. This research program evaluates the residual properties of Sc-modified Al-Zn-Mg-Cu alloy extrusions joined through friction stir welding (FSW). Mechanical and corrosion testing were performed on the baseline material and on panels friction stir welded at 175, 225, 250, 300, 350 and 400 RPM (all other weld parameters were held constant). A thermal model of friction stir welding was developed utilizing an energy-based scaling factor to account for tool slip. The proposed slip factor is derived from an empirical relationship between the ratio of the maximum welding temperature to the solidus temperature and energy per unit length of weld. The thermal model successfully predicts the maximum welding temperatures over a range of energy levels, and the mechanical and corrosion behavior is correlated to the temperature distribution predicted by the model.

Introduction

As the aerospace industry produces new and more efficient airframes, the need to provide high-strength, lightweight alloys that meet the aggressive design objectives for mechanical performance, manufacturability and service life arises. Conventional Al-Zn-Mg-Cu 7000 series alloys are widely used in aerospace applications due to their favorable mechanical properties. However, these properties degrade above 150°C due to the coarsening and/or dissolution of the strengthening phases within the microstructure. Additions of scandium (Sc) and zirconium (Zr) to 7000 series alloys can stabilize the microstructure at these elevated temperatures and can augment the mechanical performance through the formation of fine, secondary strengthening phases such as $Al_3(Sc, Zr)$ [1, 2]. Because the nanometer-sized $Al_3(Sc, Zr)$ particles also stabilize the microstructure formed during hot working operations and inhibit recrystallization during heat treatment, the potential for enhanced residual properties following joining operations, such as friction stir welding (FSW), arises.

Friction stir welded joints display excellent mechanical properties when compared to conventional fusion welds, and as such, the aerospace industry is embracing FSW technology and implementing new welding capabilities into their manufacturing sectors [3 – 8]. Over the last fifteen years, numerous investigations have sought to characterize the principles of FSW and to model the microstructural evolution. The current status of FSW research has been well summarized by Mishra and Ma [9]. The following research program characterizes the mechanical and corrosion behavior of developmental, Sc-modified Al-Zn-Mg-Cu extrusions (SSA038-T6) joined by friction stir welding.

The chemical composition of SSA038 is summarized in Table 1 along with that of aluminum 7075 for reference and comparison [1]. Whereas 7075 primarily utilizes chromium to control grain growth and recrystallization, SSA038 utilizes the synergistic combination of scandium and zirconium to stabilize the microstructure. Also the SSA038 chemistry has a higher level of zinc than 7075. The zinc level of SSA038 is more akin to that of the 7050 or 7136 alloys.

Table 1 - Chemical Composition of SSA038 with 7075 as Reference

Element	Weight Percent	
	SSA038	7075
Zn	7.11	5.60
Mg	2.14	2.50
Cu	1.56	1.60
Mn	0.25	0.30
Zr	0.17	< 0.05
Sc	0.38	--
Cr	< 0.05	0.23
Ti	< 0.05	0.20
Other, Each	0.35	1.03
Al	Balance	

Experimental Procedure

Friction Stir Welding

For this investigation, SSA038 billets were produced by UES, Inc. through direct chill casting and then extruded as 50.4 mm x 6.35 mm bars. Following extrusion, the bars were heat treated to a –T6 temper through the following schedule: (1) solution heat treat at 460°C for one hour followed by an additional hour at 480°C, (2) rapid quench in water to room temperature and (3) age at 120°C for 19 hours. After heat treatment, the bars were cut into twelve, 305 mm lengths and sent to the Edison Welding Institute to produce six friction stir welds in the configuration represented in Figure 1. As shown in the diagram, FSW occurs along the L-direction of the extrusions with a clockwise tool rotation. The diameter of the FSW tool shoulder was 17.8 mm, the pin diameter tapered linearly from 10.3 mm at the tool shoulder to 7.7 mm at the tip and the pin depth was 6.1 mm. With a constant weld velocity of 2.1 mm/s and a constant applied force of 22 kN, unique welds were produced at the following tool rotation speeds: 175, 225, 250, 300, 350 and 400 RPM.

Utilizing a Mikron M7815 Thermal Imaging Camera during welding, the temperature profile across the weld was experimentally recorded for each condition. The thermal emissivity for the infrared data was calibrated by imaging an extrusion length heated to 460°C and adjusting the emissivity value until the recorded temperature of the camera matched the reference temperature. The appropriate thermal emissivity value was determined to be 0.285. The experimental temperature data were used to verify the efficacy of the computer thermal model developed during this investigation.

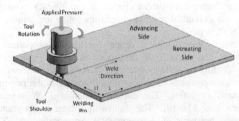

Figure 1 - Schematic of Butt-Weld Geometry and FSW Orientation

316

Mechanical Testing

Subsequent to joining, the welded panels were stored at room temperature and allowed to naturally age for at least 30 days prior to testing and investigation. From the welded panels of each tool rotation speed, full thickness (6.35 mm) tensile samples were excised perpendicular to the friction stir weld with the weld centered along the tensile specimen in the reduced section as shown in Figure 2. In this orientation, the load is applied transverse to the weld direction and across all microstructural regions associated with the welding process, i.e., weld nugget, HAZ and TMAZ. In addition to the welded tensile specimens, tensile bars of the same geometry and dimensions were also excised from an area well away from the weld region for baseline property comparison. All tensile tests were performed in accordance with ASTM E 8 utilizing an Instron 5867 screw driven test frame with a 30 kN load cell and a $0.001 - 500$ mm/min speed range. Specimen extension, crosshead deflection and load were recorded throughout the test duration. Specimen extension was measured by means of an extensometer attached to the reduced section that spanned the width of the weld. The yield strength, σ_y, was obtained by the 0.2% offset method, and the elastic modulus, E, was determined by fitting a linear regression to the elastic region of the stress-strain curve. Elongation was determined by scribing marks with known separation within the reduced section prior to testing and measuring their separation after testing.

Corrosion Testing

Coupons were then excised for exfoliation corrosion testing from the welded panels produced at each tool rotational speed. The coupons were oriented with the long edge perpendicular to the weld direction and centered on the weld to assess the corrosion behavior of all microstructural regions, as represented in Figure 2. The exfoliation corrosion behavior was evaluated at the T/10 plane in accordance with ASTM G 34 with a continuous immersion in the prescribed corrosive environment for 48 hours. The extent of exfoliation corrosion was qualitatively determined using the rating system, i.e. EA, EB, EC, etc., as outlined in the ASTM standard.

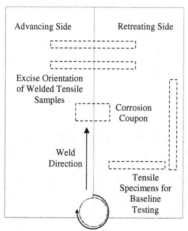

Figure 2 - Excise Location of Tensile Specimens and Corrosion Coupons from FSW Panels

317

Results and Discussion

Mechanical Properties

Six different tool rotation speeds were examined to assess the impact of changing weld parameters on the residual properties and to identify the optimum weld condition leading to the highest joint efficiency. Since the weld velocity and the applied force were held constant during FSW and only the tool rotation speed was varied, the tool RPM provides a relative measure of the energy imparted to the aluminum extrusions due to changes in the process parameters. It is, however, more appropriate to evaluate the influence of the process parameters, weld velocity (v_w), tool rotation speed (ω) and applied force (F), during friction stir welding in terms of the total heat or energy imparted to the workpieces during welding, rather than the impact of a single parameter. Since varying any of the identified parameters changes the heat input of the system, correlating the heat input with the resultant material behavior gives greater insight into the friction stir welding process.

For this reason, Khandkar's torque-based model was used to determine the energy input per unit length of weld for the Sc-modified Al-Zn-Mg-Cu alloy extrusions and to correlate this quantity with the observed material properties at various tool rotation speeds [10]. The total torque, T_{total}, can be expressed as the sum of torque contributions from the tool shoulder against the workpiece, the bottom of the tool pin against thickness material and the pin surface against thickness material. Introducing τ as the shear stress during welding, assumed to be uniform across all tool-workpiece interfaces, the total torque then becomes:

$$T_{total} = \int_{r_{i1}}^{r_o}(\tau r)(2\pi r)dr + \int_{0}^{r_{i2}}(\tau r)(2\pi r)dr + 2\pi\tau r_i^2 h \tag{1}$$

where r_o is the radius of the tool shoulder, r_{i1} is the radius of the pin at the tool shoulder, r_{i2} is the radius of the pin at the pin bottom and h is the pin height. In actuality, the magnitude of the shear stress will vary with the local welding temperature. However, for simplification, the temperature dependence is ignored, and the shear stress is estimated from the applied force during FSW. To further simplify the evaluation of Equation 1, the taper of the welding pin is ignored, i.e., $r_{i1} = r_{i2}$, thus

$$T_{total} = \tau(\frac{2}{3}\pi r_o^3 + 2\pi r_i^2 h) = 2\mu F(\frac{r_o}{3} + \frac{r_i^2}{r_o^2}h) \tag{2}$$

where F is the applied force and μ is the coefficient of friction between the tool and the extrusions. The energy per unit length of weld, E_l, is found by dividing the average power, P_{avg}, by the weld velocity to yield the expression in Equation 3:

$$E_l = \frac{P_{avg}}{v_w} = T_{total}\frac{\omega}{v_w} \tag{3}$$

Though the applied force during FSW was set to 22 kN, real time data from the welding trials revealed that the load oscillated as the machine continuously corrected the load toward the set point. Consequently, the average load during welding deviated from the desired set point. Therefore, the average load was determined from the recorded data for each weld condition and was utilized in the analysis of that condition. The recorded data verified that the weld velocity remained constant at 2.1 mm/s for all welding trials. Using a coefficient of sliding friction between aluminum and mild steel of 0.5 [11, 12], the E_l were determined for each tool rotation speed.

Table 2 summarizes the E_l values and the average mechanical properties of weld joints for different FSW conditions. The joint efficiency is defined as the ratio of the tensile strength for a given weld condition to the L-direction baseline tensile strength. After FSW at 175 RPM, the yield strength (YS), ultimate tensile strength (UTS), and elongation (e) decreased relative to

318

the baseline properties. This decrease can be associated with overaging and/or annealing due to FSW-induced heating. Increasing from 175 RPM to 250 RPM did not affect YS. However, e increased to 10% and UTS increased to 451 MPa. Increasing the rotation speed from 250 RPM to 350 RPM decreased the YS, UTS, and e to 294 MPa, 381 MPa and 5.7%, respectively. The correlation between e and UTS between 175 to 350 RPM is noteworthy. That is, as the weld joint ductility increases, the UTS and joint efficiency simultaneously increase. As a result, the FSW parameters that promote superior joint ductility also lead to higher UTS. A rise from 350 RPM to 400 RPM slightly increased YS (to 306 MPa), but considerably decreased e (to 3.0%) with little impact on the UTS.

Table 2- Summary of Mechanical Properties for Each Weld Condition

Rotation Speed (RPM)	E_l (J/mm)	UTS σ_{TS} (MPa)	YS σ_y (MPa)	EL e (%)	Joint Eff. (%)
Baseline (L)	--	663	605	12.0	--
Baseline (LT)	--	651	597	15.4	--
175	842	406	338	5.7	61
225	938	443	337	7.2	67
250	977	451	337	10.0	68
300	1331	446	309	7.9	67
350	1502	381	294	5.7	57
400	1567	383	306	3.0	58

Figure 3 displays the relationship between the strength properties and the energy per unit length of weld, and clearly shows the local maximum that is realized at 250 RPM (~980 J/mm). An analysis of variance reveals that both correlations between the tensile strength and yield strength with the applied weld energy are statistically significant.

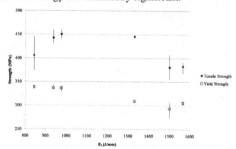

Figure 3- Ultimate and Yield Strengths as a function of the Energy per Unit Length of Weld (one standard deviation shown)

As the welding energy increases, the welding temperatures of the workpieces also increase. The SSA038 alloy is a precipitation strengthened aluminum alloy, and its strength in the −T6 temper is controlled by GPII zones and η' particles, as well as by $Al_3(Sc,Zr)$ nano-dispersoids [13, 14]. A considerable decrease in both YS and UTS after welding is a strong indication that the welding temperature was significantly higher than the aging temperature (120°C), promoting: (a) the dissolution of GPII zones, (b) coarsening of η' particles and their transformation to equilibrium η, i.e overaging, and/or (c) the partial dissolution of η' and η followed by re-precipitation after cooling. To understand the trend in mechanical behavior observed in Figure 3, a thermal model of friction stir welding was developed using NX 5.0.

Thermal Modeling

For the FSW representation and tool coordinate system shown in Figure 4, the governing heat equation for a moving heat source is given by [15]:

$$\rho c_p \nabla(uT) = -\rho c_p v_x \frac{\partial T}{\partial x} + \nabla(k\nabla T) + S_i + S_b \tag{4}$$

where u is the plastic flow velocity, v_x is the welding velocity in the x-direction, S_i is the source term for the heat generation due to the tool/workpiece interface, S_b is the source term for the heat generation rate due to plastic deformation away from the interface, and all other terms have their previous meaning. The heat flux, q, at the tool/workpiece interface may be derived from the S_i term and is given by the following expression:

$$q = \delta \mu P_N \omega r \tag{5}$$

where δ is the slip factor, P_N is the normal pressure relative to a tool face, μ is the coefficient of friction between the tool and workpiece and r is the radial distance from the tool center axis as shown in Figure 4.

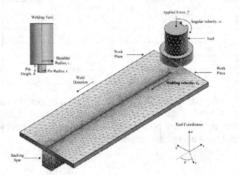

Figure 4 - Finite Element Model of FSW, Tool Coordinate System and Tool Geometry

In an earlier work, Hamilton et al. [16] proposed that a characteristic relationship exists for aluminum alloys during friction stir welding between the ratio of the maximum weld temperature to the solidus temperature of the alloy, T_{max}/T_s, and the energy per unit length of weld, E_l. From that relationship, the following energy-dependent expression for the slip factor was then derived:

$$\delta_E = \exp\left(-\frac{E_l}{(E_l)_{max}}\right) \tag{6}$$

where $(E_l)_{max}$, the maximum effective energy, is defined as the energy level for which the maximum welding temperature is equal to the solidus temperature of the alloy (i.e., $T_{max}/T_s = 1$). The solidus temperature for SSA038 is 528°C. In Equation 6, δ_E effectively represents the efficiency of frictional heat transfer, such that, when $\delta_E = 1$, sticky friction dominates. As $(E_l)_{eff}$ increases, the tool/workpiece interface softens, and the extent of slip increases, lowering the effectiveness of the heat transfer process. At the limit of $E_l = (E_l)_{max}$, $\delta_E = 0.37$.

For the tool shoulder and pin bottom, the normal pressure, P_N, is the applied pressure on the workpiece during welding. For the interface between the side of the pin and the workpiece, however, the normal pressure is the pressure that resists the tool as it traverses the workpiece during welding. For typical welding conditions, this pressure is significantly less than the applied pressure. Therefore, the frictional heat flux is primarily generated by the pin bottom and tool shoulder, and the heat flux at the pin side/workpiece interface is set equal to zero. The thermal model utilizes the average heat flux on the pin bottom, $q_{pinbottom}$, and tool shoulder, $q_{shoulder}$, to calculate the maximum welding temperatures. Assuming axisymmetric heat generation, these heat flux equations given in Equations 7 and 8 are:

$$q_{pinbottom} = \frac{\int_0^{r_i} \delta_E \mu P_N \omega r\, dr}{r_i} = \frac{\delta_E \mu P_N \omega r_i}{2}$$ (7)

and

$$q_{shoulder} = \frac{\int_{r_i}^{r_o} \delta_E \mu P_N \omega r\, dr}{r_o - r_i} = \frac{\delta_E \mu P_N \omega (r_o^2 - r_i^2)}{2(r_o - r_i)}$$ (8)

where all terms have their previous meaning.

Figure 4 shows the three-dimensional, finite element model of FSW. A backing spar, as opposed to a full backing plate, was chosen to reduce the number of equations to be solved and to shorten the computer processing time while still capturing the heat transfer characteristics between the workpiece and backing plate. The width of the backing spar is equal to the diameter of the tool, and the height is 25.4 mm. The model assumes that perfect contact exists between the two workpieces along the weld line and between the tool shoulder and the workpieces. A gap thermal conductivity of 157 W/m·K, equal to the thermal conductivity of SSA038, was applied between the pin bottom and the bottom of workpieces. For the boundaries exposed to ambient conditions, the convective coefficient was set to 15 W/m²·K. The remaining boundary conditions and thermal couplings were determined using the thermal images recorded during welding. For the tool top and spar bottom, a convection coefficient of 200 W/m²·K was applied, and for the area outside of and adjacent to the backing spar, a convective coefficient of 100 W/m²·K was used to represent the dissipation of heat into the backing plate. Lastly, a heat transfer coefficient of 250 W/m²·K was chosen to define the conductance of heat between the bottom of the workpieces and the backing spar. In this model, heat dissipation due to radiation was ignored.

Figure 5 plots the experimental temperature profile of the 400 RPM condition with the profile predicted by the thermal model for the same location. The profile extends from the weld line into the workpiece along the top surface of the FSW configuration. Although the model slightly under predicts the temperatures, the predicted profile shows good agreement with the magnitude and shape of the experimental data.

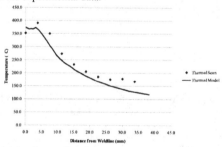

Figure 5 - Experimental Temperature Profile Compared to Temperature Profile Predicted by the Thermal Model

With the heat fluxes determined from Equations 7 and 8, the thermal model was then used to determine the maximum welding temperatures of SSA038-T6 for the six, unique effective energy levels investigated. Table 3 summarizes these temperatures. As predicted for each weld condition, the maximum welding temperature is considerably higher than the aging

321

temperature of the alloy (120°C). This over-heat explains the decrease in the tensile strength of the welded material relative to the –T6 baseline condition.

Table 3 - Predicted Maximum Welding Temperatures for each Weld Condition

Rotation Speed (RPM)	E_l (J/mm)	Predicted Max. Temp. (°C)
175	842	264
225	938	282
250	977	289
300	1331	342
350	1502	362
400	1567	386

    Figure 6 shows the transient temperature profiles that were experimentally determined for a node located on the top surface of the workpiece at the interface between the weld and HAZ. Welding conditions at 175 and 250 RPM reveal similar dwell times above the aging temperature and maximum temperatures that are only 16°C different in magnitude. On the other hand, welding at 400 RPM shows a longer dwell time above the aging temperature, and a maximum temperature that is 50°C greater than that observed at 250 RPM. The similar temperature history between welds produced at 175, 225 and 250 RPM explains the similar values of the yield strength. The decrease in YS of the welds relative to the YS of the –T6 baseline condition is most likely due to overaging, i.e., dissolution of the *GPII* zones, coarsening of η' particles and formation and growth of equilibrium η [15, 16]. Between 175 and 250 RPM, the strain hardening that occurs after yielding provides higher UTS at larger plastic strains [17], and the improved ductility within this range may simply be due to a superior weld quality produced at higher rotation speeds. This explanation, however, must be verified by microstructural examination.

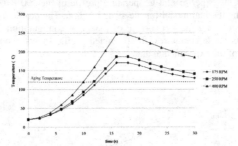

Figure 6 - Transient Temperature Profiles for a Top Surface Node at the Interface Between the Weld and HAZ

    Between the predicted maximum welding temperatures of 342°C and 386°C for 300 and 400 RPM, all metastable particles dissolve and only coarse equilibrium η phase particles and $Al_3(Sc,Zr)$ nano-dispersoids are present within the FCC Al matrix of SSA038 [1]. As a result, the yield strength values approach the YS value of the annealed alloy (~273 MPa) [1]. The slightly higher YS after 400 RPM than after 350 RPM can be due to higher super-saturation of the alloying elements after heating to a higher temperature, resulting in the re-precipitation of a larger amount of metastable particles upon cooling.

    Similar to the welding conditions at the lower rotation speeds, the UTS values are likely controlled by the plastic strain achieved after yielding and before the fracture, so that the UTS decreases with a decrease in the plastic strain (ductility). The latter is probably controlled by the weld quality, and as such, the weld parameters corresponding to 250 RPM produce the highest weld quality, and, therefore, correspond to the optimum welding condition.

Corrosion Behavior

Figure 7 displays a typical exfoliation corrosion specimen from this study, and Table 4 summarizes the exfoliation corrosion rating as prescribed by the ASTM standard for each distinct region of the corrosion coupon as indicated in the figure. For a given corrosion specimen, friction stir welding promotes a performance gradient across the coupon with the majority of corrosion occurring between the weld and HAZ, i.e., the thermo-mechanically affected zone boundary. This behavior is consistent with the corrosion sensitization seen in 7050 by other researchers [18]. The most severe corrosion occurred during weld trials at 250 RPM with the TMAZ boundary receiving an ED rating. The extent of corrosion, therefore, coincides with the trend in residual strength, that is, the maximum amount of corrosion and the maximum weld efficiency occur at 250 RPM (or ~1000 J/mm). For aluminum alloys, an inverse relationship between strength and corrosion resistance is typical. However, without microstructural verification of the welded samples, applying this reasoning to the observed FSW corrosion behavior should be done with care. For example, if the microstructure of the TMAZ is significantly strained and/or dynamically recrystallized, this will certainly influence the corrosion behavior of the region. In addition, an increase in intergranular corrosion susceptibility following FSW has been correlated to the coarsening of grain boundary precipitates and the widening of precipitate free zones [19].

Table 4 - Summary of Exfoliation Corrosion Results

Condition	Baseline	HAZ	Weld	Interface
175 RPM	EA	EA	EA	EB
225 RPM	EA	EA	EB	EB
250 RPM	EB	EA	EC	ED
300 RPM	EA	EA	EA	EB
350 RPM	EB	EA	EA	EB
400 RPM	EB	EA	EA	EB

During this investigation, the corrosion resistance within the HAZ improves for all weld conditions and receives an EA rating in each case. Nominally, the increase in the corrosion resistance in the HAZ would be prescribed to the coarsening of the strengthening precipitates in this region without significant secondary nucleation to increase the residual strength. And, as observed with the mechanical properties, the increase in corrosion at the TMAZ boundary up to 1000 J/mm may be the result of the secondary nucleation of precipitates in this area that increase the strength of the alloy. As the welding energy surpasses 1300 J/mm, the rate of precipitate coarsening dominates the precipitate kinetics, increasing the corrosion resistance, but decreasing the mechanical properties. Away from the HAZ and into the parent material, a return to the baseline corrosion resistance is observed with a rating of EA or EB, as shown in Figure 7.

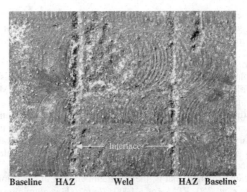
Baseline   HAZ   Weld   HAZ Baseline

Figure 7 - Typical Exfoliation Corrosion Specimen Following FSW (250 RPM)

Conclusions

Al-Zn-Mg-Cu aluminum extrusions modified with scandium were friction stir welded in a butt-weld configuration at six different tool rotation speeds (i.e., 175, 225, 250, 300, 350 and 400 RPM) and evaluated for residual properties. Mechanical testing revealed the highest joint efficiency at 250 RPM, corresponding to an energy per unit length of weld of 1000 J/mm. The tensile strength falls off sharply when the welding energies surpass 1300 J/mm. The trend in mechanical properties may be due to the competition between secondary nucleation of strengthening phases that dominates at weld energies below 1300 J/mm and precipitate coarsening that dominates at energies above 1300 J/mm. A thermal model of FSW quantifies the welding temperatures above the aging temperatures of SSA038 and the dwell times above this temperature.

The corrosion data are consistent with the trend in mechanical properties, i.e., the weld condition with the greatest susceptibility to corrosion corresponds to the weld condition that promotes the highest joint efficiency. For corrosion resistance, friction stir welding creates a performance gradient across the weld. The majority of exfoliation corrosion occurs along the TMAZ boundary, but adjacent to the weld, the corrosion resistance in the HAZ significantly improves due to the overaging of the strengthening precipitates that occurs in this region. Away from the HAZ and into the parent material, the corrosion resistance returns to the baseline behavior. The individual regions around the weld were also rated in accordance with ASTM G34, the HAZ receiving an EA and the baseline/parent material receiving an EA/EB.

Acknowledgements

The authors would like to acknowledge UES, Inc. and the Materials and Manufacturing Directorate at Wright-Patterson AFB for their support of this work.

References

1. S. V. Senkova, O. N. Senkov, D. B. Miracle, "Cryogenic and Elevated Temperature Strengths of an Al-Zn-Mg-Cu Alloy Modified with Sc and Zr, *Metallurgical and Materials Transactions A*, 57 (12) (2006) 3569 – 3575.
2. R. B. Bhat, J. D. Schloz, S. V. Senkova, O. N. Senkov, "Microstructure and Properties of Cast Ingots of Al-Zn-Mg-Cu Alloys Modified with Sc and Zr," *Metallurgical and Materials Transactions A*, 36 (8) (2005) 2115 – 2126.
3. W. M. Thomas et al., Great Britain Patent Application No. 9125978.8 (December 1991).
4. C. Dawes, W. Thomas, TWI Bulletin 6, November/December 1995, p. 124.
5. M. A. Sutton, B. C. Yang, A. P. Reynolds and J. H. Yan, "Banded microstructure in 2024-T351 and 2524-T351 aluminum friction stir welds - Part II. Mechanical characterization," *Materials Science and Engineering A-Structural Materials Properties Microstructure and Processing*, 364 (2004) 66-74.
6. B. J. Dracup, W. J. Arbegast, in: Proceedings of the 1999 SAE Aerospace Automated Fastening Conference & Exposition, Memphis, TN, October 5–7, 1999.
7. A. von Strombeck, J. F. dos Santos, F. Torster, P. Laureano, M. Kocak, in: Proceedings of the First International Symposium on Friction Stir Welding, Thousand Oaks, CA, USA, June 14–16, 1999.
8. C. Hamilton, S. Dymek, M. Blicharski, "Comparison Of Mechanical Properties For 6101-T6 Extrusions Welded By Friction Stir Welding And Metal Inert Gas Welding," *Archives of Metallurgy and Materials*, 52 (2007) 67 - 72.
9. R. S. Mishra and Z. Y. Ma, "Friction stir welding and processing," *Materials Science & Engineering R-Reports*, 50 (2005) 3-78.
10. M. Z. H. Khandkar, J. A. Khan and A. P. Reynolds, "Prediction of temperature distribution and thermal history during friction stir welding: input torque based model," *Science and Technology of Welding and Joining* 8 (3) (2003) 165 – 174.
11. O. Frigaard, O. Grong and O. T. Midling, "A process model for friction stir welding of age hardening aluminum alloys," *Metallurgical and Materials Transactions A – Physical Metallurgy and Materials Science*, 32 (5) (2001) 1189 – 1200.
12. V. Soundararajan, S. Zekovic, R. Kovacevic, Thermo-mechanical model with adaptive boundary conditions for friction stir welding of Al AA6061, *International Journal of Machine Tools and Manufacture* 45 (2005) 1577 – 1587.
13. O.N. Senkov, M.R. Shagiev, S.V. Senkova, "Effect of Sc on Aging Kinetics in a Direct Chill Cast Al-Zn-Mg-Cu Alloy," *Metallurgical and Materials Transactions A*, 39 (2008) 1034-1053.
14. O.N. Senkov, M.R. Shaghiev, S.V. Senkova, D.B. Miracle, "Precipitation of Al$_3$(Sc,Zr) Particles in a Direct Chill Cast Al-Zn-Mg-Cu-Sc-Zr Alloy during Conventional Solution Heat Treatment and Its Effect on Tensile Properties," *Acta Materialia* 2008, doi:10.1016/j.actamat.2008. 04.005.
15. R. Nandan, G. G. Roy, T. J. Lienert and T. Debroy, "Three-dimensional heat and material flow during friction stir welding of mild steel," *Acta Materialia* 55 (2007) 883 – 895.
16. C. Hamilton, S. Dymek, A. Sommers, "A thermal model of friction stir welding in aluminum alloys," *International Journal of Machine Tools and Manufacture*, 48 (2008) 1120 – 1130.
17. A.S. Argon, Strengthening Mechanisms in Crystal Plasticity, Oxford University Press, Oxford, GB, 2008.
18. J. B. Lumsden, M. W. Mahoney, G. Pollock, D. Waldron, A. Guinasso, in: The 1st International Symposium on Friction Stir Welding, California, 1999.
19. C. S. Paglia and R. G. Buchheit, "A look in the corrosion of aluminum alloy friction stir welds," *Scripta Materialia* 58 (2008) 383 – 387.

Friction Stir Welding and Processing V
Edited by: Rajiv S. Mishra, Murray W. Mahoney, and Thomas J. Lienert
TMS (The Minerals, Metals & Materials Society), 2009

Reducing Tool Axial Stresses in HSLA-65 During the Plunge

Kenneth Ross and Carl Sorensen

Brigham Young University, 435 CTB, Provo, UT, USA

Keywords: Friction Stir Welding, Stress, HSLA 65

Abstract

Using friction stir welding to join high-carbon steels would be more common if the PCBN tool had a longer tool life. PCBN tools appear to crack during the plunge due to normal loading. Reducing normal stress in the tool should increase tool life. A broad study of axial stresses during the plunge in HSLA-65 steel was conducted. Weld power and plunge heat input strongly affected stresses. Power and heat input affect the weld differently depending on whether the pin or the shoulder is engaged. Lowest stresses during pin engagement are obtained with relatively large pilot holes and rapid plunges.

Introduction

Approximately 80% of all conventional welds in industry are in steel[1]. For friction stir welding (FSW), PCBN tools are one of the tool materials used to weld steel and fail primarily due to fracture. Fracture often initiates during the plunge[2]. One approach to reduce fracture is to reduce stress during the plunge. We believe that stresses can be reduced by changing plunge parameters and using force to control the plunge speed.

Parameters studied herein include spindle speed, plunge rate, and pilot hole diameter. Plunge rate and spindle speed affect the heat generation rate and therefore the material flow stress. Also, pilot holes are believed to reduce tool stresses during the plunge. The area under the pin is heated faster and becomes hotter because there is less material under the pin to heat. Having less material at a greater temperature should produce a lower maximum stress on a tool during a plunge

The technique of using force to control plunge speed will be referred to as force control. In our approach, when the force approaches a pre-defined limit, the force control slows the plunge rate until a minimum speed is reached or the plunge is finished. A force control program was developed and tested in aluminum.

First, this manuscript describes the force control program and experimental tests in aluminum. Subsequently, the bulk of the manuscript addresses parameters that affect stress while plunging into HSLA-65. The correlations of flash formation, weld power, and plunge heat input with stress are described. The possible applications of force control in HSLA-65 are discussed.

Experimental Methods

A series of weld plunges was performed in AA7075 with convex scrolled shoulder step spiral (CS4) tools made from H13 steel to explore the effects of parameters and to develop a force control algorithm. Three series of plunges were performed in HSLA-65 with PCBN CS4[2] tools to explore the effects of plunge parameters.

Each plunge experiment was conducted on a TTI model RM-2 friction stir welding machine. An experiment consisted of a plunge at a given spindle speed and plunge feed rate. The plate either had a pilot hole of various sizes, or had no pilot hole. During the plunge, the axial force, the plunge depth, the spindle torque, the spindle speed, the motor power, and the plunge rate were measured approximately 8 times per second. The forces were converted into stresses by considering the geometry of the tool together with the plunge depth. The tool dimensions came from drawings provided by the manufacturers and was verified with an optical comparator.

Force Control

Based on previous research, a depth of 5.5 mm was used for all plunges in AA7075[3]. Following a series of welds made with various parameters, a force controlled plunge velocity algorithm was developed to minimize stresses during the plunges. The program slows the plunge rate as the plunge force approaches a predefined limit. If the plunge rate slows to 0.254 mm/min (0.01 ipm), it continues at that speed until the predefined depth is reached. In this case, the target plunge force can be exceeded. Plunges were made using the force controlled plunge algorithm to demonstrate its ability to reduce tool stresses during plunge.

HSLA 65

The three series of experiments using HSLA 65 steel plunged to a depth of 5.6mm (0.221in). The first series used predrilled pilot holes with a diameter of 2.4 mm (0.0935in), about 70% of the pin diameter. The second series used predrilled pilot holes with a diameter of 1.7mm (0.067in), 50% of the pin diameter. The third had no pilot holes. All pilot holes were drilled to a depth between 5.08 mm (0.2in) and 6.1mm (0.24in). All three conditions used plunge rates from 6.35 mm/min (0.25ipm) to 31.75 mm/min (1.25ipm) and tool rotation speeds from 450rpm to 1000rpm.

Results and Discussion

Force Control

The stress due to tool axial loads can be estimated using Equation 1,

$$\sigma = F / A \qquad (1)$$

where σ represents normal stress in the tool axial direction, F is the force in the axial direction and A represents the cross sectional area of the tool at the surface of the metal. Using equation (1), the forces measured in each plunge are converted to stress.

Figure 1 shows both force and stress vs. plunge depth during a position control friction stir weld in AA7075. The highest stress occurs at point A, just after the pin begins to plunge into the aluminum. As the tool plunges further, the surface area increases, and the stress decreases. This is seen in the constant slope from points A to B. At point B, the stress and force lines begin to drop significantly. At this point, more than half of the pin is submerged and is generating more heat by deformation causing the metal to soften significantly, explaining the dramatic drop in both the stress and force plots. During the transition region, between the shoulder and pin, engagement, the force and stress increase until point C. At point C, the shoulder of the tool is engaged. The shoulder causes more heat to be generated and provides much more surface area. The increased surface area allows the stresses to decrease despite the fact that force is increasing sharply.

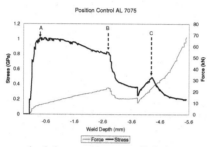

Figure 1. Force and stress vs. depth for a position controlled plunge.

A force control program was written to slow the plunge speed as the force approached a pre-defined limit. Once the speed slowed to 2.54 mm/min, it continued at that speed even if the force exceeded the pre-defined limit. The force control program reduced the maximum stress by approximately 30% as shown in Figures 2 and 3. The maximum stress in both force control and position control occur just after the pin makes contact with the material.

Figure 2 shows stresses in position control, force control, and force control with pilot holes plotted against depth. Drilling pilot holes significantly reduced stresses in the first half of the plunge. Figures 2 and 3 illustrate the dramatic reduction in stress due to pre drilled pilot holes. The maximum stress, when a pilot hole and force control are employed, is shown at point A in Figure 2. Using pilot holes and force control produces a maximum stress approximately 40% less than using only position control.

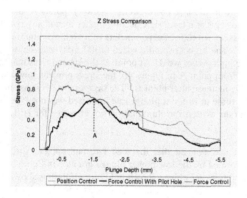

Figure 2. Maximum stresses for three plunges in AA7075 plotted against depth.

329

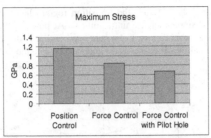

Figure 3. Maximum stresses for 3 plunges in AA7075.

HSLA-65 Steel

Three different series of experiments were completed to determine the effect of plunge rate and spindle speed on tool axial stress during the plunge in HSLA-65 steel. Spindle speeds from 450rpm to 1000rpm were tested. Plunge rates from 6.4 mm/min (0.25ipm) to 31.75 mm/min (1.25ipm) were tested. Pilot holes of 2.4mm (0.0935in) diameter were drilled for each weld in the first experiment, 1.72mm (0.067in) diameter for the second experiment and the third experiment was without pilot holes. The depth of the pilot holes was between 5.1mm (0.2in) and 6.1mm (0.24in) for the first two series of experiments.

The results for the largest pilot holes (2.4 mm diameter) show that as plunge and spindle speed are increased, the stress decreases. The power was calculated using equation 2 where Ω is RPM and M is torque. As spindle speed and plunge rate increases, power increases in HSLA-65.

$$Power = \frac{(2\pi)\Omega M}{60} \qquad (2)$$

A graph of force and stress vs. depth for the parameters of low maximum power input (3.79 kW) is shown in Figure 4a. The lowest spindle speed (450 rpm) and the lowest feed rate (6.35mm/min) were used in the low power weld. Between points A and B, the stress quickly rises to a maximum and stays at a level significantly higher than at any position after point B. A graph of force and stress vs. depth for the parameters of a high maximum power (6.23 kW) input is shown in Figure 4b. The highest spindle speed (1000 rpm) and the highest feed rate (31.75 mm/min) produced the high power weld. At point A, the stress is less than 20% of the value for the low power input. From point A in Figure 4b, the stress remains low and gradually increases until it begins to rapidly increase after point B. The decrease in stress as spindle speed increased was expected. The decrease in stress as plunge rate increased was not expected. Because of this, the force control program written for aluminum is not effective for HSLA-65 while the pin is engaged.

During the plunges, it was noticed that the feed rate and spindle speed greatly affect flash formation. The flash formed by the low power plunge graphed in Figure 4a is shown in Figures 5a and 5b and for the high power plunge shown in Figure 4b flash is shown in Figures 5c and 5d. Together, Figures 4 and 5 show that higher stresses corresponds to flash formation that curves upward off the surface of the metal and lower stresses corresponds to flat flash formation.

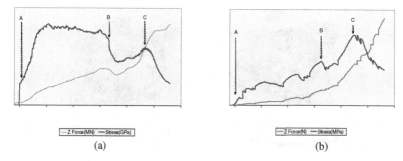

[— Z Force(MN) —Stress(GPa)]
(a)

[— Z Force(N) — Stress(MPa)]
(b)

Figure 4. (a) 3.79 kW Plunge in HSLA 65 (b) 6.23 kW Plunge in HSLA 65.

The side view shown in Figure 5b shows the flash curving upwards off the surface of the metal. The flash pushes up against the tool, significantly increasing the force required to continue the plunge. This upward curving flash caused significant wear on the locking collar of the tool as shown in Figure 6. The flash pushing up against the tool also increases the surface area of the tool contacting the material and therefore decreases the stress. This decrease in stress is not accounted for in the data presented in this work, because no technique for measuring this decrease in stress has been developed.

The flash formed by the high power input plunge is smooth and flat, Figures 5c and 5d. This is possibly because the flash was sufficiently heated and softened. Instead of pushing up against the tool, the flash was flattened. This is supported by the lack of wear on the locking collar during high power welds as shown in Figure 6. Via observation, we believe the plunge heat input corresponds more closely with the flatness of the flash than the power input. Plunge heat input was calculated using equation 3 where v is the plunge rate.

$$Plunge\ Heat\ Input. = \frac{Power}{v} \qquad (3)$$

Figures 4a and 4b are very different prior to point B. This suggests the flash shape has a greater affect before point B. To further explore this observation, weld data were divided into sections and analyzed by depth. When the pin is engaged, increased power corresponds to decreased stress. When the shoulder is engaged, increased power corresponds to increased stress. When the pin is engaged, increased plunge heat input corresponds to increased stress and when the shoulder is engaged, increased heat input corresponds to decreased stress. Stress verses power has the highest R squared values while the pin is engaged while stress verses heat input has the highest R squared values while the shoulder is engaged. The transition area between the pin and the shoulder has very low R square values for both power and heat input. Running a plunge at high power until the transition region where it will begin to run at high plunge heat input should reduce stresses. Figure 7 shows stress vs. power while the pin is engaged. Figure 8 shows stress vs. plunge heat input while the shoulder is engaged. Similar charts were created verifying these trends and have similar R squared values at all depths.

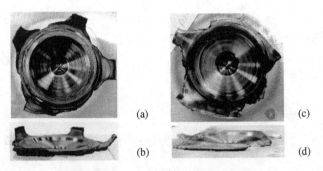

(a)

(c)

(b)

(d)

Figure 5. (a) and (b) are the top and side view of flash produced by a low power plunge. (c) and (d) are the top and side view of flash produced by a high power plunge.

Figure 6. The left side shows a tool that ran several high plunge heat input welds. The right side is a tool that ran several low plunge heat input welds. Much of the locking collar has been scraped off the tool by flash formed during the low heat input plunges.

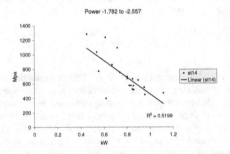

Power -1.782 to -2.557

$R^2 = 0.5199$

Figure 7. Stress vs. power at a given depth while the pin is engaged.

332

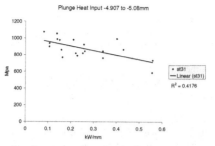

Figure 8. Stress vs. plunge heat input at a given depth while shoulder is engaged.

Based on the previous discussion, stresses may possibly be further lowered during the plunge if the plunge ran at high power while the pin engages and high plunge heat input while the shoulder engages. In Figure 9 a high power weld and high heat input weld are overlaid. The high power weld is highlighted until the transition region is reached. The high heat input weld is highlighted for the remainder of the plunge. A weld that switches from parameters that produce high power welds to parameters that produce high heat input welds might produce a stress vs. depth plot similar to the highlighted path shown in Figure 9. The change in parameters would occur while the transition region is engaged.

Stresses could be further reduced by engaging a force control program after the pin has completely engaged and before the shoulder engages. This would allow for greater plunge heat input because the plunge rate decreases as the force approaches a predefined level. Greater heat input would lower the stresses where it is the highest in Figure 9. A plunge that runs at high power while the pin engages and force control when the shoulder engages could possibly produce the lowest stresses during the plunge.

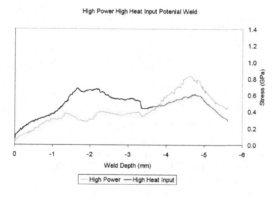

Figure 9. The path of a plunge that would switch from high power to high plunge heat input is highlighted and follows the plot of the high power weld until the shoulder is about to engage.

333

Two other series of experiments were run with the same spindle speeds and feed rates. The second series used pilot holes 1.702 mm in diameter and the third had no pilot holes. In both the second and third series increased power did not corresponded with reduced stress at any depth. Increased plunge heat input corresponds with lower stresses at almost all depths. These trends can be seen in Figure 10. We believe increased power causes lower stress only when larger pilot holes are drilled because there is a greater amount of energy per unit volume. This could be because most of the energy is on the outer edge of the pin, and most of the material directly beneath the pin has been removed.

The box plots shown in Figure 11 show that the lowest maximum stresses were produced by plunges done with the smaller diameter pre-drilled holes. Figure 12 takes the maximum stress that occurred from just after the tool makes contact with the metal to the depth where the base of the pin is at the surface of the metal for all 20 welds in each of the three sets. Larger diameter pilot holes produce the lowest stresses while the pin engages. When analyzing these box plots, the lines that extend to the maximum values can be ignored because they represent values for low plunge heat input welds. Since these stresses can be avoided by changing RPM and IPM values, they are not significant.

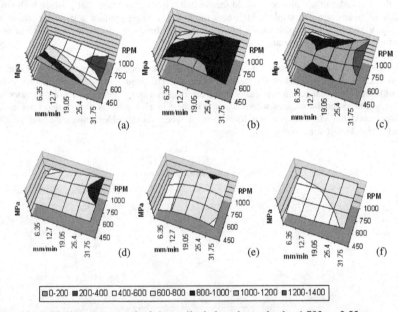

Figure 10. a) Maximum stresses for 2.4mm pilot holes, plunge depths -1.782 to -2.55mm
b) Maximum stresses for 1.702mm pilot holes, plunge depths -1.782 to -2.55mm
c) Maximum stresses without pilot holes, plunge depths -1.782 to -2.55mm
d) Maximum stresses for 2.4mm pilot holes, plunge depths -5.08 to -5.254mm
e) Maximum stresses for 1.702mm pilot holes, plunge depths -5.08 to -5.254mm
f) Maximum stresses without pilot holes, plunge depths -5.08 -5.254mm

Cracks at the base of the PCBN pin form while the pin is engaged. Therefore, reducing stresses while the pin is engaged is more important than reducing stress when the shoulder is engaged. Pilot holes should be used because they produce the lowest stresses when the pin is engaged. The median and average values for maximum stress during the entire plunge are approximately the same for all three data sets.

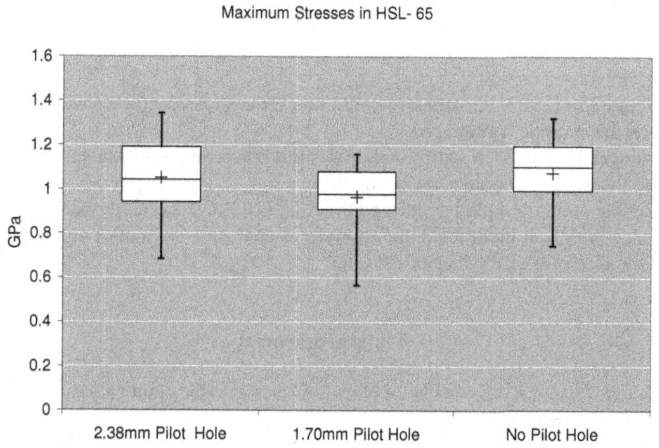

Figure 11. Box plots of maximum stress values during the plunge using.

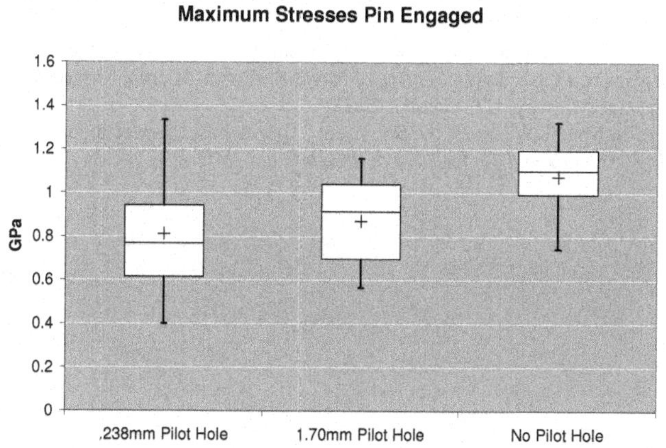

Figure 12. Box plots of maximum stresses while the pin was engaged.

Conclusions

Based on the experiments and discussion above we conclude the following:

1. Force control reduces normal axial stresses during the plunge in AA7075.
2. In HSLA-65, there exists a correlation between the shape of flash formed during the plunge and stresses during the plunge. The shape of the flash is correlated with the power/ plunge heat input into the material during the plunge.
3. During plunges in HSLA- 65 with large pilot holes, stress decreases with increased power while the pin was engaged.
4. During plunges in HSLA-65 with large pilot holes, the stress decreases with increased heat input when the shoulder is engaged.
5. During plunges in HSLA-65 with small or no pilot holes, stress decreases with heat input when both the pin and shoulder are engaged.
6. The best parameters for a plunge during HSLA 65 is to have a high power input while the pin is engaged, a high heat input while the shoulder is engaged, and a pilot hole with a diameter ~70% of the diameter of the face of the pin.

Acknowledgement

This research was funded by the Office of Naval Research under grant N00014-06-1-0501, Mr. John Deloach, program manager.

References

1. Sorensen, C.D., "Progress in Friction Stir Welding of High Temperature Materials," *Proceedings of the 14th Annual Offshore and Polar Engineering Conference*, 2004, pp. 8-14.
2. Sorensen, C.D., Nelson, T.W., Packer, S.M. and R. Kingston, "Reducing Fracture Tendencies of PCBN Tools", *Office of Naval Research Friction Stir Technologies Conference*, Midway, UT, October 2007.
3. Pew, J., Nelson, T.W., and C.D. Sorensen, "Torque based power model for friction stir welding" *Science and Technology of Welding and Joining* 12:4 (May 2007), pp. 341-347

Friction Stir Welding and Processing V
Edited by: Rajiv S. Mishra, Murray W. Mahoney, and Thomas J. Lienert
TMS (The Minerals, Metals & Materials Society), 2009

INVESTIGATION OF A DONOR MATERIAL IN FRICTION STIR WELDING

Justin. M. Rice, S. Mandal, A.A. Elmustafa

Old Dominion University
Department of Mechanical Engineering
238 Kaufman Hall, Norfolk, VA 23549

Keywords: FSW, Plunge, Donor, ABAQUS

Abstract

Excessive tool wear caused during the plunge phase of friction stir welding (FSW) is hindering the application of FSW to hard materials such as steel. This research uses a finite element model with the implementation of the Johnson-Cook material constitutive law to investigate the shear stress and axial force experienced by the tool during the plunge phase of FSW into AISI 1045. The model in this research consists of a deformable work-piece, deformable donor material, and a rigid tool. By implementing the concept of using a donor material, we are able to have localized pre-heating and minimize shear stress and axial forces, therefore reducing tool fracturing and replacement costs. The numerical simulation data supports the concept of using a donor material to reduce tool wear and the need to implement this concept to experimental work for further verification.

Introduction

Friction stir welding (FSW) is an innovative method of joining materials developed and patented by The Welding Institute in Cambridge, England. This process normally consists of three phases – the plunge stage, where ideally a non-consumable tool penetrates into the materials being joined, a dwell stage, where the tool rotates without moving in the lateral direction, and the traverse stage, where the tool moves laterally along the joint line of the materials to be welded. FSW combines frictional heating between the tool and the work-piece with intense plastic deformation of the work-piece to create solid state joining of materials and therefore making FSW the preferred method of joining materials [1-3]. Friction stir welding of aluminum has already been widely adopted as the preferred method of welding. The applications of FSW for aluminum have created a desire to use this process for joining harder materials such as composites and steels.

Currently, tool wear complications with welding harder materials have prevented the adoption of friction stir welding harder materials. Although there is research of tool wear during the traverse stage of friction stir welding, there appears to be little research studying the tool wear during the plunge stage. The plunge stage of FSW is a critical stage for tool wear, where the initial thermomechanical conditions are generated and significant material transformations occur due to high temperatures and stresses before a steady state condition is reached during the dwell phase[2]. Although there is ongoing research to study tool wear during the traverse stage, there is little research to study tool wear during the plunge stage. This research simulates the critical plunge stage using the finite element code ABAQUS/Explicit and implements the use of a donor material in order to reduce axial forces and shear stress during the plunge phase and therefore reduce the amount of tool wear.

Modeling of the plunge stage of friction stir welding poses a variety of challenges due to the high strain rates and temperatures, making it a non-linear computation. Schmidt et al [4-6] developed numerical models using ABAQUS/Explicit and the Arbitrary Lagrangian Eulerian (ALE) formulation. However, these models used a 'simplified plunge,' where the simulation starts at the point where the shoulder is already in contact with the work-piece and therefore not simulating the most difficult initial stage of contact. Goetz et al [7] developed a two dimensional model using DEFORM to simulate the metal flow around the tool and the initial tool contact, but a three dimensional model would be a better . A three dimensional model with a plunge depth of 300μm was developed by Gerlich et al [8] to simulate friction stir spot welding (FSSW) using a computational fluid dynamics approach. The difference between these models and the one presented in this research is a full, not 'simplified', three dimensional plunge of 12mm. The model in this research modifies the base model described by Mandal et al [2] by incorporating the use of a donor material to study the effect on the axial force and shear stresses.

Numerical Model

The model consists of a deformable work-piece, deformable donor, and rigid tool. In order to simplify the computational process, the threads of the FSW tool are not modeled. The work-piece and donor material consists of 1200 and 800 - 8 node brick elements, respectively with a higher mesh density around the tool plunge area. This method of meshing improves the accuracy of the solution without increased computational time. The work-piece dimensions are 100 x 100 x 15mm and the donor dimensions are 100 x 100 x 5mm. Figure 1 shows the assembly and meshing of the model with the donor material on top of work-piece. The tool dimensions are shown in Fig. 2 with the height of the overall tool being irrelevant.

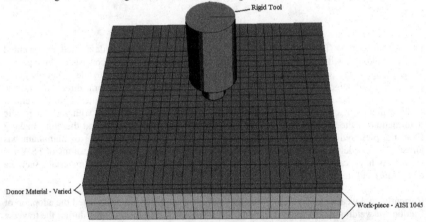

Fig. 1 Model assembly and finite element mesh

The sides of both the donor material and work-piece are constrained such that there can only be compression along the direction of the plunge and the bottom surface of the work-piece is constrained such that no bending can occur. A constant friction coefficient of 0.3 is implemented between all three surfaces [6] with the donor and work-piece being the master surface and the

338

tool being the slave surface. The tool rotation is set at 300 RPM and the tool plunge rate is set at a constant 4mm/s.

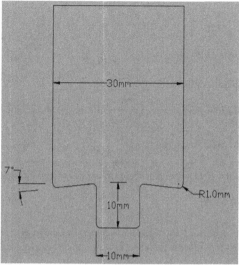

Fig 2. Tool Dimensions

Material Law and Properties

In order to model the interaction of flow stress and temperature, the temperature and strain rate dependant elastic-plastic Johnson-Cook law is used as the constitutive law. The Johnson-Cook law calculates the flow stress as a function of temperature and strain rate up to the melting point or solidus temperature at which point the stress is reduced to zero according to the equations below [10].

$$\overline{\sigma} = \left[A + B(\overline{\varepsilon}^{pl})^n \right] \left[1 + C \ln\left(\frac{\overline{\dot{\varepsilon}}^{pl}}{\dot{\varepsilon}_0} \right) \right] (1 - \hat{\theta}^m) \tag{1}$$

where $\hat{\theta}$ is a nondimensional temperature and is defined as

$$\hat{\theta} \equiv \begin{cases} 0 & \text{for} \quad \theta < \theta_{transition} \\ (\theta - \theta_{transition})/(\theta_{melt} - \theta_{transition}) & \text{for} \quad \theta_{transition} \leq \theta \leq \theta_{melt} \\ 1 & \text{for} \quad \theta > \theta_{melt} \end{cases} \tag{2}$$

Johnson-Cook strain rate dependence assumes that

$$\overline{\sigma} = \sigma^0(\overline{\varepsilon}^{pl}, \theta) R(\overline{\dot{\varepsilon}}^{pl}) \tag{3}$$

339

and

$$\bar{\dot{\varepsilon}}^{pl} = \dot{\varepsilon}_0 \exp\left[\frac{1}{C}(R-1)\right] \quad for \quad \bar{\sigma} \geq \sigma^0, \tag{4}$$

where

$\bar{\sigma}$ is the yield stress at nonzero strain rate;

$\bar{\dot{\varepsilon}}^{pl}$ is the equivalent plastic strain rate;

$\dot{\varepsilon}_0$ and C are material parameters measured at or below the transition temperature, $\theta_{transition}$;

$\sigma^0(\bar{\varepsilon}^{pl}, \theta)$ is the static yield stress; and

$R(\bar{\varepsilon}^{pl})$ is the ratio of the yield stress at nonzero strain rate to the static yield stress such that

$$R(\dot{\varepsilon}_0) = 1.0$$

$A, B, C, n,$ and m from equation 1 are material parameters that are measured at or below the transition temperature. The Johnson-Cook properties for the work-piece and donor materials are given in table 1[11-15].

Table 1. Johnson-Cook Material Properties Used in Simulation

	AISI 1045	Al 2024-T3	Al 6061-T4	Copper 10100	Units
A	553.1	369	324	440	MPa
B	600.8	684	114	150	MPa
n	0.234	0.73	0.42	0.31	NA
m	1	1.7	1.34	1.09	NA
θ_{melt}	1460	502	582	1083	°C
$\theta_{transition}$	25	25	25	25	°C
C	0.013	0.0083	0.002	0.025	NA
$\dot{\varepsilon}_0$	1	1	1	1	NA

Results and Discussion

The advantages of using a donor material are to generate localized preheating, minimize tool fracturing, reduce replacement costs, and standardization. To test the concept of using a donor material, the work-piece was assigned AISI 1045 material and Johnson-Cook properties while using three different donor materials consisting of Al 6061-T4, Al 2024-T3, and copper. The donor to work-piece thickness ratio was set at 1:3. Four simulations were performed. The first simulation consisted of an AISI 1045 work-piece and an AISI 1045 donor. Using this simulation for comparison, three more simulations were performed using the different donor materials listed above. The shear stress and axial load results are shown in Fig.3 and 4.

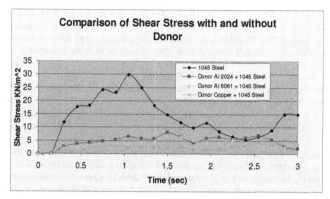

Fig. 3 Donor versus no donor shear stress comparison

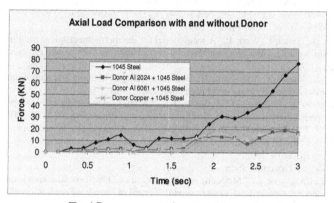

Fig. 4 Donor versus no donor axial force comparison

Comparing the data from the simulations, the use of a donor material drastically reduced both shear stresses and axial forces. The maximum shear stress of pure AISI 1045 is 30 kN/m² whereas the maximum for either of the three donor materials is 7kN/m², a reduction of more than 76%. Similarly, the maximum axial force without the use of a donor material is roughly 78kN and the maximum with either of the donor materials is 21kN, a reduction of 73%. With a significant reduction in both shear stresses and axial forces, tool wear would be reduced significantly.

Conclusion

Successful joining of aluminum using the friction stir welding process has led to the desire to weld harder materials. To do so, the tool wear must be reduced for the plunge phase of FSW. A finite element based model was developed using ABAQUS/Explicit to study the effect of using a donor material during the plunge stage. The use of a donor material resulted in a 76% reduction in shear stresses and a 73% reduction in axial forces. With this significant reduction, it is feasible to start implementing this concept to experiments to determine the interaction of materials and set industry standards for using a donor material. If experimental data produces the same results as the models, FSW of steels will become the preferred method of joining materials.

References

1. Thomas W.M., Nicholas E. D., Needham J. C., Murch M. G., Temple-Smith P. and Dawes C. J. (TWI): Improvements relating to friction welding. EP 0 615 480 B1.
2. Mandal, S., Rice, J. M. and Elmustafa, A. "A Numerical Study of the Plunge Stage in Friction Stir Welding Using ABAQUS," *Friction Stir Welding and Processing IV*, 2007, pp. 127-133.
3. Mandal, S., Williamson, K. "A thermomechanical hot channel approach to friction stir welding", *Journal of Materials Processing Technology*, 2006, pp. 190-194.
4. Schmidt, H., Hattel, J., Wert, J., "An analytical model for heat generation in friction stir welding", *Modeling and Simulation in Materials Science and Engineering*, 2004, pp. 143-157.
5. Schmidt, H., Hattel, J., "Modelling thermomechancial conditions at the tool/matrix interface in friction stir welding", *Proceedings of the 5th International Friction Stir Welding Symposium*, Metz, France, 2004.
6. Schmidt, H., Hattel, J., Wert, J., "A local model for the thermomechanical conditions in friction stir welding", *Modeling and Simulation in Materials Science and Engineering*, 2005, pp.77-93.
7. Goetz, R.L., Jata, K.V., "Modeling friction stir welding of titanium and aluminum alloys", *Friction Stir Welding and Processing*, TMS, 2001.
8. Gerlic, A., Su, P., Bendzsak, G. J., North, T.H., "Numerical modeling of FSW spot welding: preliminary results", *Friction Sir Welding and Processing III*, TMS 2005.
9. Kakarla, S. T., Muci-Kuchler, K. H., Arbegast, W. J., Allen, C. D., "Three dimensional finite element model of the friction stir spot welding process", *Friction Stir Welding and Processing III*, TMS, 2005.
10. ABAQUS v6.5 Documentation
11. Poizat, C., Campagne, L., "Modeling and Simulation of Thin Sheet Blanking Using Damage and Rupture Criteria", *International Journal of Forming Processes*. Vol. 8, 2005, pp. 29-47
12. Ozel, T., Zeren, E., "Finite Element Modeling of Stresses Induced by High Speed Machining with Round Edge Cutting Tools", *International Mechanical Engineering Congress & Exposition*. November, 2005, Orlando, FL.
13. Lesuer, D. R., Kay, G. J., LeBlanc, M. M., "Modeling Large-Strain, High-Rate Deformation in Metals", *Third Biennial Tri-Laboratory Engineering Conference on Modeling and Simulation"*, July, 2001, Pleasanton, CA.
14. "Aluminum 6061-T4", *ASM Materials Handbook*, Vol. 2, Ed. 9, 1978.
15. Gregory Kay, "Failure Modeling of Titanium 6Al-4V and Aluminum 2024-T3 With the Johnson-Cook Material Model," *U.S. Department of Transportation, Federal Aviation Administration, Final Report – DOT/FAA/AR-03/57*, September, 2003.

AUTHOR INDEX
Friction Stir Welding V

SUBJECT INDEX
Friction Stir Welding V